AF780962

VOLUME FORTY FOUR

METHODS IN GEOCHEMISTRY AND GEOPHYSICS

Principles of Electric Methods in Surface and Borehole Geophysics

METHODS IN GEOCHEMISTRY AND GEOPHYSICS

(Volumes 1–28 are out of print)

29. V.P. Dimri – Deconvolution and Inverse Theory – Application to Geophysical Problems
30. K.-M Strack – Exploration with Deep Transient Electromagnetics
31. M.S. Zhdanov and G.V. Keller – The Geoelectrical Methods in Geophysical Exploration
32. A.A. Kaufman and A.L. Levshin – Acoustic and Elastic Wave Fields in Geophysics, I
33. A.A. Kaufman and P.A. Eaton – The Theory of Inductive Prospecting
34. A.A. Kaufman and P. Hoekstra – Electromagnetic Soundings
35. M.S. Zhdanov and P.E. Wannamaker – Three-Dimensional Electromagnetics
36. M.S. Zhdanov – Geophysical Inverse Theory and Regularization Problems
37. A.A. Kaufman, A.L. Levshin and K.L. Larner – Acoustic and Elastic Wave Fields in Geophysics, II
38. A.A. Kaufman and Yu. A. Dashevsky – Principles of Induction Logging
39. A.A. Kaufman and A.L. Levshin – Acoustic and Elastic Wave Fields in Geophysics, III
40. V.V. Spichak – Electromagnetic Sounding of the Earth's Interior
41. A.A. Kaufman and R.O. Hansen – Principles of the Gravitational Method
42. A.A. Kaufman, R.O. Hansen and Robert L.K. Kleinberg – Principles of the Magnetic Methods in Geophysics
43. Michael S. Zhdanov – Geophysical Electromagnetic Theory and Methods
44. A.A. Kaufman and B.I. Anderson – Principles of Electric Methods in Surface and Borehole Geophysics

VOLUME FORTY FOUR

METHODS IN GEOCHEMISTRY AND GEOPHYSICS

Principles of Electric Methods in Surface and Borehole Geophysics

A.A. KAUFMAN
Emeritus Professor

B.I. ANDERSON
Consultant

Amsterdam • Boston • Heidelberg • London • New York • Oxford
ELSEVIER Paris • San Diego • San Francisco • Singapore • Sydney • Tokyo

Elsevier
Radarweg 29, PO Box 211, 1000 AE Amsterdam, The Netherlands
The Boulevard, Langford Lane, Kidlington, Oxford OX5 1GB, UK

Copyright © 2010 Elsevier B.V. All rights reserved

No part of this publication may be reproduced, stored in a retrieval system or transmitted in any form or by any means electronic, mechanical, photocopying, recording or otherwise without the prior written permission of the publisher

Permissions may be sought directly from Elsevier's Science & Technology Rights Department in Oxford, UK: phone (+44) (0) 1865 843830; fax (+44) (0) 1865 853333; email: permissions@elsevier.com. Alternatively you can submit your request online by visiting the Elsevier web site at http://www.elsevier.com/locate/permissions, and selecting *Obtaining permission to use Elsevier material*

British Library Cataloguing in Publication Data
A catalogue record for this book is available from the British Library

Library of Congress Cataloging-in-Publication Data
A catalog record for this book is available from the Library of Congress

ISBN: 978-0-444-63824-3
ISSN: 0076-6895

For information on all Elsevier publications
visit our website at elsevierdirect.com

10 11 12 13 14 10 9 8 7 6 5 4 3 2 1

Working together to grow
libraries in developing countries

www.elsevier.com | www.bookaid.org | www.sabre.org

ELSEVIER BOOK AID International Sabre Foundation

This book is dedicated to the outstanding Russian geophysicist
L.M. Alpin

CONTENTS

INTRODUCTION

The subject of this monograph is the physical and mathematical principles of electric methods of applied geophysics. In Chapter 1, proceeding from Coulomb's law and the principle of superposition, we introduce the concept of electric field and describe its two remarkable features observed in any medium. Special attention is paid to the system of equations of the field **E** in the integral and differential forms as well as at points of interfaces dividing media with different physical properties. Then, after the introduction of the potential, we move from the system of equations to Poisson's and Laplace's equations and discuss their fundamental solution. The theorem of uniqueness and different boundary problems are studied in detail. For illustration of the field behavior, we consider several examples, including electric dipole, double layer, electrostatic induction, polarization, and appearance of bound charges. Chapters 2 and 3 describe the electric field in a conducting and polarizable medium. First, we discuss the vector of current density, current lines, and current tubes, as well as the relation between the vector **j** and charge density and velocity. Special attention is paid to Ohm's law, principle of charge conservation, and non-Coulomb's (extraneous) force. In Chapters 2, we describe in detail the distribution of the volume and surface charges of both types, namely, free and bound ones, and demonstrate why dielectric constant does not have any influence on the electric field in a conducting medium. We consider the system of equations of the electric field and prove the theorem of uniqueness. The concept of resistance in the case of volume conductors is studied in detail. In particular, the role of resistance in different parts of the electric arrays is demonstrated. Chapter 3 is a natural continuation of Chapters 2, presenting numerous examples of field behavior. They are chosen to illustrate the field behavior in cases which are rather typical in mining, engineering, groundwater geophysics, and other application areas. In the next chapters, we discuss the physical and mathematical foundation of the main electric methods of the surface and borehole geophysics and start from geometric soundings (Chapter 4). At the beginning, we describe different arrays and introduce the concept of apparent resistivity, as well as the relation between the measured resistance and apparent resistivity. Then, the boundary-value problem for a horizontally layered medium is formulated and expressions for the potential and electric field on the earth's surface are derived. Examples of two- and three-layered curves of the apparent resistivity are described. Special attention is paid to asymptotic equations describing the field behavior at large distances from the source. We demonstrate the field behavior in a medium that is similar to the one met in marine electroprospecting, where a relatively thin and resistive layer may represent the collector. Solution of the forward and inverse problems of the geometrical soundings is the subject of Chapter 5, where initially we describe the main steps in the development of a solution of the forward problem, such as replacement of integrals by series, graphical methods of calculation of the apparent resistivity curves for a many-layered medium, numerical integration by Fillon's method, as well as the method of filters. We discuss the concept of uniqueness and solution of inverse problem for soundings, ill- and well-posed problems, stable and nonstable parameters, and equivalence and influence of geological noise. In Chapter 6, we describe profiling methods using different arrays and methods of charged body, applied in mining and groundwater geophysics, as well as the self-potential method of different origins. The third part of the monograph (Chapter 7) is dedicated to the electric methods of borehole geophysics. First, we derive expressions for the potential and electric field inside the borehole when the primary field is caused by the point source located at the borehole axis, and study the apparent resistivity curves as a function of the probe length (lateral soundings with potential and gradient probes). This part of the chapter leads us to the investigation of the radial responses of the probes and influence of the invasion zone, and then of the vertical responses of the potential and gradient probes. In addition, we describe the main features of different types of laterologs based on measuring grounding resistance and focusing of the electric field around the central part of the probes. After this, the physical and mathematical principles of the method of measuring through the casing are described in some detail. Finally, we briefly consider the mechanism of self-potential field in a medium surrounding the borehole and derive expressions for the potential at the borehole axis.

ACKNOWLEDGMENTS

We wish to express our deepest gratitude to Dr. M. Oristaglio for his contributions to all stages of preparation of this book. We appreciate very much the valuable discussions we had with him concerning all chapters of the book, insightful suggestions that helped to clarify many aspects of the text, numerous calculations, together with graphical representation of their results, and many more. We are also grateful to Dr. D. Alumbaugh, and Dr. N. Cuevas, who very carefully read Chapter 4 and made numerous corrections; as well as to Dr. S. Akselrod, Dr. V. Druskin, Dr. Y. Li, Dr. K.V. Titov, Dr. A. Zohdy, and Dr. C. Stoyer, for their very useful comments and suggestions.

LIST OF SYMBOLS

a	Major semi-axis of elliptical cylinder and spheroid
b	Minor axis of elliptical cylinder and spheroid
e	Electric charge
$\mathbf{E}$	Electric field
E_{n}	Normal component of electric field
E_{t}	Tangential component of electric field
$\mathbf{F}$	Force of interaction
g	Gravitational field
G	Green's function
h	Layer thickness
h_1, h_2, h_3	Metric coefficients
H_n	Hankel's function of **n** order
$\mathbf{i}$, $\mathbf{j}$, $\mathbf{k}$	Unit vectors in Cartesian system of coordinates
I	Current
$\mathbf{j}$	Current density vector
$\mathbf{j} = (-\mathbf{1})^{1/2}$	Imaginary unit
J_n	Bessel function of **n** order
J'_n	Derivative of the Bessel function
L_{qp}	Distance between points **q** and **p**
$\mathbf{n}$	Unit vector
$\mathbf{p}$	Dipole moment
$P_n(x)$, $Q_n(x)$	Legendre's functions of the first and second kind
R	Resistance
S	Longitudinal conductance and surface
t	Time
T	Transversal conductance
U	Scalar potential of electric field
α	Angle
δ	Volume density of charges
ε	Dielectric constant
ε, η, ϕ	Coordinates in spheroidal system of coordinates
ε_0	Constant
ε_{r}	Relative dielectric constant
ϕ	Angle
γ	Conductivity

η	Density of dipole moments
$\boldsymbol{\lambda}$	Linear density of charges and coefficient of anisotropy
$\boldsymbol{\rho}$	Resistivity
$\boldsymbol{\rho}_{\mathbf{m}}$	Mean resistivity
$\boldsymbol{\rho}_{\mathbf{n}}$	Transversal resistivity
$\boldsymbol{\rho}_{\mathbf{t}}$	Longitudinal resistivity
$\boldsymbol{\sigma}$	Surface density of charges
$\boldsymbol{\omega}$	Solid angle
$\boldsymbol{\Xi}$	Electromotive force

PART ONE

Electric Field in a Conducting Medium

CHAPTER ONE

Coulomb's Law and Stationary Electric Field

This chapter introduces the theory of stationary electric fields—that is, the theory of electric fields that do not vary with time—starting from Coulomb's law of force between electric charges and the principle of superposition. The concepts of electric field and electric potential are then developed, along with the general equations that these fields satisfy and their solutions in simple configurations. The material developed here will be used repeatedly in subsequent chapters to study the applications of stationary electric fields in surface and borehole methods of geophysics.

1.1. INTERACTION OF ELECTRIC CHARGES AND COULOMB'S LAW

The force of interaction between electric charges is, like the force of gravity, one of the fundamental concepts of physics. The French physicist Charles-Augustin de Coulomb (1736–1806) was the first to study quantitatively this remarkable phenomenon in experiments carried out at the end of eighteenth century. In this chapter, we consider these concepts only at the "macroscopic" level, which means that the dimensions of regions where "elementary" charges are located are assumed to be much smaller than the distances between these regions, while elementary charges themselves can have an arbitrary value. Coulomb's law then states that the force acting on an elementary charge located at the point p, caused by a second elementary charge located at the point q, is given by the simple expression:

$$\mathbf{F}(p) = \frac{1}{4\pi\varepsilon_0}\frac{\mathrm{d}e(q)\mathrm{d}e(p)}{L_{qp}^3}\mathbf{L}_{qp}. \qquad [1.1]$$

Here $\mathrm{d}e(p)$ and $\mathrm{d}e(q)$ are the numerical values of the elementary charges, $\mathbf{L}_{qp}$ is the vector directed from the point q to the point p, L_{qp} is the distance between these points, and ε_0 is a constant of nature. In the standard International System of units (S.I. units), force is measured in "newtons" (N),

Methods in Geochemistry and Geophysics, Volume 44
ISSN 0076-6895, DOI: 10.1016/S0076-6895(10)44001-9
© 2010 Elsevier B.V.
All rights reserved.

which according to Newton's Second Law has dimensions of mass times acceleration, or kilogram–meter per second squared (kg m s^{-2}); charge is measured in "coulombs" (C); and distance is measured in meters. The constant ε_0 then has the value

$$\varepsilon_0 = \frac{1}{36\pi} 10^{-9}\,\mathrm{F\,m^{-1}},$$

where "F" is a unit called the "farad." Its dimensions can be worked out from Eq. [1.1], but will later be expressed more simply after some further electrical units are introduced. Noting that the vector separation $\mathbf{L}_{qp}$ can be written as

$$\mathbf{L}_{qp} = L_{qp}\mathbf{L}^0_{qp},$$

where $\mathbf{L}^0_{qp}$ is a unit vector directed along line from point q to point p, and substituting into Eq. [1.1] gives the familiar "inverse square" law of force:

$$\mathbf{F}(p) = \frac{1}{4\pi\varepsilon_0}\frac{de(q)de(p)}{L^2_{qp}}\mathbf{L}^0_{qp}. \qquad [1.2]$$

In other words, the electric force of interaction between two elementary charges is directly proportional to the product of the charge strength and inversely proportional to the square of the distance between them. Unlike mass, which generates the force of gravity and is always positive, electric charge can be either positive or negative. This means that the electric force $\mathbf{F}(p)$ has the same direction as the unit vector $\mathbf{L}^0_{qp}$ when the two charges have the same sign and is therefore "repulsive"—that is, the force acts to push the charges apart. When the product of charges is negative, the force $\mathbf{F}(p)$ points in the opposite direction from $\mathbf{L}^0_{qp}$, and is "attractive" (Fig. 1.1A). Acted on

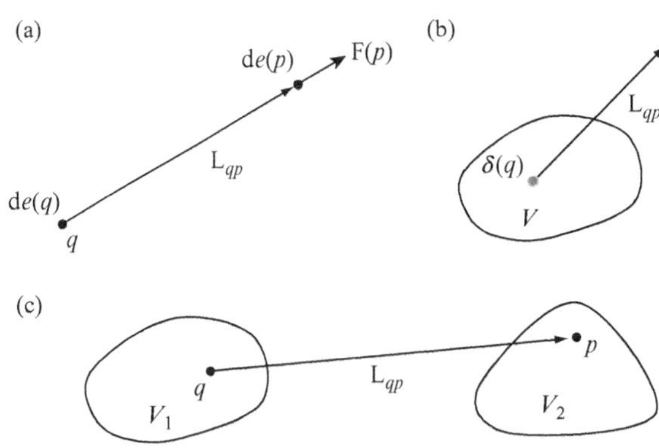

Figure 1.1 Coulomb's law and its generalization.

by the force $\mathbf{F}(p)$, the charge located at p will, unless otherwise restrained, move with an acceleration $\mathbf{a}(p)$ following Newton's second law of motion:

$$\mathbf{a}(p) = \frac{\mathbf{F}(p)}{m}, \qquad [1.3]$$

where m is a mass of the elementary charge and $\mathbf{a}$ is its vector acceleration. It is worth recalling here that the macroscopic point of view introduced earlier assumes that the charges involved in the preceding equations—which in reality occupy a finite volume of space—are located at points p and q. In effect, this is the mathematical meaning of "elementary charge." This macroscopic point of view also implies that the force has the same value at all points of the elementary volume occupied by the charge. It follows from Eq. [1.1] that the force acting on the charge at point q is the opposite of the force acting on the charge at point p, that is,

$$\mathbf{F}(q) = -\mathbf{F}(p). \qquad [1.4]$$

Thus, Newton's Third Law ("for every action, there is an equal and opposite reaction") holds for the Coulomb's force of attraction. Although the mathematical sum of the forces acting on the charges is zero,

$$\mathbf{F}(q) + \mathbf{F}(p) = 0,$$

it should be remembered that these forces act at separate points: each elementary charge feels only the force of the other charge. This latter feature is part of the macroscopic assumption, which does not allow the possibility of "self-interaction" or the interaction of an elementary charge with itself. We now consider a generalization of Coulomb's law to determine the electric force caused by a collection of elementary charges distributed throughout a volume of arbitrary size (Fig. 1.1B). The generalization invokes the principle of superposition, which holds that the force of interaction between two charges is independent of the presence of other charges. (The superposition "principle" is another assumption that can be shown to hold experimentally to a very high degree of accuracy.) Consider an elementary volume $\mathrm{d}V(q)$ surrounding the point q and containing a uniform distribution of charge, so that the total charge in the volume can be represented as the product of a charge density times the volume:

$$\mathrm{d}e(q) = \delta(q)\mathrm{d}V, \qquad [1.5]$$

where $\delta(q)$ is the volume charge density, which, by definition, is equal to the amount of charge in a unit volume of space. Its dimensions are charge per unit volume, or coulombs per cubic meter, written

$$[\delta] = \mathrm{C\,m}^{-3}.$$

In general, the density $\delta(q)$ can vary from point to point. Using the principle of superposition, as well as Eqs. [1.1] and [1.5], the force $\mathbf{F}(p)$ due to an arbitrary distribution of charge density in the volume V can be written as

$$\mathbf{F}(p) = \frac{\mathrm{d}e(p)}{4\pi\varepsilon_0} \int_V \frac{\delta(q)\mathbf{L}_{qp}}{L_{qp}^3} \mathrm{d}V. \qquad [1.6]$$

It should be obvious that, in performing the integration over the volume V, the point q varies, while the point p (where the force is being computed) remains the same. $\mathbf{F}(p)$ is thus a summation of forces having different magnitudes and directions, but all applied at the same point p. The vector force equation can be resolved into its components by introducing a system of coordinates. In Cartesian (rectangular) coordinates, with

$$q = (x_q, y_q, z_q), \quad p = (x_p, y_p, z_p),$$

the components of the force are

$$\begin{aligned} F_x(p) &= \frac{\mathrm{d}e(p)}{4\pi\varepsilon_0} \int_V \frac{\delta(q)(x_p - x_q)}{L_{qp}^3} \mathrm{d}V, \\ F_y(p) &= \frac{\mathrm{d}e(p)}{4\pi\varepsilon_0} \int_V \frac{\delta(q)(y_p - y_q)}{L_{qp}^3} \mathrm{d}V, \\ F_z(p) &= \frac{\mathrm{d}e(p)}{4\pi\varepsilon_0} \int_V \frac{\delta(q)(z_p - z_q)}{L_{qp}^3} \mathrm{d}V, \end{aligned} \qquad [1.7]$$

where

$$L_{qp} = [(x_p - x_q)^2 + (y_p - y_q)^2 + (z_p - z_q)^2]^{1/2}.$$

Coulomb's law is fundamental in that it applies in any medium; that is, the force of interaction between two charges remains the same regardless of the medium where the charges reside. Indeed, the right-hand side of Eq. [1.1] does not contain any quantity that depends on the (macroscopic) physical characteristic of matter, such as its conductivity, dielectric constant, or magnetic permeability.

1.2. SURFACE, LINEAR, AND POINT CHARGES

Equation [1.6] gives the electric force $\mathbf{F}(p)$ at any point p, which can be situated anywhere outside or inside the integration volume V (Fig. 1.1B). When the point p is inside the volume, the integral must be handled carefully because the integrand obviously becomes singular when the point q approaches p and $L_{qp} \rightarrow 0$, but is possible to show that the result is always finite. Computation with Eq. [1.6] requires volume integration, which is often cumbersome. There are three special cases when it is possible to greatly simplify the calculations by assuming that the charges are concentrated in the vicinity of a surface, a curved line, or a point.

1.2.1. Surface Distribution of Charge

Suppose that a charge distribution with volume density $\delta(q)$ is confined within a layer of arbitrary shape whose thickness $h(q)$ is much smaller than the distance between the observation point p and any point q of the volume $V(q)$ occupied by the layer (Fig. 1.2A), that is,

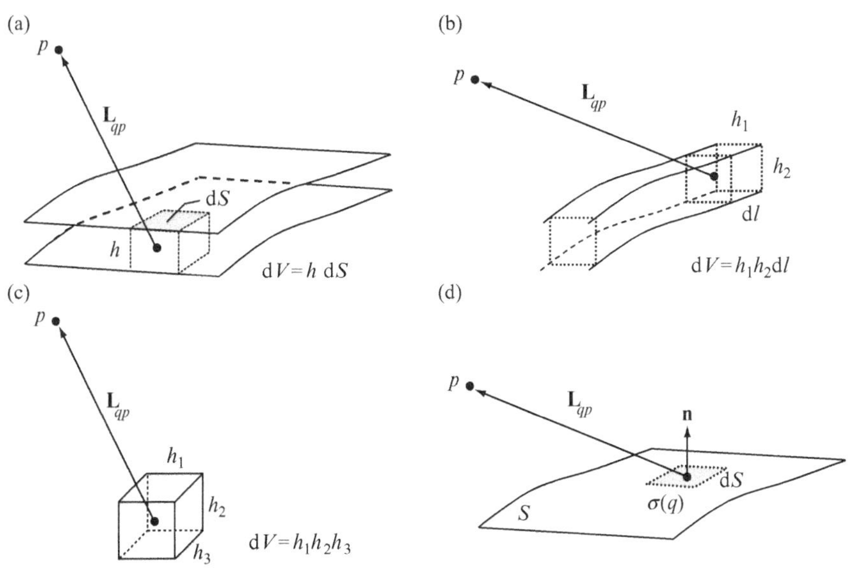

Figure 1.2 Definition of surface (A), linear (B), and point (C) charges. (D) Geometry of integral for the normal component of the electric field with a distribution of surface charge.

$$h(q) \ll L_{qp}. \qquad [1.8]$$

Assume, in addition, that the density $\delta(q)$ varies only in directions locally parallel to the layer's surface and not along its thickness (the direction locally perpendicular to the surface). An elementary volume surrounding a point in the layer can be written as $\mathrm{d}V(q) = h(q)\mathrm{d}S$, and then the total charge in the volume becomes

$$\mathrm{d}e(q) = \delta(q)h(q)\mathrm{d}S, \qquad [1.9]$$

where the point q is now only a function of two variables that define the surface. (If the surface does not fold back on itself, the coordinates of q could be represented, for example, by an expression such as $q = [x_q, y_q, z_q(x_q, y_q)]$.) In the simple case of a plane perpendicular to the z-axis, the expression would be $q = (x_q, y_q, z_q = d)$, where d is a constant. The force caused by the charge $\mathrm{d}e(p)$ on the surface becomes

$$\mathrm{d}\mathbf{F}(p) = \frac{\mathrm{d}e(p)}{4\pi\varepsilon_0}\frac{\delta(q)h(q)\mathrm{d}S}{L_{qp}^3}\mathbf{L}_{qp}, \qquad [1.10]$$

which can be rewritten as

$$\mathrm{d}\mathbf{F}(p) = \frac{\mathrm{d}e(p)}{4\pi\varepsilon_0}\frac{\sigma(q)\mathrm{d}S}{L_{qp}^3}\mathbf{L}_{qp}, \qquad [1.11]$$

where

$$\sigma(q) = \delta(q)h(q). \qquad [1.12]$$

By definition, the surface density $\sigma(q)$ is equal to the elementary charge per unit area $\mathrm{d}S$. From Eq. [1.9], it follows that decreasing the volume density and increasing the layer thickness in a way that keeps their product constant gives the same elementary charge in a volume near the surface; therefore, the force $\mathrm{d}\mathbf{F}(p)$ in Eq. [1.11] also does not change, provided of course that condition [1.8] holds—that is, the layer does not become too thick. The reverse limiting procedure leads to the concept of a pure surface charge density. In Eq. [1.12], let the thickness $h(q)$ decrease while the density $\delta(q)$ increases so that the surface density remains the same. As the thickness goes to zero, we arrive in the limit at a charge distribution confined to the surface element $\mathrm{d}S$. The charge located at this surface element

$$\mathrm{d}e = \sigma(q)\mathrm{d}S \qquad [1.13]$$

is by definition equal to the charge inside the volume $\mathrm{d}V(q) = h(q)\mathrm{d}S$. Note that in this limiting procedure, the surface density of charge $\sigma(q)$ always has a finite value, but the original volume density becomes singular, since (see Eq. [1.12])

$$\delta(q) \to \infty \text{ as } h(q) \to 0 \quad [1.14]$$

in order to keep their product constant. In reality, any macroscopic charge density is always finite, because charge occupies some finite volume. Surface charge is a mathematical concept introduced to simplify calculations when the thickness of the region in which charge resides is tiny compared to other dimensions of the problem. Applying the principle of superposition, Eq. [1.11] gives the generalized Coulomb's force law for surface distributions of charge:

$$\mathbf{F}(p) = \frac{de(p)}{4\pi\varepsilon_0}\int_S \frac{\sigma(q)}{L_{qp}^3}\mathbf{L}_{qp}\mathrm{d}S. \quad [1.15]$$

Integration over a surface S (a double integral) is generally easier than integration over a volume (a triple integral).

1.2.2. Linear Distribution of Charge

Assume next that charges are confined to lie along a curve, that is, inside a volume whose cross-sectional dimensions are much smaller that the distance between any point of the volume and an observation point p (Fig. 1.2B):

$$h_1(q) \ll L_{qp} \quad \text{and} \quad h_2(q) \ll L_{qp}. \quad [1.16]$$

Let $\mathrm{d}l$ be the increment of distance along the curve; then the elementary volume containing the charges can be written as

$$\mathrm{d}V(q) = h_1(q)h_2(q)\mathrm{d}l,$$

where h_1 and h_2 are the cross-sectional dimensions that depend on a single variable that defines the curve. The elementary charge in the volume becomes

$$de(q) = \delta(q)h_1(q)h_2(q)\mathrm{d}l, \quad [1.17]$$

provided that charge density is constant in the cross section. Writing

$$de(q) = \lambda(q)\mathrm{d}l,$$

where

$$\lambda(q) = \delta(q)h_1(q)h_2(q) \tag{1.18}$$

defines a linear density of charge, then we have in place of Eq. [1.1]

$$d\mathbf{F}(p) = \frac{de(p)}{4\pi\varepsilon_0}\frac{\lambda(q)dl}{L_{qp}^3}\mathbf{L}_{qp}.$$

The force caused by all charges in the volume V is therefore equal to

$$\mathbf{F}(p) = \frac{de(p)}{4\pi\varepsilon_0}\int_L \frac{\lambda(q)dl}{L_{qp}^3}\mathbf{L}_{qp}. \tag{1.19}$$

Clearly, the definition of a linear distribution of charge resembles that of a surface distribution. Equation [1.18] shows that in the limit as the cross-sectional dimensions shrink, the volume charge density diverges:

$$h_1(q) \to 0, \quad h_2(q) \to 0, \quad \delta(q) \to \infty.$$

Linear charge is thus another mathematical concept introduced to simplify computation, and when charge is confined to the vicinity of a curve, a triple integral becomes a line integral.

1.2.3. Point Charge

A final limiting process leads back to Coulomb's force law. Starting from Eq. [1.1] and writing

$$d\mathbf{F}(p) = \frac{de(p)}{4\pi\varepsilon_0}\frac{\delta(q)dV}{L_{qp}^3}\mathbf{L}_{qp}, \tag{1.20}$$

we now imagine that charge is uniformly distributed in a volume dV whose dimensions are much smaller than the distance L_{qp} (Fig. 1.2C). Again, let the linear dimensions of the volume decrease while the volume charge density increases so that the product

$$e(q) = \delta(q)dV \tag{1.21}$$

remains the same. In the limit, the volume charge density diverges,

$$dV \to 0 \text{ and } \delta \to \infty,$$

but we arrive at the mathematical concept of a "point charge"—charge that is localized at a point but can have an arbitrary value. It is of course physically impossible to localize charge into a region of vanishing size, but mathematically the limiting process can be defined rigorously. Substitution of Eq. [1.21] into Eq. [1.20] gives Coulomb's law (Eq. [1.1]). Although surface, linear, and point charges are characterized by divergent volume densities, the degree of singularity is different in each case. It is greatest for a point charge, diverging as the inverse cube of the linear dimension, $\delta \rightarrow h^{-3}$; whereas for line charge, the divergence goes as an inverse square h^{-2}, and for surface charges, as h^{-1}. This will translate into different behaviors of the force field in the neighborhood of different kinds of charge distribution. The generalization of Coulomb's force law allows for all four types of charge distribution (volume, surface, linear, and point):

$$\mathbf{F}(p) = \frac{\mathrm{d}e(p)}{4\pi\varepsilon_0}\left[\int_V \frac{\delta(q)}{L_{qp}^3}\mathbf{L}_{qp}\mathrm{d}V + \int_S \frac{\sigma(q)}{L_{qp}^3}\mathbf{L}_{qp}\mathrm{d}S + \int_L \frac{\lambda(q)}{L_{qp}^3}\mathbf{L}_{qp}\mathrm{d}l + \sum \frac{e(q)}{L_{qp}^3}\mathbf{L}_{qp}\right], \quad [1.22]$$

where the charge density in each term is assumed to be finite everywhere. Before moving on to applications of this law, it is worth recalling that the macroscopic theory of electric force and electric fields that follows from Coulomb's law was developed long before it was discovered that all electric charge comes in discrete units associated with subatomic "elementary" particles, such as electrons and protons. The limiting processes that lead naturally to concepts of surface, linear, and point charge are mathematical refinements of the classical macroscopic theory—these concepts are self-consistent, but independent of the microscopic physics. In reality, as physical dimensions shrink to the subatomic realm, nonclassical (quantum mechanical) effects arise that are simply not accounted for in the classical macroscopic picture.

1.3. THE ELECTRIC FIELD

We now introduce the concept of an "electric field." Like the force field $\mathbf{F}(p)$, the electric field $\mathbf{E}(p)$ is a vector attached to a point p, defined as the ratio between the electric force and the magnitude of an elementary "test charge" at p:

$$\mathbf{E}(p) = \frac{\mathbf{F}(p)}{\mathrm{d}e(p)} \quad \text{or} \quad \mathbf{F}(p) = \mathrm{d}e(p)\mathbf{E}(p). \quad [1.23]$$

The practical meaning of "test charge" is as follows: the definition of the electric field $\mathbf{E}(p)$ by Eq. [1.23] actually envisages a limiting process in which the force $\mathbf{F}(p)$ acting on charge $\mathrm{d}e(p)$ is normalized by the magnitude of the charge in the limit as the magnitude goes to zero. This limiting process implies that the field of the test charge itself becomes negligible and allows us to imagine that an electric field exists at any point p in space even if a charge is absent at this point. Clearly, that is not the case for the force of interaction given by Coulomb's law (Eq. [1.1]), which requires two distinct charges at points p and q and is essentially symmetric in the charges: if either one is removed, the force vanishes. In contrast, an electric field defined by Eq. [1.23] exists everywhere in space surrounding any distribution of charge. Charge itself can be viewed as the source of the electric field. The ideas leading from the "physical" force of interaction between two charges to the "abstract" electric field pervading all of space—a concept ultimately due to the English physicist Michael Faraday (1791–1867) in the mid-nineteenth century—would quickly lead to the full theory of the electromagnetic field, first elaborated by the Scottish physicist James Clerk Maxwell (1831–1879) and embodied in what are now called Maxwell's equations.

Under the influence of the electric field, a positive charge will move in the direction of the electric field, whereas a negative charge will move in the opposite direction. The general expression for the electric field generated by an arbitrary distribution of charge follows from Eqs. [1.22] and [1.23]:

$$\mathbf{E}(p) = \frac{1}{4\pi\varepsilon_0}\left[\int_V \frac{\delta(q)}{L_{qp}^3}\mathbf{L}_{qp}\mathrm{d}V + \int_S \frac{\sigma(q)}{L_{qp}^3}\mathbf{L}_{qp}\mathrm{d}S + \int_L \frac{\lambda(q)}{L_{qp}^3}\mathbf{L}_{qp}\mathrm{d}l + \sum \frac{e(q)}{L_{qp}^3}\mathbf{L}_{qp}\right]. \qquad [1.24]$$

Equation [1.24] can be taken as the full generalization of Coulomb's law for constant distribution of electric charges, that is, for charges that are fixed in space and do not change their magnitude in time. Moreover, under certain conditions, this expression computes the field to a very good approximation even when the electromagnetic field varies with time. This feature provides considerable physical insight into the principles and limitations of many different methods of electrical and electromagnetic surveying, including magnetotelluric sounding methods, fixed frequency and transient methods of land and marine geophysics, and logging methods for measuring the electrical properties of oil reservoirs. Several important observations follow from Eq. [1.24]:

(1) Electric charges are the ultimate sources of a constant electric field.

(2) The electric field generated by a given distribution of charge does not depend on the physical properties of the medium in which the charge resides. In other words, the electric field due to these charges remains the same whether the charge exists in free space or in a nonuniform conducting and polarizable medium. This remarkable feature follows from Eq. [1.24], because only the absolute physical constant ε_0 appears in this equation.

(3) Coulomb's law was not derived from other equations, and in this sense, it is the fundamental physical law that allows us to describe behavior of the stationary electric field. Consequently, the basic equations of the field $\mathbf{E}(p)$ will be derived from Coulomb's law as well as the principle of superposition.

(4) If a distribution of charges is given in advance, then calculation of the field (forward problem) would naturally be performed with Eq. [1.24], requiring only a simple numerical integration. Unfortunately, in nearly all practical applications, it is impossible to know the charge distribution in advance. Later, we will show how the electric fields measured in different geophysical methods can be understood as coming from net charge that appears at boundaries between media with different electrical properties.

We conclude this section with a brief discussion of units. The dimensions of electric field strength are, from its definition in Eq. [1.23], clearly those of force per unit charge, or newtons per coulomb. Recall that the newton is a derived mechanical unit with dimensions of mass times acceleration (from Newton's second law of motion), or kg m s^{-2} in S.I. units. Nevertheless, for reasons that will be clear later in the definition of the electric potential, the traditional S.I. unit of electric field strength uses another derived unit called the "volt" (V) and is given in volts per meter (V m^{-1}). The most commonly used subunits of the electric field are the following:

$$1\,\text{V m}^{-1} = 10^3\,\text{mV m}^{-1} = 10^6\,\mu\text{V m}^{-1} = 10^9\,\text{nV m}^{-1},$$

where, with the usual S.I. abbreviations, mV stands for "millivolt"; μV, for "microvolt"; and nV, for "nanovolt." Modern sensors can easily measure electric fields at the nanovolt level. It is illuminating to equate the two units of electric field, V m^{-1} and N C^{-1}, and arrive at the dimensions of volt as Nm C^{-1}, or force times distance per coulomb. Since force times distance has the units of energy, this implies that the volt is actually a measure of electrical energy per unit charge. We will see this again later in developing the concept of electric potential and voltage.

1.4. THE ELECTRIC FIELD CAUSED BY SURFACE CHARGE

We now consider a simple example to illustrate features of the field behavior caused by surface charges that appear at interfaces between different media.

1.4.1. Normal Component of the Field due to Charge on a Plane

Let charge be distributed on a finite planar surface S with the density σ (Fig. 1.2D) and consider the vector

$$\mathrm{d}\mathbf{S} = \mathbf{n}\,\mathrm{d}S,$$

where at each point $\mathbf{n}$ is a unit vector normal to the plane and directed from the back side (1) toward the front side (2). Consider the normal component of the electric field, which is directed along $\mathbf{n}$. From Eq. [1.24], each elementary charge $\sigma(q)\mathrm{d}S$ creates a field given by

$$\mathrm{d}\mathbf{E}(p) = \frac{1}{4\pi\varepsilon_0}\frac{\sigma(q)\mathrm{d}S}{L_{qp}^3}\mathbf{L}_{qp} \qquad [1.25]$$

and its normal component is

$$\mathrm{d}E_\mathrm{n}(p) = \frac{1}{4\pi\varepsilon_0}\frac{\sigma(q)\mathrm{d}S\,\mathbf{L}_{qp}\cdot\mathbf{n}}{L_{qp}^3} = \frac{\sigma(q)}{4\pi\varepsilon_0}\frac{\mathrm{d}\mathbf{S}\cdot\mathbf{L}_{qp}}{L_{qp}^3}. \qquad [1.26]$$

Since $\mathbf{L}_{qp} = -\mathbf{L}_{pq}$ we have

$$\mathrm{d}E_\mathrm{n}(p) = -\frac{\sigma(q)}{4\pi\varepsilon_0}\frac{\mathrm{d}\mathbf{S}\cdot\mathbf{L}_{pq}}{L_{pq}^3}. \qquad [1.27]$$

The ratio

$$\frac{\mathrm{d}\mathbf{S}\cdot\mathbf{L}_{pq}}{L_{qp}^3} \qquad [1.27a]$$

is actually the solid angle $\mathrm{d}\omega(p)$ subtended by the surface element $\mathrm{d}S$ viewed from the point p. Thus

$$\mathrm{d}E_\mathrm{n}(p) = -\frac{\sigma(q)}{4\pi\varepsilon_0}\mathrm{d}\omega(p). \qquad [1.28]$$

Note that according to this definition, the solid angle is positive when the point p is located "in back of" the plane. Applying the principle of superposition, we obtain for the normal component of the field caused by all surface charges:

$$E_{\mathrm{n}}(p) = -\frac{1}{4\pi\varepsilon_0}\int_S \sigma(q)\mathrm{d}\omega. \qquad [1.29]$$

In particular, if charge is distributed uniformly on the surface (σ = constant), we have

$$E_{\mathrm{n}}(p) = -\frac{\sigma}{4\pi\varepsilon_0}\omega(p), \qquad [1.30]$$

where $\omega(p)$ is the total solid angle subtended by the surface S, viewed from p. As the point p recedes from the planar surface S, the magnitude of the solid angle subtended by S decreases, and, correspondingly, the normal component of the field decreases in magnitude. As the point p approaches the plane S, the solid angle increases and, in the limit as p touches the surface, becomes equal to -2π and $+2\pi$; when the observation point p is located on either the front side (2) or back side (1) of the surface, respectively. Therefore, we have the following expressions for the normal component of the field on either side of the surface:

$$E_{\mathrm{n}}^{(2)} = \frac{\sigma}{2\varepsilon_0} \text{ and } E_{\mathrm{n}}^{(1)} = -\frac{\sigma}{2\varepsilon_0}. \qquad [1.31]$$

The normal component of the field $\mathbf{E}$ is therefore discontinuous across the surface S. This makes sense physically. For example, if positive charge is distributed over the plane, we expect intuitively that the electric field will point away from the plane as the point p approaches it from either side. To examine the field behavior in more detail, we can express E_{n} as the sum of two terms:

$$E_{\mathrm{n}}(p) = E_{\mathrm{n}}^{q}(p) + E_{\mathrm{n}}^{S-q}(p), \qquad [1.32]$$

where $E_{\mathrm{n}}^{q}(p)$ is the normal component of electric field caused by an elementary charge $\sigma(q)\mathrm{d}S$, located in the vicinity of the point q, and $E_{\mathrm{n}}^{S-q}(p)$ is the part of the normal component contributed by all of the other surface charges. It is clear that

$$E_{\mathrm{n}}^{S-q}(p) = -\frac{\sigma}{4\pi\varepsilon_0}\int_{S-q}\mathrm{d}\omega = -\frac{\sigma}{4\pi\varepsilon_0}\omega^{S-q}(p).$$

Here, $\omega^{S-q}(p)$ is the solid angle subtended by the plane surface S minus the surface element $dS(q)$. Letting the point p approach this area, the solid angle $\omega^{S-q}(p)$ tends to zero. To understand this result, recall the definition of the solid angle (Eq. [1.27a]). As the point p approaches the surface, the angle between the unit normal $\mathbf{n}$ at point q on the surface and $\mathbf{L}_{pq}$, the line connecting p to q, goes to $\pm\pi/2$ except at the point directly in front of p. At the same time, the total solid angle subtended by the surface element dS at the point q that p approaches tends to $\pm 2\pi$, because even a small surface viewed infinitely closely subtends the entire field of view:

$$\omega^{q}(p) \rightarrow \pm 2\pi \text{ if } p \rightarrow q.$$

Thus, as the surface is approached from either side, the normal component of the field is determined only by the elementary charge located at the observation point:

$$E_{\rm n}^{(2)}(q) = \frac{\sigma(q)}{2\varepsilon_0} \text{ and } E_{\rm n}^{(1)}(q) = -\frac{\sigma(q)}{2\varepsilon_0}. \qquad [1.33]$$

Thus,

$$E_{\rm n}^{(2)}(q) - E_{\rm n}^{(1)}(q) = \frac{\sigma(q)}{\varepsilon_0}, \qquad [1.34]$$

and the discontinuity of the normal component of the field is directly proportional to surface charge density.

1.4.2. Normal Component of the Field Caused by Charge on a General Surface

We now generalize the preceding results to an arbitrary surface. Starting again from Eq. [1.29] and re-expressing the integrand in terms of solid angles, we arrive at the following expressions for the normal component of the field on either side of a surface:

$$E_{\rm n}^{(2)}(p) = \frac{\sigma(q)}{2\varepsilon_0} + E_{\rm n}^{(2,S-q)}(p),\ E_{\rm n}^{(1)}(p) = -\frac{\sigma(q)}{2\varepsilon_0} + E_{\rm n}^{(1,S-p)}(p),\ \text{if } p \rightarrow q. \qquad [1.35]$$

In the case of arbitrary surface, however, it is no longer true that the total solid angle involved in the definition of the normal component $E_{\rm n}^{(S-p)}$ is necessarily zero at the point p. We can nevertheless demonstrate that this part of the field is a continuous function of p as the observation point passes through the surface. This follows from the fact that the charges generating this field are located at some distance from the point q and therefore

$$E_{\mathrm{n}}^{(1,S-q)} = E_{\mathrm{n}}^{(2,S-q)} = E_{\mathrm{n}}^{(S-q)}. \qquad [1.36]$$

Thus, Eqs. [1.35] become

$$E_{\mathrm{n}}^{(2)}(p) = \frac{\sigma(q)}{2\varepsilon_0} + E_{\mathrm{n}}^{(S-p)}(p),\ E_{\mathrm{n}}^{(1)}(p) = -\frac{\sigma(q)}{2\varepsilon_0} + E_{\mathrm{n}}^{(S-p)}(q),\ \text{if } p \to q. \qquad [1.37]$$

This means that, in general, as in the case of planar surface, a discontinuity of the normal component is described by Eq. [1.34].

1.4.3. Tangential Component of Field Near Surface Charges

The above analysis can also be applied to the tangential components of the field. Again, split the surface charge into two parts: one located on a small circular disk around the point q; the second formed by all other charges on the surface. Let the observation point p approach the point q located at the center of the disk. If surface charge is a continuous function on the surface, then the charge on the disk becomes uniform as the size of the disk shrinks. Each element of charge on the disk creates only a tangential component of the electric field (in the limit as the observation point approaches the surface). But because of circular symmetry, there are always two elements of charge on the disk that generate fields with the same magnitude and opposite directions. The total tangential field caused by the disk charge is therefore equal to zero. Now, consider the tangential component of the field near the surface caused by charges outside the small circular disk. Since the observation point hits the center of the disk as it approaches the surface, it is always a finite distance from charges outside the disk. We thus arrive at a fundamental property of electric fields: the tangential component of the electric field is a continuous function across an arbitrary surface (see section 1.12.7 for an exception to this rule):

$$\mathbf{E}_{\mathrm{t}}^{(2)}(p) = \mathbf{E}_{\mathrm{t}}^{(1)}(p). \qquad [1.38]$$

The two equations,

$$E_{\mathrm{n}}^{(2)}(q) - E_{\mathrm{n}}^{(1)}(q) = \frac{\sigma(q)}{\varepsilon_0}, \quad \mathbf{E}_{\mathrm{t}}^{(2)} = \mathbf{E}_{\mathrm{t}}^{(1)}, \qquad [1.39]$$

in fact, describe the behavior of arbitrary electromagnetic fields, including fields that vary in time. This behavior is important, because applications of electric methods in geophysics nearly always generate surface distributions of electric charge.

1.5. ELECTROSTATIC INDUCTION

We consider next a phenomenon that arises whenever a conducting body is placed in the presence of an electric field. Suppose that a conductor of arbitrary shape and size is placed in an insulating medium ("insulator") and subject to an electric field $\mathbf{E}_0$ whose sources are in the insulator. (A conductor is a medium that allows charge to flow freely in its interior or along its surface; an insulator is a medium that does not allow the flow of charge.) Under an action of the external field $\mathbf{E}_0$, positive and negative charges residing inside the conductor move in opposite directions, eventually reaching the boundary of the interface between the conductor and insulator, where they are "trapped" and accumulate. The net electric charge that appears on the conductor's surface (Fig. 1.3) in turn creates a secondary electric field $\mathbf{E}_s$, tending to oppose and cancel the primary field $\mathbf{E}_0$ inside the conductor. The total field at each point in the conductor is

$$\mathbf{E}_i = \mathbf{E}_0 + \mathbf{E}_s,$$

and it determines any further motion of charges. Eventually equilibrium is reached: the charges on the surface of the conductor distribute themselves in such a way as to completely cancel the applied field $\mathbf{E}_0$, making the total

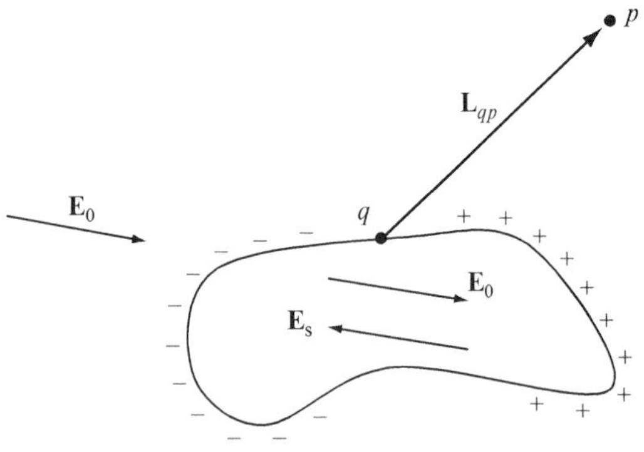

Figure 1.3 Electrostatic induction.

electric field at each point inside the conductor disappear and stopping any further motion of charge:

$$\mathbf{E}_{\mathrm{i}}(q) = 0. \qquad [1.40]$$

This phenomenon is called electrostatic induction. Two further comments are appropriate at this point:

(1) Our description of the process by which equilibrium is established—which assumes, for example, that a constant external electric field is suddenly applied to the conductor—is only approximate. In reality, "switching on" of the external electric field $\mathbf{E}_0$, as well as the accumulation and motion of surface charges, involves the propagation of electromagnetic fields within both the insulator and conductor. It is possible to show, however, that this transient stage lasts for a very short time, and after equilibrium is reached, the constant electric field is governed by Coulomb's law.

(2) The phenomenon of the electrostatic induction takes place whenever a conducting medium—even a very weak conductor—is surrounded by an insulator and the sources of the applied field $\mathbf{E}_0$ are located outside the conductor. The body can consist of any kind of conductive material such as metal, water, or porous rock; the strength of its electrical "conductivity" only plays a role in determining the time required for equilibrium to be reached, but does not change the final result or configuration of charges.

1.5.1. Secondary Electric Field Caused by Surface Charges

The secondary electric field caused by surface charges resulting from electrostatic induction is given by

$$\mathbf{E}_{\mathrm{s}}(p) = \frac{1}{4\pi\varepsilon_0}\oint_S \frac{\sigma(q)\mathrm{d}S}{L_{qp}^3}\mathbf{L}_{qp}, \qquad [1.41]$$

where $\sigma(q)$ is the surface density of charges and an integration is performed over the conductor surface. Consequently, the condition [1.40] has the form

$$\mathbf{E}_0(p) + \frac{1}{4\pi\varepsilon_0}\oint_S \frac{\sigma(q)\mathrm{d}S}{L_{qp}^3}\mathbf{L}_{qp} = 0. \qquad [1.42]$$

Here $\mathbf{E}_0(p)$ is the primary field at any point p inside the conductor, caused by charges situated in the insulator, and q is a point of the surface S. For instance, if there is only elementary charge e in the insulator located at the point r, its electric field at any point inside the conductor is

$$\mathbf{E}_0(p) = \frac{e}{4\pi\varepsilon_0} \frac{\mathbf{L}_{rp}}{L_{rp}^3}.$$

This is a consequence of the principle of superposition; as stressed before, the field caused by a specified distribution of charges does not depend on the presence of other charges—here, charges induced on the surface of the conductor. The field $\mathbf{E}_0(p)$ is, however, no longer the total electric field, which must include the field of charges that appear on the conductor. At the same time, new charge cannot be created by the process of electrostatic induction. For example, if the conductor was initially neutral (zero net charge), we still have

$$\sum e_s = 0. \qquad [1.43]$$

What the presence of the field $\mathbf{E}_0$ induces is the redistribution of positive and negative charges, so that they occupy different parts of the conductor surface, creating a local imbalance of charge—and the secondary field $\mathbf{E}_s$—even though the total charge on the conductor remains zero. The secondary field inside the conductor is known, because at each point we must have (according to the principle of electrostatic induction)

$$\mathbf{E}_s(p) = -\mathbf{E}_0(p).$$

The secondary field outside the conductor depends on the conductor's shape and dimensions, as well as dielectric properties of the insulating medium in which it resides.

1.5.2. Determination of the Field of Surface Charges

If the density of the surface charges $\sigma(q)$ is known, calculation of the secondary field $\mathbf{E}_s(p)$ outside the conductor can be easily performed from Eq. [1.41]. But the distribution of charges caused by electrostatic induction is not known in advance, because it depends on the final (equilibrium) state of the field. This "closed circle" problem—the charges determine the field, which in turn determines the configuration of charges—is present in all problems involving electric fields, and vividly demonstrates the fundamental difficulty in calculating the electric field from Eq. [1.41].

Let us illustrate the problem more directly. According to Eqs. [1.39] and [1.41], the secondary field outside a conductor and the surface charge on the conductor are related by the following equations:

$$\mathbf{E}_{\mathrm{s}}^{\mathrm{e}}(p) = \frac{1}{4\pi\varepsilon_0}\oint_S \frac{\sigma(q)}{L_{qp}^3}\mathbf{L}_{qp}\mathrm{d}S, \quad \sigma(q) = \varepsilon_0 E_{\mathrm{n}}^{\mathrm{e}}(p), \quad \text{as } p \to q, \qquad [1.44]$$

where the point p is located in the insulator and the limit is taken as it approaches the surface of the conductor. $E_{\mathrm{n}}^{\mathrm{e}}$ is the normal component of the electric field in the insulator as p touches the surface. Thus, in order to find the electric field from Coulomb's law, the function $\sigma(q)$ has to be given, but it can be calculated only if the field is known. We now develop a general approach that determines the charges and the field $\mathbf{E}_{\mathrm{s}}$, even though both are initially unknown. The key step is the derivation of a self-consistent "integral equation" that can be solved for the surface charge density $\sigma(q)$. Once the charge density is known, the field can be calculated from Eq. [1.41]. We proceed by first writing $E_{\mathrm{n}}^{\mathrm{e}}(p)$ in Eq. [1.44] as a sum of three terms:

$$E_{\mathrm{n}}^{\mathrm{e}}(p) = E_{\mathrm{n}}^{0}(p) + E_{\mathrm{n}}^{p}(p) + E_{\mathrm{n}}^{S-p}(p). \qquad [1.45]$$

Recall that this equation will be taken in the limit as the point of observation p, initially in the insulating medium, approaches a point q on the conductor. Here E_{n}^{0} is the normal component of the primary field at the point p, E_{n}^{p} is the normal component of the field caused by the elementary surface charge $\sigma(q)\mathrm{d}S$ situated in a small patch of surface at the point q (that p is approaching), and E_{n}^{S-p} is the normal component contributed by the rest of the surface charges. From Eq. [1.33], we know that, in the limit,

$$E_{\mathrm{n}}^{p}(p) = \frac{\sigma(q)}{2\varepsilon_0}, \qquad [1.46]$$

where $\mathbf{n}$ is the unit vector directed from the conductor surface into the insulating medium. In addition, by the generalized Coulomb's law, we have

$$\mathbf{E}^{S-p}(p) = \frac{1}{4\pi\varepsilon_0}\oint_{S-p} \frac{\sigma(q)}{L_{qp}^3}\mathbf{L}_{qp}\mathrm{d}S \quad \text{if } q \neq p.$$

Taking the normal component of this equation gives

$$E_{\mathrm{n}}^{S-p}(p) = \frac{1}{4\pi\varepsilon_0}\oint_{S-p} \sigma(q)\frac{(\mathbf{L}_{qp}\cdot\mathbf{n})}{L_{qp}^3}\mathrm{d}S \quad \text{if } q \neq p, \qquad [1.47]$$

where $\mathbf{n}$ is the unit vector at the point p of the surface S and $\mathbf{L}_{qp}\cdot\mathbf{n}$ is the dot product at the point p near the surface S:

$$\mathbf{L}_{qp}\cdot\mathbf{n} = L_{qp}\cos(\mathbf{L}_{qp},\mathbf{n}).$$

Substituting for all terms of Eq. [1.45] and taking the limit gives

$$\frac{\sigma(p)}{\varepsilon_0} = E_{\mathrm{n}}^0 + \frac{\sigma(p)}{2\varepsilon_0} + \frac{1}{4\pi\varepsilon_0}\oint_{S-p}\sigma(q)K(p,q)\mathrm{d}S,$$

or, rearranging

$$\sigma(p) = 2\varepsilon_0 E_{\mathrm{n}}^0 + \frac{1}{2\pi}\oint_{S-p}\sigma(q)K(p,q)\mathrm{d}S, \qquad [1.48]$$

where p is now a point of the surface S, and the term $E_{\mathrm{n}}^0(p)$ and

$$K(p,q) = \frac{\mathbf{L}_{qp}\cdot\mathbf{n}(p)}{L_{qp}^3} = \frac{\cos(\mathbf{L}_{qp},\mathbf{n})}{L_{qp}^2}$$

are known functions. Equation [1.48] is called an integral equation because the unknown function, the surface charge density $\sigma(q)$, appears under the integral. In the mathematical terminology used for such equations, it is a Fredholm integral equation of the second kind, because the unknown function occurs both outside and inside the integral. As with differential equations, there are analytical methods for solving Eq. [1.48], but these are limited to special configurations. We can, however, obtain a numerical solution in nearly all cases by representing the surface of the conductor by a collection of small cells within which the charge density is assumed to be constant. The integral Equation [1.48] can thus be rewritten approximately as

$$\sigma(p) = 2\varepsilon_0 E_{\mathrm{n}}^0(p) + \frac{1}{2\pi}\sum_{n=1}^{N}\sigma(q)K(p,q)\Delta S \quad \text{if } q \neq p. \qquad [1.49]$$

Writing this system once for every cell, we obtain a system of N linear equations with N unknowns; its solution gives the distribution of charge $\sigma(q)$. Applying Eq. [1.41] gives the secondary electric field outside the conductor. This example shows how the determination of the field becomes much more complicated when the charge distribution is not known in advance.

1.5.3. Role of Electrostatic Induction in Electrical Methods of Geophysics

The principle of electrostatic induction is important to a full understanding of electrical methods of geophysics. Consider, for example, an Earth model consisting of a sequence of conducting formations with different electric properties. Free electrical charges are always present above the Earth's surface; they arise from atmospheric processes and other reasons, but their precise distribution is unknown. If it were necessary to account for the field of these charges in electrical methods of geophysics, the situation would be hopeless. But because of electrostatic induction, atmospheric charge induces a charge on the Earth's surface, and this surface charge in turn creates a field inside the Earth that exactly compensates the field caused by atmospheric charge. In other words, because the Earth is a conductor, unknown constant charges situated outside of it (in the atmosphere or anywhere in space) have no effect on the stationary electric field within the Earth. A constant electric field in a conducting medium can be caused only by charges located inside this medium or at its surface. Fortunately, because of electrostatic induction there is no need to take into account the influence of unknown sources in the atmosphere, and a constant electric field can be used to survey inside the Earth. Charges induced on the Earth's surface do, however, create an electric field in the atmosphere that does not compensate the primary field of free atmospheric charge. For instance, the constant atmospheric electric field near the Earth's surface has (only) a vertical component equal to about 100 V m^{-1}. This is much greater than the horizontal electric field created and measured in electrical methods.

1.6. VOLTAGE OF THE ELECTRIC FIELD

We now describe other concepts and methods that enable determination of the electric field, even though some charges may initially be unknown. Starting again from Coulomb's law, we develop two remarkable features of the electric field in any medium and introduce the concept of voltage.

1.6.1. Voltage of the Field E

Consider an elementary displacement dl (Fig. 1.4A) and assume that the field $\mathbf{E}(q)$ is constant at its points. The dot product

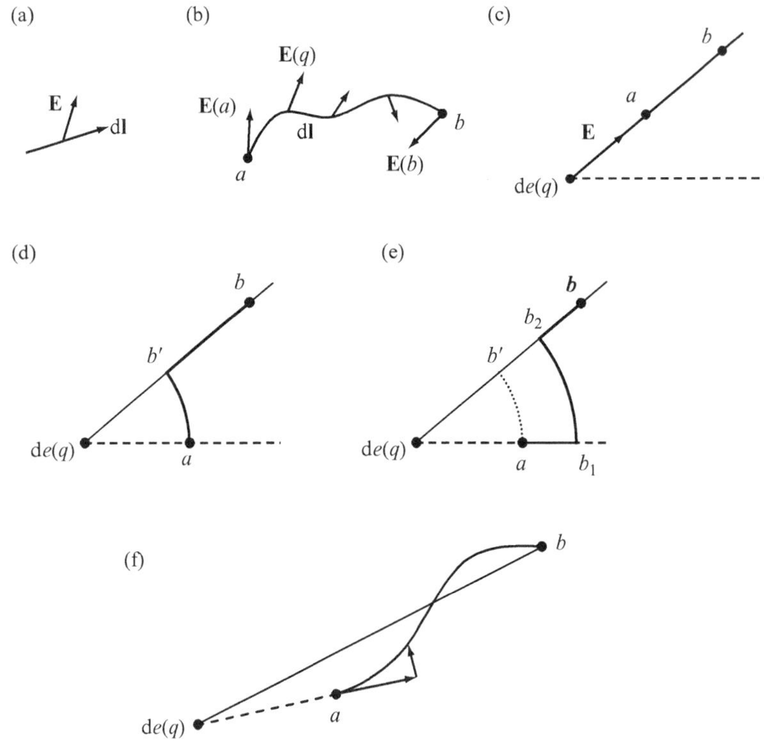

Figure 1.4 Path independence of voltage.

$$dV = \mathbf{E} \cdot d\mathbf{l} = E\,dl \cos(\mathbf{E}, d\mathbf{l}) \quad [1.50]$$

defines the voltage of the field **E** along the small path dl. Clearly, the voltage can be positive, negative, or zero depending on the angle between the field and the path. The voltage between two points a and b, caused by the electric field along an arbitrary path L (Fig. 1.4B) is defined to be

$$V = \int_a^b \mathbf{E} \cdot d\mathbf{l} = \int_a^b E(q) \cos(\mathbf{E}, d\mathbf{l})\, dl. \quad [1.51]$$

Here $(\mathbf{E}, d\mathbf{l})$ is the angle between the electric field and the displacement vector **dl** at each point of the path. The product **E**·**dl** is actually the work performed by the electric field in moving a unit positive charge along the path **dl**. This product has dimension of work per unit charge, which defines the unit "volt":

$$1\,\text{V} = 10^3\,\text{mV} = 10^6\,\mu\text{V} = 10^9\,\text{nV}.$$

The integral in Eq. [1.51] represents the work in carrying a unit positive charge between two end points *a*, *b* of some path *L*. The familiar device called a "voltmeter" measures this work and, according to Eq. [1.51], certainly is a very "clever" instrument which, in principle, divides a path into many elements, calculates elementary work along each, and performs the integral.

1.6.2. Independence of Voltage on a Path of Integration

Starting from Coulomb's law, we now prove a remarkable feature of the electric field voltage caused by charges—namely, that this voltage between any two points is independent of the path taken between the points. At the beginning, we assume that the source of the field is a single elementary charge d*e*; then its electric field is

$$\mathbf{E}(p) = \frac{1}{4\pi\varepsilon_0}\frac{\mathrm{d}e}{L_{qp}^3}\mathbf{L}_{qp}.$$

Consider first the simple case when the end points *a* and *b* of the path are situated along the same radius vector $\mathbf{L}_{qp}$ extending outward from the charge. The path of integration in Eq. [1.51] is then along this radius (Fig. 1.4C) and the voltage is easily calculated as a simple integral,

$$V = \int_a^b \mathbf{E}\cdot\mathrm{d}\mathbf{l} = \frac{\mathrm{d}e}{4\pi\varepsilon_0}\int_a^b \frac{\mathrm{d}\mathbf{l}\cdot\mathbf{L}_{qp}}{L_{qp}^3} = \frac{\mathrm{d}e}{4\pi\varepsilon_0}\int_a^b \frac{\mathrm{d}l}{L_{qp}^2},$$

because

$$\mathrm{d}\mathbf{l}\cdot\mathbf{L}_{qp} = \mathrm{d}l\,L_{qp}\cos 0 = L_{qp}\mathrm{d}l.$$

Performing the integral gives

$$V = \frac{\mathrm{d}e}{4\pi\varepsilon_0}\left(\frac{1}{L_{qa}} - \frac{1}{L_{qb}}\right). \qquad [1.52]$$

Consider next the case when points *a* and *b* lie at the end of different radius vectors $\mathbf{L}_{qa}$ and $\mathbf{L}_{qb}$ (Fig. 1.4D), and assume that point *b* is farther away from the point charge than *a*. To compute the integral, we choose a path L_1 that consists of two pieces. The first piece is a circular arc ab' that extends from the tip of $\mathbf{L}_{qa}$ until it intersects $\mathbf{L}_{qb}$; the second piece proceeds along the radius vector $\mathbf{L}_{qb}$ until it reaches *b*. For this path, the voltage is

$$V = \frac{\mathrm{d}e(q)}{4\pi\varepsilon_0}\left[\int_{\mathrm{arc}\,ab'} \frac{\mathrm{d}\mathbf{l}\cdot\mathbf{L}_{qp}}{L_{qp}^3} + \int_{b'}^{b} \frac{\mathrm{d}\mathbf{l}\cdot\mathbf{L}_{qp}}{L_{qp}^3}\right].$$

The first term, the integral along the arc ab', vanishes because the dot product $\mathbf{dl}\cdot\mathbf{L}_{qp}$ is zero at everyone point along the path (the radius vector is perpendicular to a circular arc). Thus, the voltage between points a and b is again equal to

$$V = \frac{de(q)}{4\pi\varepsilon_0}\left(\frac{1}{L_{qa}} - \frac{1}{L_{qb'}}\right) = \frac{de(q)}{4\pi\varepsilon_0}\left(\frac{1}{L_{qa}} - \frac{1}{L_{qb}}\right) \qquad [1.53]$$

because $L_{qb} = L_{qb'}$. In other words, the voltage depends only on the radial distances between the two points and the charge. Consider now a slightly more complicated path ab_1b_2b (Fig. 1.4E) that first proceeds outward along the radius vector $\mathbf{L}_{qa}$ to the point b_1, then swings along a circular arc from b_1 to the point b_2 on the radius vector $\mathbf{L}_{qb}$, and then proceeds outward along this vector to point b. The integral along the circular arc is zero, as before, and we have

$$V = \frac{de(q)}{4\pi\varepsilon_0}\left[\left(\frac{1}{L_{qa}} - \frac{1}{L_{qb_1}}\right) + \left(\frac{1}{L_{qb_1}} - \frac{1}{L_{qb}}\right)\right] = \frac{de(q)}{4\pi\varepsilon_0}\left(\frac{1}{L_{qa}} - \frac{1}{L_{qb}}\right) \qquad [1.54]$$

because points b_1 and b_2 are located at the same radial distance from the charge. We can now generalize this result to an arbitrary path between the points a and b (Fig. 1.4F). Any element $\mathbf{dl}$ of this path can be represented as

$$\mathbf{dl} = \mathbf{dl}_1 + \mathbf{dl}_2,$$

where $\mathbf{dl}_1$ is a displacement along the radius vector $\mathbf{L}_{qp}$, and $\mathbf{dl}_2$ is a circular arc at the same radius. Thus, we have for the elementary voltage

$$\mathbf{E}\cdot\mathbf{dl} = \mathbf{E}\cdot\mathbf{dl}_1 + \mathbf{E}\cdot\mathbf{dl}_2 = \mathbf{E}\cdot\mathbf{dl}_1. \qquad [1.55]$$

Using this approach, we represent an arbitrary path as a sum of circular arcs and displacements along radius vectors. All integrals along arcs are zero, while the sum of integrals along the radius vectors is an alternating sum that collapses to two terms:

$$V = \int_a^b \mathbf{E}\cdot\mathbf{dl} = \frac{de(q)}{4\pi\varepsilon_0}\left(\frac{1}{L_{qa}} - \frac{1}{L_{qb_1}} + \frac{1}{L_{qb_1}} - \frac{1}{L_{qb_2}} + \frac{1}{L_{qb_2}} - \cdots - \frac{1}{L_{qb}}\right)$$
$$= \frac{de(q)}{4\pi\varepsilon_0}\left(\frac{1}{L_{qa}} - \frac{1}{L_{qb}}\right).$$

We have established the first fundamental feature of a stationary electric field given by Coulomb's law: the voltage between two points is independent of the path along which integration is carried out, and depends only on its end points. This result can be written as

$$\int_{L_1} \mathbf{E} \cdot \mathrm{d}\mathbf{l} = \int_{L_2} \mathbf{E} \cdot \mathrm{d}\mathbf{l} = \int_{L_3} \mathbf{E} \cdot \mathrm{d}\mathbf{l} = \cdots = \int_{L_n} \mathbf{E} \cdot d\mathbf{l}, \qquad [1.56]$$

where L_i are different paths with the same end points. This feature is not obvious from the definition of the voltage. In fact, as the path changes, its length and shape change, as well as the direction and magnitude of the electric field at its points. Thus, the elementary dot products $\mathbf{E}\cdot\mathrm{d}\mathbf{l}$ will vary considerably between different paths, but their sum remains the same for all possible paths. It is an amazing property of the electric field generated by Coulomb's law. By the principle of superposition, we conclude that this property is valid for any distribution of charges and holds in any medium with arbitrary electrical properties. It should be stressed that this "path independence" of the voltage holds for an electric field caused by constant charges, and in general does not apply to time-varying fields. It is also worth noting that the line integral of the gravitational field **g** produced by a mass distribution

$$\int_a^b \mathbf{g} \cdot \mathrm{d}\mathbf{l}$$

is also path independent.

1.6.2.1 Example 1

To illustrate the relationship between voltage and electric field, assume that the probes of a voltmeter are connected to the points N and M by a straight path, short enough so that the electric field is constant at its points (Fig. 1.5A). Then the voltage is

$$V = \int_N^M \mathbf{E} \cdot \mathrm{d}\mathbf{l} = \int_N^M E_l \mathrm{d}l = E_l MN, \qquad [1.57]$$

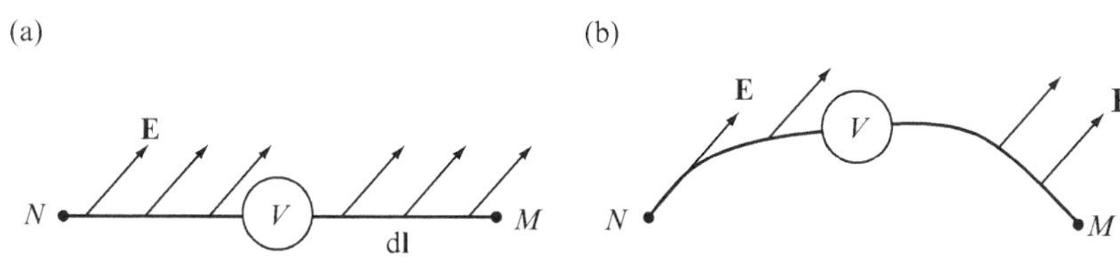

Figure 1.5 Measurement of the electric field **E** by a voltmeter.

where E_l is the component of the field along the line directed from N to M and MN is the path length. In this configuration, the voltmeter in effect determines this component of the field:

$$E_l = \frac{V}{MN}. \quad [1.58]$$

A negative voltage means, for example, that the electric field and the direction of the displacement **dl** forms an angle exceeding 90°. If, however, the shape and length of the path connecting the voltmeter to the same terminal points (Fig. 1.5B) change, the voltage stays the same, but the electric field along the path can no longer be determined by Eq. [1.58], since the tangential component of the field varies along the path.

1.6.3. The Voltage along a Closed Path

The path independence of voltage can be expressed in another form. Consider a closed contour L (Fig. 1.6A), which can be represented as a sum of two contours: L_{acb} and L_{bda}.

From Eq. [1.56], we have

$$\int_{acb} \mathbf{E} \cdot d\mathbf{l} = \int_{adb} \mathbf{E} \cdot d\mathbf{l}. \quad [1.59]$$

In these integrals, each element **dl** is directed from a to b. Reversing the direction of the integral on the right-hand side of Eq. [1.59] gives

$$\int_{acb} \mathbf{E} \cdot d\mathbf{l} = -\int_{bda} \mathbf{E} \cdot d\mathbf{l};$$

that is,

$$\int_{acb} \mathbf{E} \cdot d\mathbf{l} + \int_{bda} \mathbf{E} \cdot d\mathbf{l} = 0, \quad \text{or} \quad \oint \mathbf{E} \cdot d\mathbf{l} = 0. \quad [1.60]$$

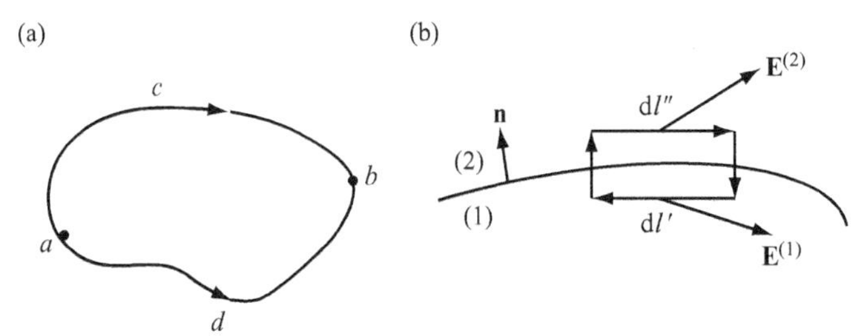

Figure 1.6 (A) Circulation of the field **E** and (B) illustration of Eq. [1.64].

The voltage along an arbitrary closed path is zero. The integral $\oint \mathbf{E} \cdot d\mathbf{l} = 0$ is called the "circulation" of the electric field or the electromotive force:

$$\Xi = \oint_L \mathbf{E} \cdot d\mathbf{l} = 0. \qquad [1.61]$$

The path L can have an arbitrary shape and it can intersect media with various electric parameters. The electromotive force caused by the electric charges is zero, and this fact indicates that a Coulomb's field cannot alone generate a steady electric current. "Non-Coulomb" electromotive forces have to be introduced to generate a constant current flow. This phenomenon will be studied in the next chapter.

1.7. THE FIRST EQUATION OF THE ELECTRIC FIELD

Equation [1.60]

$$\oint \mathbf{E} \cdot d\mathbf{l} = 0$$

is called the first equation for the electric field in integral form. It holds for stationary electric fields along arbitrary paths passing through points at any distance from each other. Since this equation does not contain derivatives, it can be applied at any point of a medium. First, consider regular points where all components of the field are continuous functions. It may be appropriate to notice that in a reality all points are regular, but due to introduction of point, linear, and surface charges, some or all components of the field demonstrate singularities. First, consider regular points. Applying Stoke's theorem, we have

$$\oint_L \mathbf{E} \cdot d\mathbf{l} = \int_S \operatorname{curl} \mathbf{E} \cdot d\mathbf{S} = 0,$$

where curl has its usual meaning as a vector differential operator (defined below). Because this equation holds for any surface S bounded by the path L, we conclude

$$\operatorname{curl} \mathbf{E} = 0. \qquad [1.62]$$

Equation [1.62] is the differential form of the first equation for the stationary electric field, which holds in the vicinity of a regular point. For instance, in the Cartesian system of coordinates:

$$\begin{vmatrix} \mathbf{i} & \mathbf{j} & \mathbf{k} \\ \frac{\partial}{\partial x} & \frac{\partial}{\partial y} & \frac{\partial}{\partial z} \\ E_x & E_y & E_z \end{vmatrix} = 0$$

or

$$\frac{\partial E_z}{\partial y} = \frac{\partial E_y}{\partial z}, \quad \frac{\partial E_x}{\partial z} = \frac{\partial E_z}{\partial x}, \quad \frac{\partial E_y}{\partial x} = \frac{\partial E_x}{\partial y}, \qquad [1.63]$$

that is, Eq. [1.62] is a system of three partial differential equations of the first order. At points where first derivatives of electric field are regular, the system of Equations [1.62] or [1.63] has exactly the same content as the integral form (1.60): the electric field, which obeys Coulomb's law, contains no net circulation or vortices. To obtain a differential form of Eq. [1.60] near surface charges where the normal component E_n is discontinuous, we apply this equation for the elementary path, shown in Fig. 1.6B. Since elements $\mathbf{dl}'$ and $\mathbf{dl}''$ are separated by a distance dh that tends to zero, in place of Eq. [1.60] we obtain

$$\mathbf{E}^{(2)} \cdot \mathbf{dl}'' + \mathbf{E}^{(1)} \cdot \mathbf{dl}' + \mathbf{E} \cdot \mathbf{dh} = 0$$

or

$$E_t^{(2)} dl - E_t^{(1)} dl = 0.$$

Finally, it gives

$$E_t^{(2)} - E_t^{(1)} = 0 \text{ or } \mathbf{n} \times \left(\mathbf{E}^{(2)} - \mathbf{E}^{(1)}\right) = 0. \qquad [1.64]$$

Here $\mathbf{t}$ and $\mathbf{n}$ are the unit vectors tangential and normal to the surface S, respectively. This shows again that the tangential component of the field is continuous across a surface distribution of charge. Later, we will demonstrate that there is a special case—namely, a nonuniform double layer of surface charge—where the tangential component of the electric field is discontinuous. We have derived three forms of the first equation based on Coulomb's law and the principle of superposition:

$$\oint \mathbf{E} \cdot \mathbf{dl} = 0, \quad \operatorname{curl} \mathbf{E} = 0, \quad E_t^{(1)} = E_t^{(2)}. \qquad [1.65]$$

Each expresses the same fact: the electromotive force caused by electric charges is zero. In other words, the voltage between two arbitrary points does not depend on the path of integration. Let us make two comments about these equations:

(1) The first two equations do not generally hold when the field is time varying, because they do not take into account the second "source" of electric fields—a change of the magnetic field with time. On the other hand, the third equation is valid for a time-varying electromagnetic field. This is because the derivation of this surface boundary condition relies on the fact that the area surrounded by a closed integration path straddling a surface goes to zero, and therefore the flux of the time-varying magnetic field through this area also vanishes.

(2) Even though the equations

$$\oint \mathbf{E} \cdot d\mathbf{l} = 0 \quad \text{and} \quad \operatorname{curl} \mathbf{E} = 0$$

are not in general true when the fields vary with time, there are cases when Coulomb's law can still be used to explain the behavior of the time-varying electric field to a very good approximation. These situations include the magnetotelluric sounding method and its modifications, as well as other methods.

1.8. FLUX OF THE ELECTRIC FIELD

Now we demonstrate a second remarkable feature of the electrical field related to the concept of electric "flux." As before, consider first an elementary charge $de(q)$ at the point q and its field:

$$\mathbf{E}(p) = \frac{de(q)}{4\pi\varepsilon_0} \frac{\mathbf{L}_{qp}}{L_{qp}^3}.$$

By definition, its flux through an elementary surface $dS(p)$ is (Fig. 1.7A) defined to be

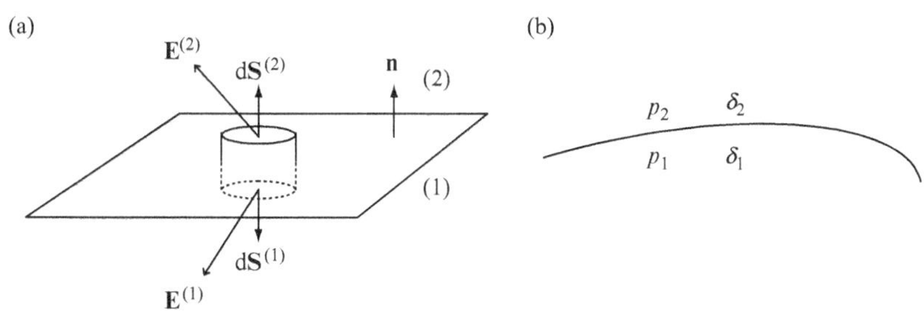

Figure 1.7 (A) Configuration for derivation of Eq. [1.73] at the interface between two media and (B) discontinuity of divergence of the field **E**.

$$\mathbf{E}(p)\cdot \mathrm{d}\mathbf{S}(p) = \frac{\mathrm{d}e(q)}{4\pi\varepsilon_0}\frac{\mathbf{L}_{qp}\cdot \mathrm{d}\mathbf{S}(p)}{L_{qp}^3},$$

or

$$\mathbf{E}\cdot \mathrm{d}\mathbf{S} = \frac{\mathrm{d}e(q)}{4\pi\varepsilon_0}\mathrm{d}\omega(q), \qquad [1.66]$$

where $\mathrm{d}\omega(q)$ is the solid angle subtended by the elementary surface $\mathbf{dS}(p)$ viewed from the point q. There is no obvious physical meaning to flux given by Eq. [1.66]. This is simply a dot product of two different vectors: the electric field $\mathbf{E}(p)$ and $\mathrm{d}\mathbf{S} = \mathrm{d}S\mathbf{n}$. It can be positive or negative or zero; the latter happens when the field is tangential to the surface. Because any surface can be represented as a collection of elementary surface elements, the total flux through an arbitrary surface is equal to the integral:

$$\int_S \mathbf{E}\cdot \mathrm{d}\mathbf{S} = \frac{\mathrm{d}e(q)}{4\pi\varepsilon_0}\omega(q). \qquad [1.67]$$

Here $\omega(q)$ is the total solid angle subtended by the surface S. In particular, if the surface S is closed and surrounds the charge $\mathrm{d}e(q)$, Eq. [1.67] reduces to

$$\oint_S \mathbf{E}\cdot \mathrm{d}\mathbf{S} = \frac{\mathrm{d}e(q)}{\varepsilon_0},$$

since the solid angle $\omega(q)$ subtended by any closed surface viewed by a point in its interior is 4π. When a charge is located outside the volume enclosed by this surface,

$$\oint_S \mathbf{E}\cdot \mathrm{d}\mathbf{S} = 0,$$

because in this case the total solid angle $\omega(q)$ is zero. The principle of superposition generalizes this equation to an arbitrary distribution of charges:

$$\oint_S \mathbf{E}\cdot \mathrm{d}\mathbf{S} = \frac{1}{\varepsilon_0}\left[\int_V \delta\,\mathrm{d}V + \int_S \sigma\,\mathrm{d}S + \int_L \lambda\,\mathrm{d}l + \sum_i e_i\right], \qquad [1.68]$$

where δ, σ, λ, and e_i are, respectively, volume, surface, linear, and point charges located in the volume V enclosed by the surface S. Charges located outside the volume V also generate an electric field inside V, but their electric field has no net flux through the surface S. Let us write Eq. [1.68] in the form

$$\oint_S \mathbf{E}_i \cdot d\mathbf{S} = \frac{e}{\varepsilon_0}, \quad \oint_S \mathbf{E}_e \cdot d\mathbf{S} = 0$$

or [1.69]

$$\oint_S \mathbf{E} \cdot d\mathbf{S} = \frac{e}{\varepsilon_0},$$

where $\mathbf{E}_i$ and $\mathbf{E}_e$ are the electric fields caused by charges located inside and outside the volume V, e is the total charge inside the volume, and

$$\mathbf{E} = \mathbf{E}_i + \mathbf{E}_e.$$

Equations [1.69] illustrate the second remarkable feature of the electric field caused by constant charges and hold in any medium. Each charge outside a given volume creates an electric field at points of the surface S enclosing the volume, but the total flux through the surface, caused by these charges, is equal to zero. This result does not depend on the shape and dimensions of the surface, or on the spatial distribution of charges. Similarly, any change of position of charges inside the volume V, or variation of the surface S, has no effect on the total flux, even though the fluxes through elementary surfaces vary. According to Eq. [1.68], the flux is defined only by the amount of charges inside the volume V.

1.9. THE SECOND EQUATION OF THE ELECTRIC FIELD

1.9.1. Second Equation at Regular Points

First, consider regular points where only volume charges can be located. In such places, Eq. [1.68] is simplified, and we have

$$\oint_S \mathbf{E} \cdot d\mathbf{S} = \frac{1}{\varepsilon_0} \int_V \delta \, dV. \qquad [1.70]$$

Equation [1.70] is traditionally called the second equation for the electric field in integral form. It applies at regular points—where field values are nonsingular, in fact, over any surface where the integral can be defined—and shows that the fluxes of $\mathbf{E}$ through all patches of the closed surface S are related in such way that their sum defines the total charge inside. At the same time, it provides no information about the distribution of charge. For instance, the volume may contain equal amounts of positive and negative charges so that

$$\oint \mathbf{E} \cdot d\mathbf{S} = 0$$

in spite of the presence of isolated charges inside the volume V. To overcome this problem and characterize the distribution of charges pointwise, we use Gauss' theorem to convert the surface integral on the left-hand side to a volume integral:

$$\oint_S \mathbf{E}\cdot d\mathbf{S} = \int_V \operatorname{div}\mathbf{E}\, dV = \frac{1}{\varepsilon_0}\int_V \delta\, dV,$$

where div is the usual divergence operator of vector analysis (defined below). Since this equation holds for any volume, the integrands in the two volume integrals must be equal, or

$$\operatorname{div}\mathbf{E} = \frac{\delta}{\varepsilon_0}. \qquad [1.71]$$

Equation [1.71] is called the second equation for the electric field in differential form. It applies at regular points where the divergence is defined, and equates the divergence of the electric field at any point to the charge density at the same point. For instance, in Cartesian coordinates, we have

$$\frac{\partial E_x}{\partial x} + \frac{\partial E_y}{\partial y} + \frac{\partial E_z}{\partial z} = \frac{\delta}{\varepsilon_0}, \qquad [1.72]$$

which is another partial differential equation of the first order. In spite of the presence of the derivatives, the divergence of the electric field does not characterize its rate of change, but rather the volume density of charge for instance, the field $\mathbf{E}$ may change very rapidly near points where charges are absent and $\operatorname{div}\mathbf{E} = 0$.

1.9.2. Second Equation Near Surface Charges

Earlier, we investigated the behavior of the field in the vicinity of surface charges. We now obtain the same result proceeding from Eq. [1.70]. As already shown, the normal component of the electric field is discontinuous in passing through a surface charge. Therefore, the derivative $\partial E_{\mathrm{n}}/\partial n$ does not exist and Eq. [1.72] does not hold. However, its integral form (Eq. [1.70]) can be applied to an elementary cylindrical surface enclosing a small patch of surface charge (Fig. 1.7A). The total charge in the cylinder is the charge $\sigma\, dS$ on the surface patch, and we have

$$\mathbf{E}_2 \cdot d\mathbf{S}_2 + \mathbf{E}_1 \cdot d\mathbf{S}_1 = \frac{\sigma\, dS}{\varepsilon_0} \quad \text{or} \quad E_{\mathrm{n}}^{(2)} - E_{\mathrm{n}}^{(1)} = \frac{\sigma}{\varepsilon_0}. \qquad [1.73]$$

Here we took into account that

$$d\mathbf{S}_2 = -d\mathbf{S}_1, \quad dS_2 = dS_1 = dS,$$

and that the flux of the electric field through the lateral surface of the cylinder vanishes because its area vanishes as $h \to 0$. Equation [1.73] is the surface analogy of Eq. [1.71], which was earlier derived in a different way. Thus, we obtained three forms of the second equation:

$$\oint \mathbf{E} \cdot d\mathbf{S} = \frac{e}{\varepsilon_0}, \quad \operatorname{div} \mathbf{E} = \frac{\delta}{\varepsilon_0}, \quad E_n^{(2)} - E_n^{(1)} = \frac{\sigma}{\varepsilon_0}, \qquad [1.74]$$

which describe the field in any medium.

1.10. SYSTEM OF EQUATIONS OF THE ELECTRIC FIELD

We can now write two systems of equations for the time-invariant (stationary) electric field that follow from Coulomb's law and the principle of superposition:

1.10.1. Integral Form

$$1.\ \oint \mathbf{E} \cdot d\mathbf{l} = 0, \quad 2.\ \oint \mathbf{E} \cdot d\mathbf{S} = \frac{e}{\varepsilon_0}. \qquad [1.75]$$

1.10.2. Differential Form

$$1.\ \operatorname{curl} \mathbf{E} = 0, \quad 2.\ \operatorname{div} \mathbf{E} = \frac{\delta}{\varepsilon_0},$$

and [1.76]

$$1.\ \mathbf{n} \times (\mathbf{E}_2 - \mathbf{E}_1) = 0, \quad 2.\ \mathbf{n} \cdot (\mathbf{E}_2 - \mathbf{E}_1) = \frac{\sigma}{\varepsilon_0}.$$

Some comments are appropriate at this point:

(1) Any electric field caused by constant charges satisfies Eqs. [1.75] and [1.76]; that is, the field given by the generalized Coulomb's law is a solution of the system of field equations in integral and differential form.

(2) These equations contain the same information as Coulomb's law, but make it possible to solve for the field in cases where direct use of Coulomb's law is not practical (because the distribution of charge is not known in advance).

(3) When there are only volume distributions of charges ($\sigma = 0$), we have

$$\text{curl}\,\mathbf{E} = 0, \quad \text{div}\,\mathbf{E} = \frac{\delta}{\varepsilon_0}$$

and [1.77]

$$E_t^{(2)} = E_t^{(1)}, \quad E_n^{(2)} = E_n^{(1)};$$

that is, the normal and tangential components of the field are everywhere continuous functions, including at points where the volume density is discontinuous. However, in such places the first derivatives of the electric field can experience a discontinuity (Fig. 1.7B) and

$$\text{div}\,\mathbf{E}_1(p_1) \neq \text{div}\,\mathbf{E}_2(p_2) \quad \text{if} \ p_1 \to p, \ p_2 \to p.$$

(4) Systems [1.75] and [1.76] are generally valid and correctly describe the electric field, not only in a free space but also in any polarizable and conducting medium. In particular, the stationary electric field, measured in all electrical methods of geophysics, is a solution of these systems. (As was pointed out, there is one exceptional situation where continuity of the tangential component of the electric field (Eq. [1.76]) does not hold across a surface, even though the first equation of the set (Eq. [1.75]) does hold.)

(5) Comparison with the equations for the gravity field $\mathbf{g}$ in differential form

$$\text{curl}\,\mathbf{g} = 0, \quad \text{div}\,\mathbf{g} = -4\pi\gamma\delta$$

and

$$\mathbf{n} \times (\mathbf{g}_2 - \mathbf{g}_1) = 0, \quad \mathbf{n}\cdot(\mathbf{g}_2 - \mathbf{g}_1) = -4\pi\gamma\sigma$$

shows that the two systems are identical. This happens because the stationary electric field and the gravity field are generated by sources (of charge and mass) and obey "inverse square" laws.

(6) There is nevertheless one fundamental difference between the two systems of equations for electric and gravity fields when studying their solutions in geophysical applications. The source of the gravity (attraction) field—that is, the right-hand side of the second equation, the mass density distribution—is usually fixed in advance, whereas the density of electric charges is generally unknown. When the field of a given distribution of charges is considered in free space, in the absence of any conducting and polarizable medium, we can use Coulomb's law to calculate the field $\mathbf{E}$ in the same manner as

Newton's law is used to compute the attraction field of a given mass distribution. But this case has no practical value for electric methods in geophysics. When electric fields are generated in the presence of conducting and polarizable bodies, such as the Earth, we cannot know the density of charges before the field **E** is calculated. Earlier, we used the example of the electrostatic induction to illustrate this problem.

(7) The system of field equations at regular points (Eqs. [1.77]) consists of four partial differential equations of the first order with respect to four unknowns, namely, three components of the field **E** and the density of charge. As is well known, each differential equation can have an infinite number of solutions; therefore, this system of equations must be supplemented by additional information about the field behavior to determine the field.

(8) The system of Equations [1.76] for the field **E** is complicated, and it is natural to simplify it as much as possible. This subject will be discussed in the next section.

1.11. POTENTIAL OF THE ELECTRIC FIELD

1.11.1. Relation Between the Field E and Potential *U*

To reduce the system of equations of the electric field, we now introduce the concept of a scalar potential $U(p)$. We start with one of the fundamental identities of vector analysis. For any function $\phi(p)$ with second partial derivatives, we have

$$\operatorname{curl}(\operatorname{grad}\phi) = 0.$$

It is clear from this identity (which follows simply from the equality of mixed partial derivatives) that the first equation of the set [1.76] will hold identically if we define the electric field as the gradient of a scalar function:

$$\mathbf{E} = -\operatorname{grad} U. \qquad [1.78]$$

The function $U(p)$ is called the potential of the electric field. The gradient of a scalar field is a vector field that characterizes at each point the maximum rate of change of the function; that is, it points in the direction of maximum change and its magnitude is equal to the rate of increase along this direction. Because of the minus sign in Eq. [1.78], the electric field **E** points in direction of maximum decrease of the potential. Its component along any direction l is

$$E_l = -\frac{\partial U}{\partial l}. \qquad [1.79]$$

In Cartesian coordinates, Eq. [1.78] becomes

$$E_x = -\frac{\partial U}{\partial x}, \quad E_y = -\frac{\partial U}{\partial y}, \quad E_z = -\frac{\partial U}{\partial z}$$

and

$$\operatorname{grad} U = \frac{\partial U}{\partial x}\mathbf{i} + \frac{\partial U}{\partial y}\mathbf{j} + \frac{\partial U}{\partial z}\mathbf{k}.$$

There are several reasons why it is useful to introduce the potential U. Before describing these in the following sections, several comments are appropriate:

(1) There are an infinite number of functions $U(p)$ that give the same electric field; they differ from each other by a constant function, because

$$\operatorname{grad} U = \operatorname{grad}(U + C).$$

This indicates that only difference in the potential $U(p)$ can have physical meaning.

(2) From Eq. [1.78], we conclude that the potential has the same dimensions as voltage because its gradient has the same dimensions as the electric field (V m^{-1}).

(3) The minus sign at the right-hand side of Eq. [1.78] is conventional. It can be omitted, and the new potential will differ by sign only.

1.11.2. Relation Between Voltage and Potential

First, let us express the voltage in terms of potential. It is clear that along an elementary path the voltage can be represented as

$$\mathbf{E}\cdot d\mathbf{l} = -\operatorname{grad} U \cdot d\mathbf{l} = -\frac{\partial U}{\partial l} dl = -dU, \qquad [1.80]$$

where dU is a change of the potential within the element dl. Note that by definition

$$\frac{\partial U}{\partial l} = \mathbf{l}_0 \cdot \operatorname{grad} U,$$

and $\mathbf{l}_0$ is the unit vector of the displacement $d\mathbf{l}$. Then integrating along any path, we obtain

$$V = \int_p^b \mathbf{E} \cdot \mathrm{d}\mathbf{l} = -\int_p^b \mathrm{d}U = U(p) - U(b). \qquad [1.81]$$

Thus, the voltage of the electric field along any path with end points p and b can be written as the difference of potentials between these points. Certainly, such procedure is much simpler than a calculation of the dot product $\mathbf{E}\cdot\mathrm{d}\mathbf{l}$ at each point of the path and their summation. This is one of the reasons for the introduction of the potential. It may be proper to emphasize again that a voltmeter measures the voltage $\int_p^b \mathbf{E}\cdot\mathrm{d}\mathbf{l}$, and the latter is expressed in terms of the potential difference.

1.11.3. Relation Between Potential and Charges

Next we use Eq. [1.81], rewritten as

$$U(p) = U(b) + \int_p^b \mathbf{E} \cdot \mathrm{d}\mathbf{l} \qquad [1.82]$$

to define the potential caused by an arbitrary distribution of charge. At great distances from charges, the field $\mathbf{E}$ vanishes, and the potential tends to a constant since

$$\frac{\partial U}{\partial l} \to 0.$$

Then, letting b go to infinity and assuming that the potential at large distances from charges goes to zero,

$$U(\infty) = 0, \qquad [1.83]$$

we have

$$U(p) = \int_p^\infty \mathbf{E} \cdot \mathrm{d}\mathbf{l}. \qquad [1.84]$$

The force acting on the elementary charge $\mathrm{d}e$ is equal to

$$\mathrm{d}\mathbf{F} = \mathbf{E}\,\mathrm{d}e,$$

so we can say that the potential $U(p)$ characterizes the work performed by the field $\mathbf{E}$ to move a unit positive charge from the point p to "infinity." This work is path independent. To find the relationship between the potential and charges, we use of Eq. [1.84] and assume first that there is only a single charge at the point q. Then

$$\mathbf{E}(a) = \frac{de}{4\pi\varepsilon_0} \frac{\mathbf{L}_{qa}}{L_{qa}^3},$$

and Eq. [1.84] gives

$$U(p) = \int_p^\infty \mathbf{E}(a) \cdot \mathbf{dl}.$$

Here a is any point of the path of integration. Bearing in mind that the voltage does not depend on the path, we choose the latter along the straight line L_{qa} connecting points: q, p, and a. In such a case, $\mathbf{L}_{qa} \cdot \mathbf{dl} = L_{qa} \mathrm{d}l$, and we have

$$U(p) = \frac{de}{4\pi\varepsilon_0} \int_p^\infty \frac{\mathrm{d}L_{qa}}{L_{qa}^2}.$$

Thus,

$$U(p) = \frac{de}{4\pi\varepsilon_0 L_{qp}}, \qquad [1.85]$$

and the potential due to elementary charge is inversely proportional to the distance L_{qp} between the charge and the observation point. Making use of the principle of superposition for an arbitrary distribution of the volume, surface, linear, and point charges, we arrive at the following expression for the potential:

$$U(p) = \frac{1}{4\pi\varepsilon_0} \left[\int_V \frac{\delta \,\mathrm{d}V}{L_{qp}} + \int_S \frac{\sigma \,\mathrm{d}S}{L_{qp}} + \int_L \frac{\lambda \,\mathrm{d}l}{L_{qp}} + \sum \frac{e(q)}{L_{qp}} \right]. \qquad [1.86]$$

It is worth emphasizing that Eq. [1.86] is valid in any medium, characterized by any distribution of conductivity and dielectric constant. Comparison with Eq. [1.24] shows that the potential $U(p)$, as a scalar function, is much more simply related to charge than the vector electric field $\mathbf{E}(p)$. This is the second reason for using the potential.

1.11.4. Poisson's and Laplace's Equations for Potential

We now describe a third and perhaps most important reason for introducing the potential $U(p)$. Recall that the motivation for using the potential was that the first equation of the field, curl $\mathbf{E}=0$, would be satisfied identically by setting

$$\mathbf{E} = -\operatorname{grad} U.$$

Substituting this expression into the second equation of the system [1.77], we find that at regular points, the potential satisfies the equation:

$$\operatorname{div}(\operatorname{grad} U) = -\frac{\delta}{\varepsilon_0} \quad \text{or} \quad \nabla^2 U = -\frac{\delta}{\varepsilon_0}. \qquad [1.87]$$

Equation [1.87] is called Poisson's equation. At points where charge is absent, it reduces to Laplace's equation:

$$\nabla^2 U = 0. \qquad [1.88]$$

In Cartesian coordinates, these two equations become explicitly

$$\frac{\partial^2 U}{\partial x^2} + \frac{\partial^2 U}{\partial y^2} + \frac{\partial^2 U}{\partial z^2} = -\frac{\delta}{\varepsilon_0} \quad \text{and} \quad \frac{\partial^2 U}{\partial x^2} + \frac{\partial^2 U}{\partial y^2} + \frac{\partial^2 U}{\partial z^2} = 0. \qquad [1.89]$$

The steps leading to these equations are illustrated below. Thus, we have replaced a system of equations for individual field components with one equation for the scalar function U, which is a great simplification. As pointed out before, the system [1.77] does not define the field **E** uniquely; similarly, the potential field equations are differential equations of the second order, which have an infinite number of solutions. Let us explore this nonuniqueness in detail. We demonstrated that at each point

$$U(p) = \frac{1}{4\pi\varepsilon_0} \int_V \frac{\delta(q)}{L_{qp}} \, \mathrm{d}V. \qquad [1.90]$$

This function is called the fundamental solution of Poisson's equation. In other words, substituting Eq. [1.90] into the first equation of the set [1.89], we find that a sum of the second derivatives equals $-\delta/\varepsilon_0$ at the same point. If net charge is absent at this point, this sum is zero. We can say that the sum of second derivatives of the potential with respect

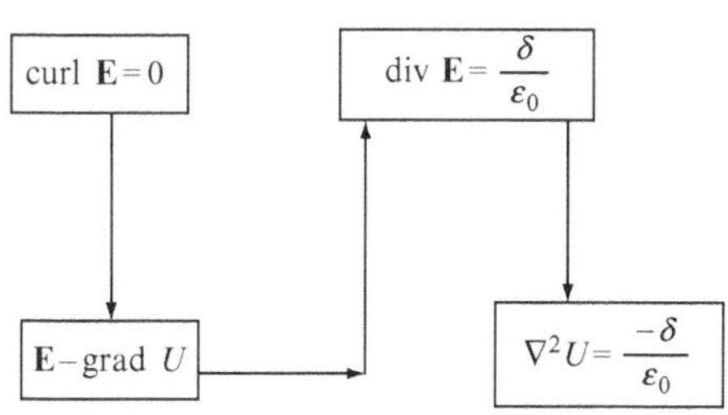

to x, y, and z defines the charge density pointwise, and, consequently, this sum is insensitive to charge at other locations. The analogy with the gravitational field is obvious and, as pointed out earlier, follows because both fields obey similar physical laws. Now, imagine that functions U_1 and U_2 are solutions of Poisson's and Laplace's equations, respectively:

$$\nabla^2 U_1 = -\frac{\delta}{\varepsilon_0} \text{ and } \nabla^2 U_2 = 0.$$

Then the function

$$U_3 = U_1 + U_2$$

is also a solution of Poisson's equation. In fact, we have

$$\nabla^2 U_3 = \nabla^2 U_1 + \nabla^2 U_2 = -\frac{\delta}{\varepsilon_0},$$

which means that solutions of Poisson's equation differ from each other by functions that satisfy Laplace's equation. Such functions may be interpreted as potentials caused by charges located outside an observation point. With this in mind, a comment concerning Eq. [1.90] is appropriate. When the observation point p is located outside a charge distribution, the distance L_{qp} is always nonzero and the integrand always has a finite value. When the observation point is inside a charge distribution, the denominator L_{qp} tends to zero, but it can be shown that integrand in Eq. [1.90] does not become singular (for any volume or surface distribution of charge). This holds because the numerator (an elementary volume) decreases more rapidly when $L_{qp} \to 0$. Thus, the potential has everywhere a finite value provided that the point and linear charges are not considered.

1.11.5. Behavior of Potential Near Surface Charges

Equations [1.87] and [1.88] are valid at regular points only, so it is interesting to consider also the behavior of the potential in the vicinity of the surface charges where the normal component of the field is discontinuous. From Eq. [1.81], it follows that the potentials at points p_1 and p_2 on opposite sides of the surface are

$$U(p_1) = \int_{p_1}^{\infty} \mathbf{E} \cdot \mathbf{dl} \text{ and } U(p_2) = \int_{p_2}^{\infty} \mathbf{E} \cdot \mathbf{dl}$$

and

$$U(p_2) - U(p_1) = \int_{p_2}^{p_1} \mathbf{E} \cdot \mathbf{dl}. \qquad [1.91]$$

Because the electric field has everywhere a finite value and the distance between points p_1 and p_2 is vanishingly small, we conclude that the potential difference tends to be zero; that is, the potential of the electric field is continuous across any surface, carrying a charge with a density σ:

$$U_1(p) = U_2(p), \qquad [1.92]$$

where U_1 and U_2 are the potential on opposite sides of the surface.

1.11.6. System of Equations for Potential

Now we are ready to replace the system of equations for the electric field by another system for the potential, also based on Coulomb's law and the principle of superposition and valid in any medium. This set of equations is given below, making use of the equality $E_n = -\partial U/\partial n$.

$$\boxed{\nabla^2 U = \frac{-\delta}{\varepsilon_0}}$$

$$\boxed{\text{A. } U_2 - U_1 = 0} \qquad \boxed{\text{B. } \frac{\partial U_2}{\partial n} - \frac{\partial U_1}{\partial n} = -\frac{\sigma}{\varepsilon_0}} \qquad (1.93)$$

As usual, several comments about this system are appropriate:

(1) System (1.93) consists of three parts; the first is Poisson's equation which describes the potential at regular points where its second derivatives exist. The other two parts characterize the potential in the vicinity of a surface; in essence, they are the surface analogues of Poisson's equation. There are in general an infinite number of functions U which obey all three conditions.

(2) We will demonstrate later that there is a special charge distribution, called a double layer, across which the potential can be discontinuous.

(3) Replacing the derivative by a finite difference, Poison's equation can be approximately represented as

$$\frac{6}{h^2}\left[U(p) - U^{av}(p)\right] = \frac{\delta}{\varepsilon_0}. \qquad [1.94]$$

Here $2h$ is the length of the side of an elementary cubic volume surrounding the point p. $U(p)$ and $U^{av}(p)$ are the values of the potential at the middle point and its average value at this point, respectively. As follows from Eq. [1.94], in the absence of a charge at some point p

$$U(p) = U^{av}(p) \quad \text{or} \quad \nabla^2 U = 0, \qquad [1.95]$$

that is, the potential is a harmonic function at such points. In the one-dimensional case, its analogue is a linear function. In the vicinity of points where the charge is positive, the potential exceeds the average value: $U(p) > U^{av}(p)$, while at places with negative charge the opposite relation holds: $U(p) < U^{av}(p)$.

(4) Conditions A and B of the system (1.93) ensure continuity of the tangential component and discontinuity of the normal component of the electric field across the surface charges.

(5) This system is identical to that for the potential of the gravitational field, except for the coefficient in front of the volume and surface charges.

(6) In the vicinity of point and linear charges, the potential tends to infinity. Such singularities must be taken into account in solving forward problems for the electric field.

(7) Transition from the electric field to the scalar potential often greatly simplifies the determination of the field, that is, the solution of the forward problem, and this fact constitutes the third reason that justifies the use of the system (1.93).

(8) Introduction of the potential itself and, correspondingly, the set (1.93), does not immediately solve the problem of determining the field, which is caused in general by the lack of information about charge distribution. Later, we will describe how to overcome this difficulty by solving a boundary-value problem for the potential.

1.12. THE ELECTRIC DIPOLE AND DOUBLE LAYER

Now we will consider several examples of the potential and electric fields caused by different charge distributions. Special attention is paid to two cases: the electric dipole and double layer, because they often play an important role in the theory and practice of electric methods.

1.12.1. Vector Lines of the Electric Field and Equipotential Surfaces

In order to visualize a field behavior, it is convenient to use the vector lines of the field **E** and equipotential surfaces of the function U. By definition, the field **E** is tangential to the vector line at each of its points, and

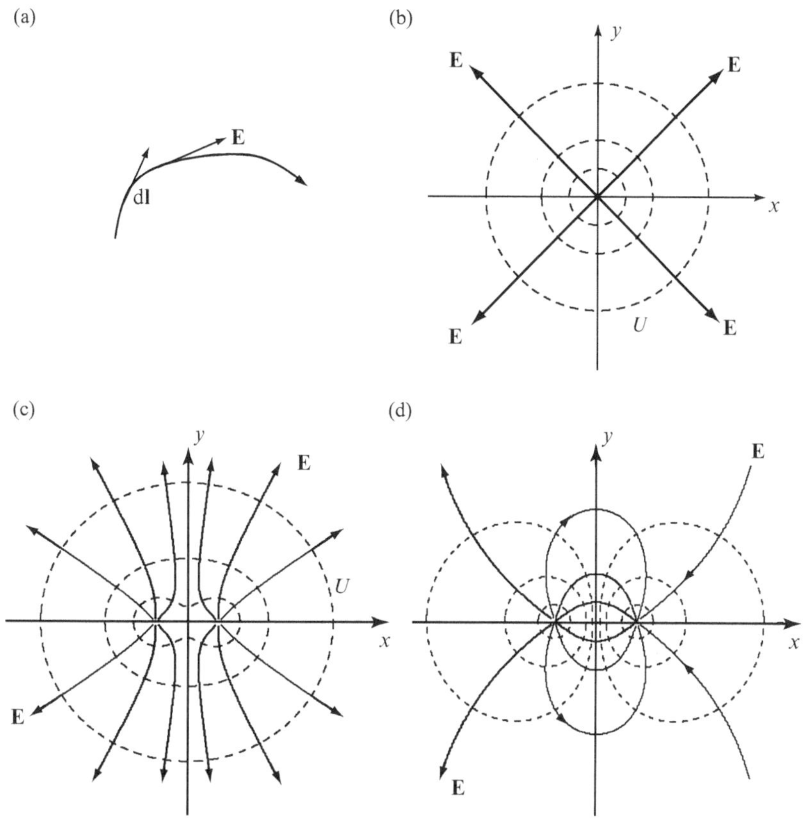

Figure 1.8 (A) Field line of the electric field vector, (B) field lines (solid) and equipotential surfaces (dashed) surrounding an elementary charge at the origin, (C) field lines and equipotential surfaces of two charges with the same magnitude and sign, and (D) field lines and equipotential surfaces of two charges with the same magnitude and opposite sign.

they have the same direction (Fig. 1.8A). Correspondingly, the equation of the vector line is

$$\cos(\mathbf{E}, \mathbf{dl}) = 1, \quad [1.96]$$

where $\mathbf{dl}$ is an elementary displacement of this line. Let us choose a curvilinear orthogonal system of coordinates x_1, x_2, and x_3. Then, we have

$$\mathbf{E} = E_1\mathbf{i}_1 + E_2\mathbf{i}_2 + E_3\mathbf{i}_3 \text{ and } \mathbf{dl} = \mathrm{d}l_1\mathbf{i}_1 + \mathrm{d}l_2\mathbf{i}_2 + \mathrm{d}l_3\mathbf{i}_3, \quad [1.97]$$

where components of the displacement along coordinate axes are

$$\mathrm{d}l_1 = h_1\mathrm{d}x_1, \quad \mathrm{d}l_2 = h_2\mathrm{d}x_2, \quad \mathrm{d}l_3 = h_3\mathrm{d}x_3$$

and h_1, h_2, h_3 are metric coefficients. Inasmuch as vectors **dl** and **E** form the same angle with the coordinate axes, we can write at each point of the coordinate line:

$$\frac{E_1}{E}=\frac{dl_1}{dl},\quad \frac{E_2}{E}=\frac{dl_2}{dl},\quad \frac{E_3}{E}=\frac{dl_3}{dl}.$$

Thus,

$$\frac{dl_1}{E_1}=\frac{dl_2}{E_2}=\frac{dl_3}{E_3}\quad \text{or}\quad \frac{h_1 dx_1}{E_1}=\frac{h_2 dx_2}{E_2}=\frac{h_3 dx_3}{E_3}, \qquad [1.98]$$

and the latter represents the system of differential equations describing the vector lines of the electric field. For instance, in the Cartesian system of coordinates, Eqs. [1.98] give

$$\frac{dx}{E_x}=\frac{dy}{E_y}=\frac{dz}{E_z}$$

or

$$\frac{dy}{dx}=\frac{E_y}{E_x}\quad \text{and}\quad \frac{dz}{dx}=\frac{E_z}{E_x}. \qquad [1.99]$$

Thus, knowing components of the electric field, we can find the slope of the vector line at each its point and, applying numerical methods, draw the family of vector lines. As follows from the first equation of the electric field, curl $\mathbf{E}=0$, all vector lines are open and terminate at points where charges are located. To describe the scalar field of the potential, the equipotential surfaces are usually introduced. By definition, at each point of such surface the potential remains the same and its equation is

$$U(x_1,x_2,x_3)=C. \qquad [1.100]$$

Here C is some constant. As we know, the gradient of the field U and its directional derivative, $\partial U/\partial l$, along some line l are related as

$$\frac{\partial U}{\partial l}=|\operatorname{grad} U|\cos(\mathbf{i}_l,\nabla U). \qquad [1.101]$$

In particular, $\mathbf{i}_l$ is the unit vector tangent to the equipotential surface. This shows that $\operatorname{grad} U$ is normal to the equipotential surface; otherwise, we would observe a change of the potential at this surface. From the relation $\mathbf{E}=-\operatorname{grad} U$, it follows that at regular points the electric field, as well as its vector lines, is perpendicular to the equipotential surfaces.

1.12.2. Example 1: The Elementary Charge

We start from the simplest possible case and suppose that the elementary charge is located at the origin of coordinates (Fig. 1.8B). Because of symmetry, it is sufficient to consider fields only in the plane XY, and it is obvious that components of the electric field are

$$E_x = \frac{e}{4\pi\varepsilon_0} \frac{x}{\left(x^2 + y^2\right)^{3/2}}, \quad E_y = \frac{e}{4\pi\varepsilon_0} \frac{y}{\left(x^2 + y^2\right)^{3/2}}.$$

Consequently, the equation of the vector lines is

$$\frac{dx}{x} = \frac{dy}{y}. \qquad [1.102]$$

Here dx and dy are projections of the elementary displacement $\mathbf{dl}$ of the vector line on the coordinate axes. Integration of Eq. [1.102] gives

$$\ln y = \ln x + C_1 \text{ or } \frac{y}{x} = C, \qquad [1.103]$$

that is, along the vector line, the ratio of coordinates remains constant. In other words, the angle between the vector line and the x-axis does not change. Thus, these lines are straight and all of them have terminal points at the origin. Of course, the same result directly follows from Coulomb's law:

$$\mathbf{E}(\mathbf{R}) = \frac{e}{4\pi\varepsilon_0} \frac{\mathbf{R}_0}{R^2},$$

where $\mathbf{R}_0$ is the unit vector of the radius vector. Assuming that the density of vector lines is directly proportional to the field magnitude, we see that it decreases with an increase of the distance from the charge. This means that these lines allow us to characterize both the magnitude and direction of the field $\mathbf{E}$. In particular, it is stronger where the density of vector lines is higher, and this simple rule will be used in much more complicated cases.

Since

$$U(\mathbf{R}) = \frac{e}{4\pi\varepsilon_0 R},$$

the equation of the equipotential surfaces is

$$R = \text{constant or } x^2 + y^2 + z^2 = C, \qquad [1.104]$$

and it describes the family of spherical surfaces perpendicular to the vector lines. Usually, these surfaces are plotted in such way that the difference of potentials of the neighboring surfaces is the same. Then the density of these surfaces becomes directly proportional to the electric field. This simple example is very useful for understanding of many electric methods.

1.12.3. Example 2: Two Elementary Charges

Assume that two charges with equal magnitude but either the same or opposite signs are located at the x-axis at points $x = \pm a$. Potential $U(x, y, z)$ caused by these charges is

$$U(p) = \frac{e}{4\pi\varepsilon_0}\left(\frac{1}{R_1} \pm \frac{1}{R_2}\right), \qquad [1.105]$$

and components of the electric field are

$$\begin{aligned} E_x(x, y, z) &= \frac{e}{4\pi\varepsilon_0}\left(\frac{x-a}{R_1^3} \pm \frac{x+a}{R_2^3}\right), \\ E_y(x, y, z) &= \frac{e}{4\pi\varepsilon_0}\left(\frac{y}{R_1^3} \pm \frac{y}{R_2^3}\right), \\ E_z(x, y, z) &= \frac{e}{4\pi\varepsilon_0}\left(\frac{z}{R_1^3} \pm \frac{z}{R_2^3}\right). \end{aligned} \qquad [1.106]$$

Here $e > 0$ and

$$R_1 = \left[(x-a)^2 + y^2 + z^2\right]^{1/2}, \quad R_2 = \left[(x+a)^2 + y^2 + z^2\right]^{1/2}.$$

The positive charge is located at the point $(a, 0, 0)$, while the sign $\pm$ corresponds to the cases when at the point $(-a, 0, 0)$ either the positive or negative charge is placed. Because of the symmetry with respect to the x-axis, we consider the fields U and $\mathbf{E}$ only in the XY-plane, and assume first that both charges are positive. Figure 1.8C shows the equipotential lines, which represent the intersection of the equipotential surfaces by this plane. Near each charge, they are almost circular, since the influence of the second charge is small. With an increase of the distance from charges, the shape of lines becomes more complicated, and they enclose both charges. Then, the equipotential lines again tend to circles. It happens because at great distances, the potential of the two charges behaves as that of the single charge $2e$ placed at the origin. At the point $(x=0,\ y=0,\ z=0)$

the electric field is equal to zero, that is, $\text{grad}\, U = 0$. This is actually a "saddle point" of the potential, where it increases in the two directions along the x-axis going toward the charges, but decreases in opposite directions pointing along the y-axis. By drawing lines perpendicular to the equipotential lines, or solving Eq. [1.99], we obtain the family of the vector lines of the field **E** (Fig. 1.8C). All vectors lines start from charges, while the other terminal points are located at infinity. It is useful to notice that in the vicinity of the plane $x=0$ and not far from the origin the vector lines are almost parallel to this plane. Such "focusing" of the field due to the presence of two charges is often used in borehole geophysics. Figure 1.8D shows the behavior of the potential and electric field caused by charges of equal magnitude and opposite sign. In this case, each vector line has the same terminal points, $x = \pm a$, and everywhere $\text{grad}\, U \neq 0$; that is, the field is regular at all points except at the two point charges.

1.12.4. Behavior of Potential *U* and Field E far away from Charges

Suppose that a point of observation p is far away from charges distributed in some volume V with density $\delta(q)$. This means that distances L_{qp} from the point p to any point q of the volume is nearly constant. Correspondingly, Eq. [1.90] can be written as

$$U(p) = \frac{1}{4\pi\varepsilon_0 L_{qp}} \int_V \delta(q) \mathrm{d}V = \frac{e}{4\pi\varepsilon_0 L_{qp}} \quad \text{if } L_{qp} \to \infty. \qquad [1.107]$$

Here q is any point inside of the volume and e is total charge. Correspondingly, for the electric field we have

$$\mathbf{E}(p) = \frac{e\mathbf{L}_{qp}}{4\pi\varepsilon_0 L_{qp}^3} \quad \text{if } L_{qp} \to \infty. \qquad [1.108]$$

Thus, regardless of the distribution of charges, the potential and electric field behave as those of the point charge at large distances, provided the total charge differs from zero. Otherwise, these fields decrease more rapidly.

1.12.5. The Electric Dipole

As before, assume that there are two elementary charges equal by magnitude but different in sign (Fig. 1.9A). Then, the potential due to these charges is

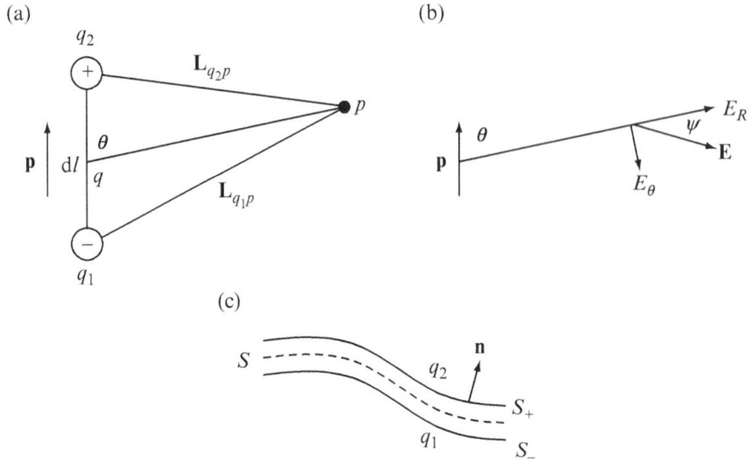

Figure 1.9 (A) Geometry of the electric dipole with positive and negative charges of equal magnitude q separated by a distance dl. The product q dl remains finite as dl goes to zero. (B) Electric field components in spherical coordinates. (C) Double layer of surface charge.

$$U(p) = \frac{e}{4\pi\varepsilon_0}\left(\frac{1}{L_{q_2p}} - \frac{1}{L_{q_1p}}\right), \qquad [1.109]$$

where q_1 and q_2 are the points of charge location. We will consider the field only at distances that greatly exceed the separation dl between charges:

$$\mathrm{d}l \ll L_{qp}. \qquad [1.110]$$

Here q is the middle point of charges $+e$ and $-e$. In this case, the system of charges is called an electric dipole, and it plays an important role for understanding phenomenon of polarization, as well as in the theory of electric methods. Taking into account the inequality [1.110], we can expect that a change of the function $1/L_{qp}$ between points q_1 and q_2 is very small, and it can be represented as

$$\frac{1}{L_{q_2p}} - \frac{1}{L_{q_1p}} = \mathrm{d}l\frac{\partial}{\partial l}\frac{1}{L_{qp}}, \qquad [1.111]$$

where $(\partial/\partial l)(1/L_{qp})$ is the directional derivative along the line connecting both charges. In essence, we expanded the left-hand side of Eq. [1.111] in a power series and discarded all terms except the first one. By definition of gradient, Eq. [1.109] can be represented in the form

$$U(p) = \frac{e\,dl}{4\pi\varepsilon_0}\frac{\partial}{\partial l}\frac{1}{L_{qp}} = \frac{e}{4\pi\varepsilon_0}\mathbf{dl}\cdot \overset{q}{\mathrm{grad}}\frac{1}{L_{qp}}. \qquad [1.112]$$

Here the index q means that only this point varies. Since

$$\overset{q}{\mathrm{grad}}\frac{1}{L_{qp}} = \frac{\mathbf{L}_{qp}}{L_{qp}^3},$$

Eq. [1.112] becomes

$$U(p) = \frac{1}{4\pi\varepsilon_0}\frac{\mathbf{p}\cdot\mathbf{L}_{qp}}{L_{qp}^3} \qquad [1.113]$$

and

$$\mathbf{p} = e\mathbf{dl} \qquad [1.114]$$

is called the moment of the electric dipole. This vector shows the direction of the displacement of the positive charge with respect to negative one, and its magnitude equals the product of the positive charge and the distance between them, dl. The vector $\mathbf{p}$ is the parameter characterizing the electric dipole. It may be proper to note that Eq. [1.109] correctly describes the potential everywhere, but Eq. [1.113] is valid only at relatively large distances from charges (when the condition [1.110] holds). From Eq. [1.113], it follows that

$$U(p) = \frac{1}{4\pi\varepsilon_0}\frac{p\cos\theta}{L_{qp}^2}. \qquad [1.115]$$

Here θ is the angle between the vector $\mathbf{p}$ and the radius vector from the dipole center to an observation point (Fig. 1.9A). It is clear that the potential of the electric dipole decays with distance more rapidly than the potential caused by elementary charge, and also depends on the angle θ. In particular, if $\theta<\pi/2$, the potential is positive, since charge $+e$ is closer to the observation point; at the equipotential plane $\theta=\pi/2$, it is equal to zero because the distance from both charges to any point of this plane is the same. Finally, the potential is negative when $\theta>\pi/2$. To study the electric field, it is convenient to use a spherical system of coordinates, R, θ, ϕ, with the z-axis directed along the dipole moment (Fig. 1.9B). Then, Eq. [1.115] becomes

$$U(R,\theta,\phi) = \frac{p}{4\pi\varepsilon_0}\frac{\cos\theta}{R^2}.$$

Since

$$\mathbf{E} = -\operatorname{grad} U,$$

in spherical coordinates, we have

$$E_R = -\frac{\partial U}{\partial R}, \quad E_\theta = -\frac{1}{R}\frac{\partial U}{\partial \theta}, \quad E_\phi = -\frac{1}{R\sin\theta}\frac{\partial U}{\partial \phi},$$

and the components of the field **E** are

$$E_R = \frac{2p\cos\theta}{4\pi R^3}, \quad E_\theta = \frac{p\sin\theta}{4\pi R^3}, \quad E_\phi = 0. \qquad [1.116]$$

Thus, the electric field of the dipole is confined in the meridian planes ($E_\varphi = 0$), and decreases with R more rapidly than that of the elementary charge. In agreement with Coulomb's law, at all points of the z-axis the field **E** is directed parallel to the dipole moment, regardless of whether the angle θ is equal to 0 or π. In the equatorial plane, $\theta = \pi/2$, the field has a direction opposite to that of the dipole moment:

$$E_R = 0, \quad E_\theta = \frac{p}{4\pi R^3}.$$

It is useful show that a direction of the vector **E** along a given radius vector, θ = constant, remains the same. In fact, Eqs. [1.116] give

$$\tanh\psi = \frac{E_\theta}{E_R} = \frac{1}{2}\tan\theta. \qquad [1.117]$$

Here ψ is the angle between the radius vector and the electric field. Thus, with the increase of the distance from the dipole (θ = constant), both components of the field decrease, but the orientation of the field **E** does not change. In deriving formulas for the electric dipole, we made two assumptions: (1) the sum of two charges is equal to zero and (2) the field is considered only at distances that greatly exceed the separation between charges. This suggests that Eqs. [1.115] and [1.116] can also describe the potential and electric field of an arbitrary neutral system of charges, for which $\sum e_i = 0$, provided that observation points are located far away with respect to dimensions of the volume where charges are located. For instance, during electrostatic induction, charges of both signs appear on a conductor surface. With an increase of the distance, the potential U and the field **E** of these charges tend to those of the electric dipole and such behavior occurs for any shape and size of a conductor.

1.12.6. System of Dipoles and the Vector of Polarization P

Now suppose that there is a continuous distribution of the electric dipoles in a volume V with the dipole moments $\mathbf{p}(q)$, and our goal is to find the potential at any point inside and outside this volume. First, consider the potential caused by dipoles inside an elementary volume dV. By definition, such volume is so small that all dipole moments have the same magnitude and direction, and they are located at the same distance from an observation point. This means that each dipole inside an elementary volume creates the same potential, and their superposition also gives the potential of the electric dipole. We thus obtain for the potential caused by this volume

$$dU(p) = \frac{1}{4\pi\varepsilon_0}\frac{d\mathbf{M}\cdot\mathbf{L}_{qp}}{L_{qp}^3}, \qquad [1.118]$$

where $d\mathbf{M}$ is the sum of the dipole moments—that is, we have replaced the system of electric dipoles by one equivalent dipole with the moment $d\mathbf{M}$. By analogy with the volume density of charges, let us characterize a distribution of the dipole moments by vector $\mathbf{P}$, defined by

$$d\mathbf{M} = \mathbf{P}\,dV \qquad [1.119]$$

and called the electric "polarization vector." Later we will explain the origin of this name, but for now it is enough to understand that it can be viewed as a density of dipoles in an elementary volume. Applying the principle of superposition, we now obtain an expression for the potential caused by all dipoles of the volume V,

$$U(p) = \frac{1}{4\pi\varepsilon_0}\int_V \frac{\mathbf{P}(q)\cdot\mathbf{L}_{qp}}{L_{qp}^3}\,dV, \qquad [1.120]$$

and it is valid everywhere, including points inside a volume and at its surface. Performing a differentiation with respect to the point p, it is a simple matter to find components of the electric field.

1.12.7. Double Layer and its Potential

Finally, we consider a special distribution of surface charge that has unusual properties. Imagine that positive and negative charges with density $\sigma(q_2)$ and $\sigma(q_1)$ are distributed on two surfaces S_+ and S_-, respectively, and that the separation between them dl is much smaller than the distance from an observation point to these surfaces (Fig. 1.9C). Every opposing pair of surface elements has charges of the same magnitude, but opposite sign.

Such a system of charges is called a "double layer," and is widely used as a model in the theory of self-potential and induced polarization methods of geophysics, as well as in marine electrical prospecting for highly resistive zones within an oil reservoir. Of course, the electric force between the two layers tends to pull the two layers of charge toward each other and neutralize the total charge. Therefore, as in the case of the electric dipole, there must be some force with opposite direction that preserves the double layer. For convenience, choose some surface located between surfaces S_+ and S_- with the normal **n** (Fig. 1.9C). Now we are ready to find the potential caused by this system of charges. Consider a point q of this surface and the two elementary charges located at the distance $\mathrm{d}l$ from each other:

$$e_2 = e = \sigma(q)\mathrm{d}S \text{ and } e_1 = -e = -\sigma(q)\mathrm{d}S.$$

These charges form an electric dipole with the moment $\mathrm{d}\mathbf{P}$ equal to

$$\mathrm{d}\mathbf{P} = e\,\mathrm{d}\mathbf{l} \text{ or } \mathrm{d}\mathbf{P}(q) = \sigma(q)\mathrm{d}S\,\mathrm{d}l\,\mathbf{n},$$

since $\mathrm{d}\mathbf{l}$ and $\mathbf{n}$ have the same direction. Thus,

$$\mathrm{d}\mathbf{P} = \eta\,\mathrm{d}\mathbf{S}, \qquad [1.121]$$

where the function

$$\eta(q) = \sigma(q)\mathrm{d}l \qquad [1.122]$$

characterizes the density of the dipole moments. Correspondingly, $\eta(q)$ is called the double-layer density. As follows from Eq. [1.118], the potential caused by this elementary dipole is

$$\mathrm{d}U(p) = \frac{\eta(q)}{4\pi\varepsilon_0}\frac{\mathrm{d}\mathbf{S}\cdot\mathbf{L}_{qp}}{L_{qp}^3}$$

or

$$\mathrm{d}U(p) = -\frac{\eta(q)}{4\pi\varepsilon_0}\frac{\mathrm{d}\mathbf{S}\cdot\mathbf{L}_{pq}}{L_{qp}^3}, \qquad [1.123]$$

where $\mathrm{d}\mathbf{S} = \mathrm{d}S\mathbf{n}$ is the element of the middle surface S. In essence, we have replaced a real distribution of charges located at two different surfaces S_- and S_+ by the mathematical model of the double layer placed on the surface S. As with an individual dipole, it is obvious that this replacement does not change the field if the observation points are located far away. It is useful to represent Eq. [1.123] in the form

$$dU(p) = -\frac{\eta(q)}{4\pi\varepsilon_0}d\omega(p). \qquad [1.124]$$

Here $d\omega(p)$ is the solid angle over which we see the elementary surface dS from the point p. Applying the principle of superposition, we have for the potential caused by the double layer

$$U(p) = -\frac{1}{4\pi\varepsilon_0}\int_S \eta(q)d\omega. \qquad [1.125]$$

For instance, if the double layer is uniform (η=constant), we have

$$U(p) = -\frac{\eta}{4\pi\varepsilon_0}\omega(p), \qquad [1.126]$$

where $\omega(p)$ is the solid angle subtended by the double-layer surface S when viewed from the point p. Calculation of the solid angle allows us to easily find the potential and electric field for certain simple configurations. In the case when the uniform double layer is closed, we have for points inside and outside of S

$$U^{\mathrm{i}} = -\frac{\eta}{\varepsilon_0} \text{ and } U^{\mathrm{e}} = 0, \qquad [1.127]$$

respectively. Therefore, outside a uniform and closed double layer the electric field is zero. Consider now the behavior of the potential and field $\mathbf{E}$ near the double layer when the density $\eta(q)$ is not uniform. The potential at an observation point p as it approaches a point q on the double layer ($p \to q$) can be represented at each point as the sum of two terms:

$$U(p) = U^{q}(p) + U^{S-p}(p). \qquad [1.128]$$

Here $U^q(p)$ is the potential caused by the element of the double layer, $\eta(q)dS(q)$, and $U^{S-q}(p)$ is the potential is due to the rest part of the double layer. When the observation point p approaches the point q ($p \to q$), the solid angle subtended by the surface $dS(q)$ tends to $\pm 2\pi$, and correspondingly for the potential caused by this element, we have

$$U(p) = \pm\frac{\eta(p)}{2\varepsilon_0} \quad \text{if } p \to q.$$

Therefore, the potential U at the front and back sides of the double layer is

$$U_+(p) = \frac{\eta(p)}{2\varepsilon_0} + U_+^{S-p}(p)$$

and $\qquad , \text{ if } p \to q. \qquad$ [1.129]

$$U_-(p) = -\frac{\eta(p)}{2\varepsilon_0} + U_-^{S-p}(p)$$

Because the charges that contribute to the potential U^{S-p} are located away from the point q, this part of the potential is a continuous function at q. That is,

$$U_+^{S-p}(p) = U_-^{S-p}(p) \quad \text{as } p \to q.$$

Thus, the difference of the potential across the double layer is simply

$$U_+(p) - U_-(p) = \frac{\eta(p)}{\varepsilon_0} \quad \text{if } p \to q. \qquad [1.130]$$

The potential caused by the double layer is therefore discontinuous in crossing the layer, and the size of the discontinuity is proportional to the density $\eta(q)$ at the point of observation. This behavior of the potential is exceptional in two ways. First, as we saw earlier, potential fields satisfying Coulomb's law are generally continuous everywhere. Here, the discontinuity arises from the limiting process that squeezes the two surfaces of opposite charge S_- and S_+ into a single, idealized dipole layer. It we consider the potential between the surfaces—that is inside the double layer—the discontinuity disappears. Perhaps even more interesting is the second way in which Eq. [1.30] is unusual. Taking the derivative on both sides of this equation in a direction tangent to the double layer gives

$$\frac{\partial U_+}{\partial t} - \frac{\partial U_-}{\partial t} = \frac{1}{\varepsilon_0}\frac{\partial \eta}{\partial t},$$

where t stands for a coordinate limited to the surface of the double layer. Thus,

$$E_t^+(p) - E_t^-(p) = -\frac{1}{\varepsilon_0}\frac{\partial \eta(p)}{\partial t} \quad \text{if } p \to q \qquad [1.131]$$

and the tangential component of the electric field is also discontinuous across a double layer at points where the density η varies along the layer. It is important to remember that Eq. [1.130] describes the potential at points within the double layer and, therefore, that the derivative must be tangent to the layer. This equation cannot be used to calculate the derivative perpendicular to the layer, that is, the derivative along the normal **n**, since that requires knowledge of the potential outside the layer. Let us now consider the normal components of the field. Recall that the normal component of the field **E** in the vicinity of a small element of surface charge is

$$E_{\mathrm{n}} = \pm \frac{\sigma}{2\varepsilon_0}.$$

Therefore, two elementary surfaces with charges $\sigma(p)\mathrm{d}S$ and $-\sigma(p)\mathrm{d}S$ create a field

$$E_{\mathrm{n}}^{p}(p) = \frac{\sigma(p)}{\varepsilon_0},$$

in the thin space between the two surface, that is, within the double layer. This field is directed from the positive to the negative charge. At the same time, the normal component of the field of charges outside S_- and S_+ is equal to zero when $p \to q$. This allows us to conclude that the normal component of the electric field is continuous across the double layer:

$$E_{\mathrm{n}}^{+}(p) = E_{\mathrm{n}}^{-}(p) \quad \text{or} \quad \frac{\partial U_+}{\partial n} = \frac{\partial U_-}{\partial n}. \qquad [1.132]$$

It is helpful to explain in a little more detail why the double layer has such unusual properties. By definition, the voltage between surfaces of the double layer is

$$U_+(p) - U_-(p) = \int_+^- \mathbf{E}^{\mathrm{i}} \cdot \mathrm{d}\mathbf{l}, \qquad [1.133]$$

where $\mathbf{E}^{\mathrm{i}}$ is the electric field in the region inside the double layer, that is, between the two surfaces S_- and S_+. When we take the mathematical model of a double layer and squeeze the two surfaces into a single layer S, the distance **dl** in Eq. [1.133] goes to zero, which means that the field $\mathbf{E}^{\mathrm{i}}$ between the surfaces tends to infinity, so that the voltage at the right-hand side of Eq. [1.133] has a finite value. As pointed out earlier, there must be an external field $\mathbf{E}^{\mathrm{ext}}$ of some sort that supports the double layer (in opposition to the Coulomb's field of the charges themselves). The external force is

$$\Xi^{\mathrm{ext}} = \int_-^+ \mathbf{E}^{\mathrm{ext}} \cdot \mathrm{d}\mathbf{l}.$$

Taking into account Eqs. [1.130] and [1.133], we have

$$\Xi^{ext} = U_+ - U_- = \frac{\eta}{\varepsilon_0}. \qquad [1.134]$$

1.13. SOLUTIONS OF POISSON'S EQUATION AND UNIQUENESS

Earlier we showed that Poisson's equation has an infinite number of solutions within any region of space, each differing by a function that satisfies Laplace's equation. These latter functions may be interpreted as potentials caused by distributions of charge located outside the region where Poisson's equation is given. Thus, determination of the potential, and in turn the electric field, for a given configuration requires additional information about the field behavior. Usually, this additional information comes in the form of values for the potential, or field, on the surface or boundary of the region where the field is to be determined (Fig. 1.10A). We will now formulate conditions on a surface S that uniquely define the electric field $\mathbf{E}$ inside the volume V bounded by the surface. The derivation starts from Gauss' theorem, which is the natural "bridge" between field values inside the volume and those at the boundary surface:

$$\int_V \mathrm{div}\,\mathbf{X}\,\mathrm{d}V = \oint_S \mathbf{X}\cdot \mathrm{d}S = \oint_S X_n \mathrm{d}S. \qquad [1.135]$$

Here $\mathbf{X}$ is a continuous function within the volume V, and X_n is its component along the unit normal vector $\mathbf{n}$ directed outward from the surface S. Let U_1 and U_2 be two particular solutions of Poisson's equation in the volume V (there are an infinite number):

$$\nabla^2 U_1 = -\frac{\delta}{\varepsilon_0} \text{ and } \nabla^2 U_2 = -\frac{\delta}{\varepsilon_0} \qquad [1.136]$$

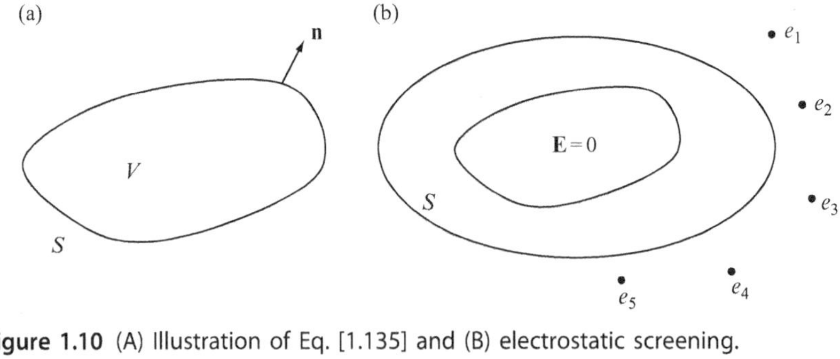

Figure 1.10 (A) Illustration of Eq. [1.135] and (B) electrostatic screening.

and take their difference

$$U_3(p) = U_2(p) - U_1(p). \quad [1.137]$$

Because Poisson's equation is linear,

$$\nabla^2 U_3 = \nabla^2(U_2 - U_1) = \nabla^2 U_2 - \nabla^2 U_1 = 0$$

the function $U_3(p)$ is a solution Laplace's equation:

$$\nabla^2 U_3 = 0. \quad [1.138]$$

We now construct a vector function $\mathbf{X}(p)$ from $U_3(p)$ as follows:

$$\mathbf{X} = U_3 \operatorname{grad} U_3 = U_3 \nabla U_3. \quad [1.139]$$

Substitution of Eq. [1.139] into Eq. [1.135] gives

$$\int_V \operatorname{div}(U_3 \nabla U_3) \mathrm{d}V = \oint_S U_3 \operatorname{grad}_\mathrm{n} U_3 \mathrm{d}S, \quad [1.140]$$

where $\operatorname{grad}_\mathrm{n} U_3$ is the component of the gradient along the normal $\mathbf{n}$, that is,

$$\operatorname{grad}_\mathrm{n} U_3 = \frac{\partial U_3}{\partial n}. \quad [1.141]$$

Note that the boundary surface S surrounding the volume may consist of several different surfaces. Expanding the integrand on the left-hand side of Eq. [1.140] gives

$$\operatorname{div}(U_3 \nabla U_3) = U_3 \nabla^2 U_3 + \nabla U_3 \cdot \nabla U_3 = (\nabla U_3)^2, \quad [1.142]$$

because, by construction, $\nabla^2 U_3 = 0$. Substituting Eqs. [1.141] and [1.142] into Eq. [1.140] gives

$$\int_V (\nabla U_3)^2 \mathrm{d}V = \oint_S U_3 \frac{\partial U_3}{\partial n} \mathrm{d}S. \quad [1.143]$$

This identity, which in essence is the Gauss' theorem, allows us to formulate the most important boundary conditions for a unique solution of Poisson's equation. Notice first that the integrand on the left-hand side is never negative because $(\nabla U_3)^2 \geq 0$, and that the function U_3 was defined as the difference of two solutions of Poisson's equation $(U_2 - U_1)$. We now formulate boundary conditions for the potential of the electric field $\mathbf{E}$ that uniquely define this field inside the volume V.

Suppose that the surface integral on the right-hand side of Eq. [1.143] vanishes. Then,

$$\int_V (\nabla U_3)^2 \mathrm{d}V = 0, \qquad [1.144]$$

and, because the integrand cannot be negative, we conclude that at every point of the volume

$$\operatorname{grad} U_3 = 0. \qquad [1.145]$$

This means that the derivative of the function U_3 in any direction l is zero:

$$\frac{\partial U_3}{\partial l} = \frac{\partial (U_2 - U_1)}{\partial l} = 0.$$

Thus, if the surface integral in Eq. [1.143] vanishes, then the derivatives of the two solutions of Poisson's equation (U_1 and U_2) used to construct U_3 are equal to each other in any direction:

$$\frac{\partial U_1}{\partial l} = \frac{\partial U_2}{\partial l}.$$

This means that these solutions can differ only by a constant:

$$U_2(p) = U_1(p) + C, \qquad [1.146]$$

and C is the same for all points of the volume V including the surface S. Next we will define conditions under which the surface integral

$$\oint_S U_3 \frac{\partial U_3}{\partial n} \mathrm{d}S \qquad [1.147]$$

vanishes and, correspondingly, Eq. [1.146] becomes valid. There are at least three such conditions which are described below, and they often play an important role in the theory of electric methods.

1.13.1. Case 1

Suppose that the potential $U(p)$ is known on the boundary surface, that is,

$$U(p) = \varphi_1(p) \text{ on } S, \qquad [1.148]$$

and we are looking for a solution of Poisson's equation that satisfies the condition [1.148]. Let us assume that there are two different solutions of this equation inside the volume: $U_1(p)$ and $U_2(p)$, that coincide on the boundary surface:

$$U_1(p) = U_2(p) = \varphi_1(p) \text{ on } S.$$

Then their difference is a function that is equal to zero:

$$U_3(p) = 0 \text{ on } S$$

and, consequently, the surface integral in Eq. [1.147] vanishes. It follows from Eq. [1.146] that the function U_3 must be constant in the volume V, and this constant must be zero, because U_3 equals zero on the boundary S. We have thus proved that Poisson's equation and Eq. [1.148], that is,

$$\nabla^2 U = -\frac{\delta}{\varepsilon_0} \text{ in } V \text{ and } U(p) = \varphi_1(p) \text{ on } S \qquad [1.149]$$

uniquely define the potential U and the electric field $\mathbf{E}$, since $\mathbf{E} = -\nabla U$. The pair of Equations [1.149] is called a "Dirichlet" boundary-value problem (after the German mathematician Johann Dirichlet, 1805–1859) and can be interpreted as follows. The electric field is a sum of two terms: one is the field caused by charges in the volume V, which can be calculated by Coulomb's law since the density of charges is known; the second term is the field generated by charges located outside V, and the boundary condition on the surface S replaces the absence of knowledge about a distribution of these charges. A consequence of Eq. [1.148] is that the component of the electric field E_t along any direction t tangential to the boundary surface is known, since $E_t = -\partial U/\partial t$. Therefore, this boundary-value problem can be written in terms of the field $\mathbf{E}$ as

$$\text{curl}\,\mathbf{E} = 0, \qquad \text{div}\,\mathbf{E} = \frac{\delta}{\varepsilon_0}$$

and [1.150]

$$E_t = -\frac{\partial \varphi_1}{\partial t} \text{ on } S.$$

1.13.2. Case 2

Assume that two different solutions of Poisson's equation $U_1(p)$ and $U_2(p)$ have the same normal derivative on the surface S, that is,

$$\frac{\partial U_1}{\partial n} = \frac{\partial U_2}{\partial n} = \varphi_2(p) \text{ on } S. \qquad [1.151]$$

It is clear that the normal derivative of the difference of these solutions vanishes on the boundary S:

$$\frac{\partial U_3}{\partial n} = 0 \text{ on } S.$$

Therefore, the surface integral in Eq. [1.143] as in the previous case equals zero, and, correspondingly, inside the volume, we have

$$\nabla U_3 = 0.$$

This means that any solutions of Poisson's equation can differ from each other in the volume V only by some constant if their normal derivatives coincide on the boundary surface S. Thus, a second boundary-value problem that uniquely defines the field $\mathbf{E}$ can be written as

$$\nabla^2 U = -\frac{\delta}{\varepsilon_0}$$

and [1.152]

$$\frac{\partial U}{\partial n} = \phi_2(p) \text{ on } S$$

or

$$\operatorname{curl} \mathbf{E} = 0, \quad \operatorname{div} \mathbf{E} = \frac{\delta}{\varepsilon_0}$$

and [1.153]

$$E_n = -\phi_2(p).$$

This formulation is called Neumann's boundary-value problem (after the German mathematician Carl Neumann, 1832–1925). Unlike the previous case, Eq. [1.152] defines the potential only to within a constant C, but of course the field is determined uniquely, since $\operatorname{grad} C = 0$.

1.13.3. Case 3

Next we suppose that the boundary surface S is equipotential; that is,

$$U(p) = C \text{ on } S, \qquad [1.154]$$

where C is unknown. In addition, it is assumed that the value of the integral

$$\oint_S \frac{\partial U}{\partial n} \mathrm{d}S = \phi_3 \qquad [1.155]$$

is given.

We will show now that solutions of Poisson's equation, $U_1(p)$ and $U_2(p)$, satisfying conditions [1.154] and [1.155] can differ from each other by a constant only as in the case of Neumann's boundary-value problem. Consider again the surface integral in Eq. [1.143]:

$$\oint_S U_3 \frac{\partial U_3}{\partial n} \mathrm{d}S.$$

Because the boundary surface S is an equipotential surface for both the fields, U_1 and U_2, their difference

$$U_3(p) = U_2(p) - U_1(p)$$

is also constant at this surface. Consequently, we have

$$\oint_S U_3 \frac{\partial U_3}{\partial n} \mathrm{d}S = U_3 \oint_S \frac{\partial U_3}{\partial n} \mathrm{d}S.$$

Equation [1.155] then gives

$$\oint_S U_3 \frac{\partial U_3}{\partial n} \mathrm{d}S = U_3 \oint_S \frac{\partial U_3}{\partial n} \mathrm{d}S = U_3 \left[\oint_S \frac{\partial U_2}{\partial n} \mathrm{d}S - \oint_S \frac{\partial U_1}{\partial n} \mathrm{d}S \right] = 0$$

and, in accordance with Eq. [1.143],

$$\nabla U_3 = 0 \text{ or } U_2(p) = U_1(p) + C.$$

Thus boundary conditions [1.154] and [1.155] define the solution of Poisson's equation within the volume V up to a constant and, correspondingly, a third boundary-value problem can be posed as

$$\nabla^2 U = -\frac{\delta}{\varepsilon_0}$$

and [1.156]

$$\oint_S \frac{\partial U}{\partial n} \mathrm{d}S = \phi_3 \text{ on } S,$$

where S is an equipotential surface, or

$$\operatorname{curl} \mathbf{E} = 0, \quad \operatorname{div} \mathbf{E} = \frac{\delta}{\varepsilon_0}$$

and [1.157]

$$\oint_S E_\mathrm{n} \mathrm{d}S = -\phi_3.$$

Here E_n is the normal component of the field, and it coincides with the total field since at the equipotential surface the tangential component vanishes. There are many cases when the potential can be found by solving the third boundary-value problem. For instance, the determination of the electric field outside a conductor is reduced to solution of this problem provided that the total charge is known at its equipotential surface. In fact, as follows from Gauss' formula, we have

$$\oint_S E_n \mathrm{d}S = \frac{e}{\varepsilon_0},$$

that is, the given value of ϕ_3 differs from the charge on the conductor surface by a constant only.

1.13.4. Boundary-Value Problems

Let us summarize the results, which are as follows:

(1) Three types of the boundary-value problems were formulated corresponding to three kinds of boundary conditions that were found to uniquely define the electric field inside the volume.

(2) As was already mentioned, the volume V can be surrounded by several surfaces, and at every point on them, one of these conditions has to be specified.

(3) The procedure of determination of these conditions, based on the use of Gauss' theorem, is called the theorem of uniqueness.

(4) In general, boundary surfaces may have an arbitrary shape, as well as location, and here it is appropriate to distinguish three cases. In the first case, the boundary surfaces are located in the vicinity of given charges, which are usually sources of the so-called primary field. In essence, boundary conditions on such surfaces replace information about parameters of these sources (single charge, electric dipole, double layer, and others). In the second case, the boundary surface is located far away from all sources and for this reason it is natural to assume that the field is very small at points of this surface. Moreover, from an analysis of the specific problem, it is usually possible to predict the manner in which a field decreases at great distances from charges. This information defines the boundary condition at infinity. And finally, one more case deserves our attention: the boundary surface is often the surface where measurements of the electric field are performed.

(5) In accordance with the theorem of uniqueness, a solution of the forward problem, that is, a determination of the electric field, includes the following steps:

(a) We have to solve Poisson's equation.

(b) We have to select among those solutions of such functions that satisfy boundary conditions.

(c) If there are interfaces, where Poisson's equation does not hold, we have to choose among the last group a function that also obeys at these surfaces two conditions:

$$U_1 = U_2 \text{ and } \frac{\partial U_2}{\partial n} - \frac{\partial U_1}{\partial n} = -\frac{\sigma}{\varepsilon_0},$$

provided that double layer is absent. Here σ is the surface density of charges and the normal $\mathbf{n}$ is directed from the back to front side of the interface.

(6) It is convenient to distinguish two kinds of boundary-value problems. One is called the internal boundary-value problem, where the volume V is surrounded by the surface S. The other is an external boundary-value problem when an observation point is located between the surface S and at infinity. For illustration of application of the theorem of uniqueness, consider one example.

1.13.5. Example: Electrostatic Shielding

The uniqueness theorem sheds light on the phenomenon of electrostatic induction and the shielding effect. First, suppose that some volume V is surrounded by a conducting surface S, for instance, a metal foil, and sources of the field $\mathbf{E}_0$ are located outside this volume (Fig. 1.10B). As we know, due to electrostatic induction, positive and negative charges arise on the external surface of the conductor, and they are distributed in such a way that the potential does not change within a conductor. Our task is to find a field $\mathbf{E}$ in the volume V. Bearing in mind that the surface S is an equipotential surface, we have for each point of the volume V

$$\nabla^2 U = 0 \text{ and } U = C, \text{ on } S, \qquad [1.158]$$

where C is unknown constant. As follows from the theorem of uniqueness we formulated, Dirichlet's boundary-value problem and, correspondingly, Eq. [1.158] uniquely define the electric field as well as its potential. We will look for a solution, applying a trial-and-error method. Suppose that within the volume V, the potential is also constant, the same constant C, Eq. [1.158]; that is,

$$U(p) = C \text{ in } V. \qquad [1.159]$$

It is obvious that this function automatically satisfies the boundary condition and Laplace's equation since even the first derivative of a constant is zero. Thus, we have found the potential inside the volume, and there are no other solutions of this boundary-value problem. Inasmuch as the potential is constant, the electric field vanishes in the volume surrounded by an arbitrary conducting surface:

$$\mathbf{E} = 0. \tag{1.160}$$

In other words, we have described the effect of electrostatic screening and proved it by using the uniqueness theorem. Note that, in this case, the medium outside and inside the conductor is an insulator, and our goal was to determine the field surrounded by an internal surface of the conductor. Next assume that sources of the field $\mathbf{E}_0$ are enclosed by a conducting surface S, but we are interested in the behavior of the field outside this surface where charges are absent. Again, due to the electrostatic induction, charges of both signs appear at the external and internal surfaces of the conductor and they acquire the same potential. Taking into account that the field $\mathbf{E}$ is zero within a conductor, the total flux of this field through any surface S_i inside a conductor is equal to zero:

$$\oint_S \mathbf{E} \cdot d\mathbf{S} = 0.$$

This means that the induced charges at the internal surface of the conductor are equal in magnitude and opposite in sign to the charges located in the volume V. Consequently, at the external surface the same charge as that in the volume V appears since the total charge of the conductor is zero. It is obvious that this system of charges creates a field outside the conductor. This shows the electrostatic shielding does not work in both directions. Let us demonstrate it using again the theorem of uniqueness. In this case, the potential outside a conductor satisfies the following conditions:

(1) At regular points

$$\nabla^2 U = 0.$$

(2) At points of the external surface of a conductor

$$U = C.$$

(3) At infinity, the potential tends to zero:

$$U \to 0.$$

Certainly, any constant potential does not simultaneously obey both boundary conditions, provided that $C \neq 0$, and therefore the potential changes from point to point. We see that the electric field differs from zero although charges in the volume V are surrounded by a conductor. At the same time, according to the uniqueness theorem, if the potential of the conductor becomes equal to zero (e.g., due to grounding), then it vanishes at all points outside the conductor:

$$U(p) = 0. \qquad [1.161]$$

It also follows from this same reasoning that, if we surround a region of space by two conducting surfaces and connect them, the electric field between the surfaces disappears and we have shielded the region from the field caused by external charges even without grounding. This happens because the contact between the two surfaces bring them both to the same potential, and inside the volume a potential that is constant and equal to that of the boundary surfaces satisfies Laplace's equation uniquely.

1.14. POLARIZATION OF A MEDIUM

Earlier we presented two equations of the electric field that describe its behavior at regular points and interfaces:

$$\operatorname{div}\mathbf{E} = \frac{\delta}{\varepsilon_0} \text{ and } E_{2n} - E_{1n} = \frac{\sigma}{\varepsilon_0}. \qquad [1.162]$$

It is natural to raise the following question: what kinds of charges are included in the functions δ and σ? It turns out that in actual materials it is useful to distinguish two kinds of charge: free and bound (or polarization) charge,

$$\delta = \delta_0 + \delta_b \text{ and } \sigma = \sigma_0 + \sigma_b, \qquad [1.163]$$

where subscripts 0 and b denote free and bound charge, respectively. What is the purpose of this distinction? Earlier we showed, for example, that a medium placed in an electric field produces a change of the field because of the redistribution of charge on the conductor. This happens because every element of a conductor contains (in principle) an unlimited amount of charge that is free to move under the action of an applied electric field. As described in the section on electrostatic induction, eventually equilibrium is reached in which free charge moves to the surface of the conductor and cancels the electric field inside the conductor. In this section, we will describe a second type of charge, called bound charge, which can in principle appear in any medium. This description is based on a microscopic model that considers the material to be composed, for example, of molecules in a crystal structure that consist of a neutral collection of positive and negative charges bound to each other by atomic forces. The application of an external electric field causes the charges to change their positions slightly, separating positive and negative charge; but unlike free charges, bound charges do not leave their elementary volumes. This phenomenon is called "polarization," and it characterizes dielectric properties of a medium. Before deriving expressions for the electric potential and field that arise from polarization, we first make three comments:

(1) In reality every medium displays both conductivity and polarization.

(2) In some substances, atoms have a dipole moment even in the absence of the external field, but in most materials, the dipole moment appears only under action of the field, which produces some kind of deformation of an atom.

(3) Because of thermal effects, electric dipoles at the atomic scale are subject to random motions and are therefore randomly aligned.

Under the action of the external electric field, some fraction of the dipoles become oriented along the resultant field; that is, an ordered distribution is observed. To determine the field caused by these dipoles, we first consider an elementary volume dV which contains an equal amount of positive and negative charges. They are slightly displaced from each other and therefore have a dipole moment $d\mathbf{M}=\mathbf{P}\,dV$, where $\mathbf{P}$ is the vector of polarization introduced earlier when we considered the system of dipoles. This is the dipole moment of a unit volume; that is, it is the density of dipole moments $\mathbf{P} = d\mathbf{M}/dV$. The greater the amount of charge of one sign and the greater the distance between the positive and negative charges, the greater the polarization vector directed toward the positive charges. In most cases, we can assume that the polarization vector $\mathbf{P}$ is directly proportional to the field $\mathbf{E}$; that is,

$$\mathbf{P} = \alpha\mathbf{E}, \qquad [1.164]$$

where α is called the "polarizability." It characterizes (macroscopically) the ability of a material to support polarization of charge. The field $\mathbf{E}$ on the right-hand side of Eq. [1.164] is the total field caused by all charges, including dipoles in neighboring parts of the medium. Correspondingly, it consists of the fields caused by free charge and by polarization:

$$\mathbf{E} = \mathbf{E}_0 + \mathbf{E}_p, \qquad [1.165]$$

Here $\mathbf{E}_0$ and $\mathbf{E}_p$ are electric fields caused by free and polarization charges, respectively. Note that the macroscopic concept of vector $\mathbf{P}$ allows us to perform in every elementary volume a transition from the micro- to the macroscale: in place of a system of atomic dipoles, we deal with one equivalent dipole with the moment d$\mathbf{M}$. Because all charge within an elementary volume $dV(q)$ creates a field, we can use Eq. [1.118] for the potential of an elementary dipole to calculate the field created by an elementary volume of polarization:

$$dU_p(p) = \frac{\mathbf{P}(q)\cdot\mathbf{L}_{qp}}{4\pi\varepsilon_0 L_{qp}^3}\,dV, \tag{1.166}$$

where L_{qp} is the distance from the point q that defines the position of the element dV to the observation point p. All parts of the volume V generate an electric field $\mathbf{E}_p$ whose potential (Eq. [1.120]) is therefore given by the integral:

$$U_p(p) = \frac{1}{4\pi\varepsilon_0}\int_V \frac{\mathbf{P}(q)\cdot\mathbf{L}_{qp}}{L_{qp}^3}\,dV. \tag{1.167}$$

1.15. THE POTENTIAL AND ELECTRIC FIELD CAUSED BY POLARIZATION

Equation [1.167] cannot be used directly for calculation because it suffers from the same "closed circle" problem described earlier: on the one hand, calculation of the potential U_p from this equation requires that the polarization vector $\mathbf{P}$ is known; on the other hand, the polarization vector itself is determined by the total field (see Eq. [1.164]) including the field associated with the potential, $\mathbf{E}_p = -\operatorname{grad} U_p$, caused by the polarization effect. This problem arises because the system of Equations [1.76] in effect contains three unknowns: the electric field $\mathbf{E}$, free charges (δ_0, σ_0), and polarization vector $\mathbf{P}$. Using the polarizability α, we will now modify this system to create a new system with only two unknowns, namely, the electric field $\mathbf{E}$ and density of free charge. We start by performing simple transformations on the right-hand side of Eq. [1.167] and introduce a new vector called the electric induction field. The potential of the electric field caused by free charges

$$U_0(p) = \frac{1}{4\pi\varepsilon_0}\left[\int_V \frac{\delta_0(q)dV}{L_{qp}} + \oint_S \frac{\sigma_0(q)}{L_{qp}}dS\right] \tag{1.168}$$

will be studied in detail in the next chapter; here we consider only the field caused by the polarization. Let us represent this potential as

$$U_{\mathrm{p}}(p) = \frac{1}{4\pi\varepsilon_0}\int_V \mathbf{P}(q)\cdot \overset{q}{\nabla}\frac{1}{L_{qp}}\,\mathrm{d}V, \qquad [1.169]$$

where ∇^q denotes that the gradient operator to the q coordinates of the function L_{qp},

$$\overset{q}{\nabla}\frac{1}{L_{qp}} = \frac{\mathbf{L}_{qp}}{L_{qp}^3}.$$

Using the identity

$$\mathbf{P}(q)\cdot\overset{q}{\nabla}\frac{1}{L_{qp}} = \overset{q}{\nabla}\cdot\frac{\mathbf{P}}{L_{qp}} - \frac{1}{L_{qp}}\overset{q}{\nabla}\cdot\mathbf{P},$$

we obtain from Eq. [1.169]

$$U_{\mathrm{p}}(p) = \frac{1}{4\pi\varepsilon_0}\int_V \frac{\mathrm{div}\,\mathbf{P}(q)}{L_{qp}}\mathrm{d}V - \frac{1}{4\pi\varepsilon_0}\int_V \frac{\mathrm{div}\,\mathbf{P}}{L_{qp}}\mathrm{d}V. \qquad [1.170]$$

We next simplify this expression using Gauss' theorem, replacing the first volume integral by a surface integral. In carrying out this step, we must allow for the possibility that either the polarizability α or the field $\mathbf{E}$, or both, are discontinuous functions—for example, across an interface between two different materials—and therefore the vector of polarization $\mathbf{P}$ is also discontinuous. Assume then that in the volume V, surrounded by the surface S, there is some interior surface S_i where $\mathbf{P}$ is discontinuous. Enclosing S_i by a surface S_0 (Fig. 1.11A), we obtain the volume V_1 bounded by the surfaces S (the boundary of the original volume) and S_0,

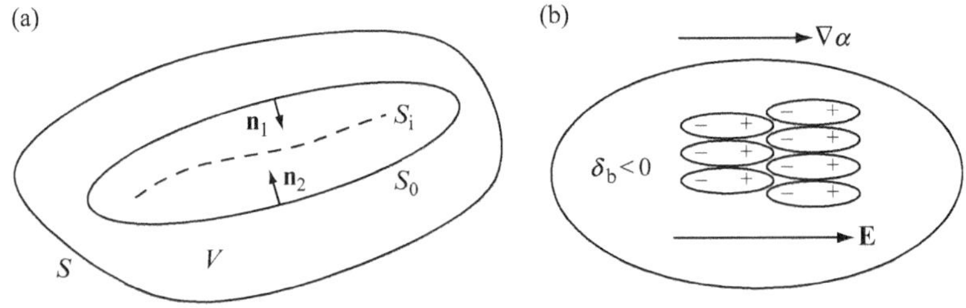

Figure 1.11 (A) Surfaces involved in the derivation of Eq. [1.171] for the field of polarization charge and (B) appearance of bound charge when the electric field and the gradient of the polarizability are aligned.

where all fields are regular and Gauss' theorem can be applied. We then take the limit as the surface S_0 shrinks around the discontinuity and eventually coincides with the front and back sides of S_i. The volume V_1 tends to the original volume V, but the surface of integration includes the surface S and both sides of S_i, where the vector $\mathbf{P}$ is discontinuous. Applying Gauss' theorem to this configuration in the first integral in Eq. [1.170], we obtain

$$U_p(p) = \frac{1}{4\pi\varepsilon_0}\int_{S_1}\frac{\mathbf{P}^{(1)}\cdot\mathrm{d}\mathbf{S}}{L_{qp}} + \frac{1}{4\pi\varepsilon_0}\int_{S_2}\frac{\mathbf{P}^{(2)}\cdot\mathrm{d}\mathbf{S}}{L_{qp}} + \frac{1}{4\pi\varepsilon_0}\int_{S}\frac{\mathbf{P}\cdot\mathrm{d}\mathbf{S}}{L_{qp}} - \frac{1}{4\pi\varepsilon_0}\int_{V}\frac{\mathrm{div}\,\mathbf{P}}{L_{qp}}\mathrm{d}V, \qquad [1.171]$$

where $\mathbf{P}^{(1)}$ and $\mathbf{P}^{(2)}$ are the values of the vector of polarization at the back (S_1) and front (S_2) sides of the surface S_i, respectively. The first two terms can be collapsed by recognizing that the normal has to be directed outside the volume (Fig. 1.11A); we have

$$\int_{S_1}\frac{\mathbf{P}^{(1)}\cdot\mathrm{d}\mathbf{S}}{L_{qp}} + \int_{S_2}\frac{\mathbf{P}^{(2)}\cdot\mathrm{d}\mathbf{S}}{L_{qp}} = \int_{S_i}\frac{\mathbf{P}^{(1)}\cdot\mathbf{n}_1 + \mathbf{P}^{(2)}\cdot\mathbf{n}_2}{L_{qp}}\mathrm{d}\mathbf{S} = \int_{S_i}\frac{P_{\mathrm{n}}^{(1)} - P_{\mathrm{n}}^{(2)}}{L_{qp}}\mathrm{d}S, \qquad [1.172]$$

since $\mathbf{n}_1 = \mathbf{n}$ and $\mathbf{n}_2 = -\mathbf{n}$, and $\mathbf{n}$ is the normal directed from the back to the front side of the surface S_i. Now we will show that the integral over the surface S (Eq. [1.171]), which envelops all possible media, vanishes. At infinity, the electric field and therefore polarization vector $\mathbf{P}$ decreases at a rate inversely proportional to the square of the distance. Consequently, taking a spherical surface S of very large radius R and making use of the mean value theorem we have

$$\oint_S \frac{\mathbf{P}\cdot\mathrm{d}\mathbf{S}}{L_{qp}} \to \frac{C}{R^3}4\pi R^2 \to 0 \text{ as } R \to \infty,$$

where C is some constant. Thus, we have derived a new expression for the potential caused by a polarization of a medium:

$$U_p(p) = \frac{1}{4\pi\varepsilon_0}\int_{V}\frac{-\mathrm{div}\,\mathbf{P}}{L_{qp}}\mathrm{d}V + \frac{1}{4\pi\varepsilon_0}\int_{S_i}\frac{-\mathrm{Div}\,\mathbf{P}}{L_{qp}}\mathrm{d}S, \qquad [1.173]$$

where

$$\mathrm{Div}\,\mathbf{P} = P_{\mathrm{n}}^{(2)} - P_{\mathrm{n}}^{(1)} \qquad [1.174]$$

is the difference of the normal component of the polarization vector on either side of the surface of discontinuity S_i. Letting the volume V be all of space, we can include all different media, including those parts of the volume where the polarization effect vanishes ($\alpha = 0$); obviously, such regions give no contribution to the volume integral.

1.16. VECTOR OF ELECTRIC INDUCTION AND SYSTEM OF EQUATIONS OF ELECTRIC FIELD

1.16.1. Bound Charges

Although Eqs. [1.167] and [1.173] give the same potential created by polarization of a medium, the second equation turns out to be more useful for understanding the charge distribution within a medium as well as for the formulation of boundary-value problems. First, comparison of Eqs. [1.168] and [1.173] shows the striking resemblance of these equations, and it suggests that, due to polarization, volume and surface charges appear, the density of which is defined by

$$\delta_b = -\operatorname{div}\mathbf{P} \quad \text{and} \quad \sigma_b = -\operatorname{Div}\mathbf{P} = P_n^{(1)} - P_n^{(2)}. \qquad [1.175]$$

The presence of these charges was established analytically by the transformation of Eq. [1.167], but we can arrive at the same conclusion from the physical point of view. Indeed, if the polarizability varies from point to point, the density of dipole moments also changes, and correspondingly we can expect the appearance of a net charge. This phenomenon is shown in Fig. 1.11B, in the case where negative charges arise because the parameter α increases in the direction of the field $\mathbf{E}$. If the electric field decreases in a direction where the polarizability also decreases, positive charge appears. This effect is especially obvious at interfaces with different dielectric properties. By definition, these charges cannot flow freely through the volume (unlike free charge), but only slightly change their position, which is why they are called bound charge. The different extent of movement of the free and bound charge is the sole difference between them; physically, both can create an electric field according to Coulomb's law. Thus, from Eqs. [1.168], [1.173], and [1.175], we see that, due to the polarization, bound charges arise at regular points of a medium, as well as at interfaces, and these charges create an electric field $\mathbf{E}_p$ that obeys Coulomb's law. Correspondingly, the potential and electric field caused by the free and bound charges are

$$U(p) = \frac{1}{4\pi\varepsilon_0}\int_V \frac{\delta(q)}{L_{qp}}\mathrm{d}V + \frac{1}{4\pi\varepsilon_0}\int_S \frac{\sigma(q)}{L_{qp}}\mathrm{d}S$$

and [1.176]

$$\mathbf{E}(p) = \frac{1}{4\pi\varepsilon_0}\int_V \frac{\delta(q)\mathbf{L}_{qp}}{L_{qp}^3}\mathrm{d}V + \frac{1}{4\pi\varepsilon_0}\int_S \frac{\sigma(q)\mathbf{L}_{qp}}{L_{qp}^3}\mathrm{d}S,$$

where δ and σ are the total density of the free and bound charges (Eqs. [1.163]). These equations were derived earlier from Coulomb's law, but now we know the origin of the charges. There are places in a medium where either, or both, types of charges can be present.

1.16.2. Vector of Electric Induction

We now recast the system of equations of the electric field in a polarizable and conducting medium in slightly different form. Splitting the total charge into free and bound components, we have for regular points

$$\operatorname{curl}\mathbf{E} = 0, \quad \operatorname{div}\mathbf{E} = \frac{1}{\varepsilon_0}(\delta_0 + \delta_b). \qquad [1.177]$$

We also assume that relations [1.164] and [1.175] hold:

$$\mathbf{P} = \alpha\mathbf{E} \text{ and } \delta_b = -\operatorname{div}\mathbf{P},$$

and that the polarizability α is known. We can now eliminate the unknown bound charges from the second equation. Substituting the expression δ_b into the second Equation [1.177] gives

$$\operatorname{div}\mathbf{E} = \frac{\delta_0}{\varepsilon_0} - \frac{\operatorname{div}\mathbf{P}}{\varepsilon_0} \text{ or } \operatorname{div}\left(\mathbf{E} + \frac{\mathbf{P}}{\varepsilon_0}\right) = \frac{\delta_0}{\varepsilon_0}. \qquad [1.178]$$

We introduce the auxiliary vector

$$\mathbf{D} = \varepsilon_0\mathbf{E} + \mathbf{P} \qquad [1.179]$$

and obtain

$$\operatorname{div}\mathbf{D} = \delta_0, \qquad [1.180]$$

since ε_0 is constant. The vector $\mathbf{D}$ is called the vector of "electric induction." Using the polarizability (Eq. [1.164]), we can express it in terms of the electric field:

$$\mathbf{D} = \varepsilon_0\mathbf{E} + \alpha\mathbf{E} = \varepsilon_0\left(1 + \frac{\alpha}{\varepsilon_0}\right)\mathbf{E} = \varepsilon_0(1 + \beta)\mathbf{E}, \qquad [1.181]$$

where $\beta = \alpha/\varepsilon_0$ is the dielectric susceptibility of a medium. We also define the "dielectric constant":

$$\varepsilon = \varepsilon_0(1 + \beta) = \varepsilon_r\varepsilon_0. \qquad [1.182]$$

Equation [1.181] becomes simply

$$\mathbf{D} = \varepsilon\mathbf{E}. \qquad [1.183]$$

The dielectric constant connects the vector of electric induction with the electric field and *vice versa.* In this relation, it is appropriate to note the following. Equation [1.183] is simple and gives the impression that both vectors **E** and **D** have the same physical meaning since they differ only by the scalar ε which is often uniform within a medium. The vector of the electric induction, however, is the sum of two completely different vectors; namely, the electric field **E** and the polarization **P**, which represents a density of dipole moments within a medium. Moreover, the name "dielectric constant," traditionally used for ε, is an unfortunate terminology because this quality will generally vary from point to point in a medium. From the identity

$$\operatorname{curl}\mathbf{D} = \operatorname{curl}(\varepsilon\mathbf{E}) = \varepsilon \operatorname{curl}\mathbf{E} + (\operatorname{grad}\varepsilon)\times\mathbf{E} = (\operatorname{grad}\varepsilon)\times\mathbf{E},$$

it follows that the field **D** is caused by free charge and vortices. In particular, at the interface of media with different dielectric constants, the tangential component of the electric field is continuous: $E_t^{(1)} = E_t^{(2)}$, which implies that the tangential component of the vector **D** has a discontinuity:

$$\varepsilon_1 E_t^{(1)} \neq \varepsilon_2 E_t^{(2)},$$

which indicates the presence of surface vortices. Because volume and surface vortices do not exist, the vector of the electric induction **D** has little or no physical meaning for constant fields and is introduced mainly to simplify equations of the field **E**. To complete the transformations, we have for points at the interface between two media (Eq. [1.162])

$$E_n^{(2)} - E_n^{(1)} = \frac{\sigma_0}{\varepsilon_0} + \frac{P_n^{(1)} - P_n^{(2)}}{\varepsilon_0} = \frac{\sigma}{\varepsilon_0}$$

or

$$\left(E_n^{(2)} + \frac{P_n^{(2)}}{\varepsilon_0}\right) - \left(E_n^{(1)} + \frac{P_n^{(1)}}{\varepsilon_0}\right) = \frac{\sigma_0}{\varepsilon_0} \text{ or } D_n^{(2)} - D_n^{(1)} = \sigma_0. \qquad [1.184]$$

Thus, the normal component of the vector **D** can be discontinuous, but this discontinuity is defined only by the density of free charge at the interface (not total charge, as with the electric field).

1.16.3. The System of Equations of the Electric Field and Potential

We now write down the full system of field equations in a form that involves only the density of free charge as sources on the right-hand side:

$$1.\ \operatorname{curl}\mathbf{E} = 0, \quad 2.\ \operatorname{div}\mathbf{D} = \delta_0$$

and [1.185]

$$1.\ E_t^{(2)} - E_t^{(1)} = 0, \quad 2.\ D_n^{(2)} - D_n^{(1)} = \sigma_0.$$

We were able to do eliminate bound charge from the original set of equations, because of the following assumptions: (1) there is a linear relation between the polarization vector and the field **E** and (2) the polarizability is known. As for the density of free charges, it is still unfortunately unknown in the general case of a conducting and polarizable medium, except for special cases. Next, we derive the system of equations for the potential. As before, from the first of the Equations [1.185], we obtain $\mathbf{E} = -\nabla U$, and substituting this into the second equation gives

$$\operatorname{div}(\varepsilon \operatorname{grad} U) = -\delta_0 \text{ or } \nabla\cdot(\varepsilon\nabla U) = -\delta_0. \qquad [1.186]$$

This is not Poisson's equation, but does reduce to it in regions where a medium is uniform:

$$\varepsilon\nabla^2 U = -\delta_0 \text{ or } \nabla^2 U = -\frac{\delta_0}{\varepsilon}. \qquad [1.187]$$

If, in addition, free charge is absent, then the potential becomes a harmonic function satisfying Laplace's equation:

$$\nabla^2 U = 0.$$

Expanding the left-hand side of Eq. [1.186] gives

$$\varepsilon\nabla^2 U + \nabla U\cdot\nabla\varepsilon = -\delta_0 \text{ or } \nabla^2 U = -\frac{\delta_0}{\varepsilon} - \frac{\nabla U\cdot\nabla\varepsilon}{\varepsilon}, \qquad [1.188]$$

and, as will be shown in the next section, the right-hand side of this equation characterizes the density of free and bound charges. The system of equations for the potential is given below (Eqs. [1.189]). Here we took into account that continuity of the tangential components of the electric field follows from continuity of the potential and that

$$D_n = \varepsilon E_n = -\varepsilon \frac{\partial U}{\partial n}.$$

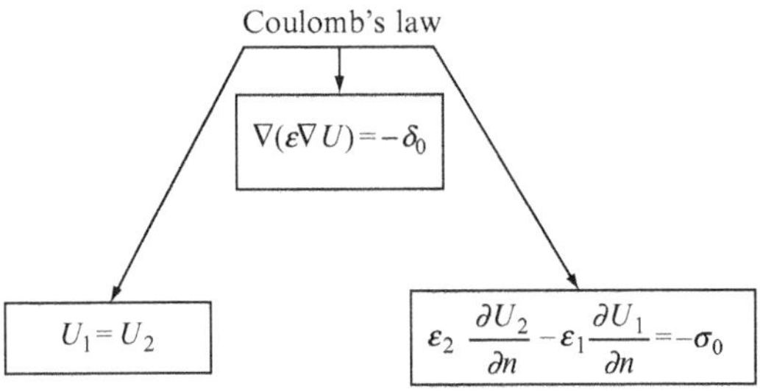

(1.189)

This system of equations for the potential describes the behavior of the electric field at regular points and at interfaces of any polarizable and conducting medium. In addition, let us make one comment. The dielectric constant varies for different types of rocks. Generally, these values are given in the terms of the "relative" dielectric constant, defined by $\varepsilon = \varepsilon_r \varepsilon_0$. In particular, the relative dielectric constant ε_r for water is 81, and for oil, it is about 2 to 3. Such differences can, in principle, be used to distinguish formations saturated by oil and water, but because rocks have conductivity it is impossible to measure their dielectric constant by measuring a constant electric field in the presence of conduction current. In the next chapter we will show that, when constant current flows in a conducting medium, the dielectric constant has no influence on the behavior of the field, despite the presence of bound charges.

1.17. DISTRIBUTION OF BOUND CHARGES

To better understand the field behavior and its sources, we conclude this chapter by studying the distribution of bound charges.

1.17.1. Volume Density of Bound Charge

According to Eq. [1.183],

$$\operatorname{div}(\varepsilon \mathbf{E}) = \varepsilon \operatorname{div} \mathbf{E} + \mathbf{E} \cdot \operatorname{grad} \varepsilon = \delta_0$$

and since

$$\operatorname{div} \mathbf{E} = \frac{\delta_0 + \delta_b}{\varepsilon_0}$$

we obtain

$$\delta_0 = \frac{\varepsilon}{\varepsilon_0}(\delta_0 + \delta_b) + \mathbf{E}\cdot \operatorname{grad}\varepsilon$$

or

$$\delta_b = \frac{1-\varepsilon_r}{\varepsilon_r}\delta_0 - \mathbf{E}\cdot \frac{\operatorname{grad}\varepsilon}{\varepsilon_r}. \qquad [1.190]$$

It is convenient to distinguish two types of bound charges:

$$\delta_b = \delta_b^{(1)} + \delta_b^{(2)},$$

where

$$\delta_b^{(1)} = \frac{1-\varepsilon_r}{\varepsilon_r}\delta_0 \text{ and } \delta_b^{(2)} = -\mathbf{E}\cdot \frac{\nabla\varepsilon}{\varepsilon_r}. \qquad [1.191]$$

The first type of charges arises in the vicinity of points where free charge is present; and since $\varepsilon_r > 1$, these bound charges always have a sign opposite to that of free charges. The second type of charges can arise in places where the dielectric "constant" ε varies. When electric field has a projection in the direction of the vector grad ε, negative charge appears; if these vectors form an angle that exceeds $\pi/2$, positive charge arises. In particular, when the electric field is perpendicular to the direction of maximum change of ε, this type of bound charge is absent.

1.17.2. Surface Density of Bound Charge

We next consider surface bound charge, proceeding from equation

$$D_n^{(2)} - D_n^{(1)} = \sigma_0.$$

This equality can be rewritten as

$$\begin{aligned} D_n^{(2)} - D_n^{(1)} &= \varepsilon_2 E_n^{(2)} - \varepsilon_1 E_n^{(1)} \\ &= \frac{1}{2}[(\varepsilon_2+\varepsilon_1)(E_n^{(2)} - E_n^{(1)}) + (\varepsilon_2 - \varepsilon_1)(E_n^{(2)} + E_n^{(1)})] = \sigma_0. \end{aligned}$$

Because

$$E_n^{(2)} - E_n^{(1)} = \frac{\sigma}{\varepsilon_0},$$

we have

$$\frac{1}{2}\left[(\varepsilon_1+\varepsilon_2)\frac{(\sigma_0+\sigma_b)}{\varepsilon_0}+(\varepsilon_2-\varepsilon_1)\left(E_n^{(2)}+E_n^{(1)}\right)\right]=\sigma_0$$

or

$$\varepsilon_r^{av}(\sigma_0+\sigma_b)+(\varepsilon_2-\varepsilon_1)E_n^{av}=\sigma_0. \qquad [1.192]$$

Here,

$$\varepsilon_r^{av}=\frac{\varepsilon_1+\varepsilon_2}{2\varepsilon_0}=\frac{\varepsilon_{1r}+\varepsilon_{2r}}{2} \text{ and } E_n^{av}=\frac{E_n^{(1)}+E_n^{(2)}}{2}.$$

Therefore,

$$\sigma_b=\frac{1-\varepsilon_r^{av}}{\varepsilon_r^{av}}\sigma_0-\frac{\varepsilon_2-\varepsilon_1}{\varepsilon_r^{av}}E_n^{av}, \qquad [1.193]$$

and, as with a volume distribution, we have distinguished two types of bound surface charge.

CHAPTER TWO

Electric Field and Steady Current Flow in Conducting Media

In Chapter 1, we found out that, due to polarization, bound charges appear in a medium, and in accordance with Coulomb's law, they create the electric field. However, it turns out, surprisingly, that the influence of bound charges vanishes when a steady current flows in the medium under an applied electric field. The flow of current can, therefore, be calculated without any knowledge of a medium's dielectric properties. In this chapter, we study this remarkable feature in some detail, and formulate appropriate boundary-value problems that determine the electric field. First, we will define more precisely what constitutes "electric current" and investigate its relationship to the motion of free charges in a conductor.

2.1. ELECTRIC CURRENT

2.1.1. Ordered Motion of Charge

In considering electrostatic induction, we have established that the electric field, caused by charges outside of a conductor, induces charges on its surface in such a way that the total field vanishes at every point within the conducting medium. This means that the field due to external charges is not able to cause an ordered motion of either electrons or ions, and consequently, they move randomly as if these external charges were absent. Next, suppose that with help of other charges located inside a conducting medium and, in particular, on its surface, we create a constant electric field which may vary from point to point but does not change with time. Of course, as before, this field obeys Coulomb's law. If we know the distribution of these charges, then the electric field at each point of the conducting and polarizable medium can be defined as

$$\mathbf{E}(p) = \frac{1}{4\pi\varepsilon_0}\int_V \frac{\delta(q)}{L_{qp}^3}\mathbf{L}_{qp}\mathrm{d}V + \frac{1}{4\pi\varepsilon_0}\int_S \frac{\sigma(q)}{L_{qp}^3}\mathbf{L}_{qp}\mathrm{d}S.$$

Methods in Geochemistry and Geophysics, Volume 44
ISSN 0076-6895, DOI: 10.1016/S0076-6895(10)44002-0
© 2010 Elsevier B.V.
All rights reserved.

Later, we will formulate precise conditions that guarantee the existence of a steady electric field within a conducting medium. For now, let us accept that this is possible and study the movement of free charges through the medium, a phenomenon called "electric current." This is the third phenomenon, after electrostatic induction and polarization, that we are going to consider in detail. It is obvious that an elementary charge subjected to a steady electric field will, unless restrained, move under the action of the electric force. The thinking of this process helps to distinguish two time intervals. The first, or transient, interval must deal with electromagnetic fields that vary with time, because it is impossible to instantly create a steady field. In the transient period, electromagnetic energy propagates through the medium to establish the field, and this process is inevitably accompanied by a transformation of electrical energy into heat (diffusion). It can be shown that the transient interval lasts for a very short time for media and distances involved in the electrical methods of geophysics. The second interval deals with fields that have reached a steady state and they do not change with time. Since **E** is constant in this state, an ordered motion of the charges is observed. This ordered motion of charges is called electric current, or simply current. In metals, electric current is the motion of electrons, whereas in most rocks, it is the motion of ions dissolved in fluids filling the rocks' pores. Consequently, it is useful to distinguish between "electronic" and "ionic" conductivity. In both cases, however, a complex, random movement of microscopic charge accompanies an ordered motion in some direction, which is defined by the electric field. It is worth noting that the motion of each charge is determined by the magnitude and direction of the electric field in the vicinity of the point where the charge is located, and that, in this sense, its motion is independent of the field at other locations. The movement of charges in real materials is complicated by various factors, such as the complexity of a rock's pore space or the metal's crystal structures, the presence of fractures or defects, and the interaction between different charge carriers. To simplify the treatment, we imagine that in any elementary volume, all microscopic charges of each sign move with the same velocity. This assumption allows a simple transition from the micro- to macroscale and enables us to treat current as the motion of charge distributed continuously in a medium.

2.1.2. Vector of Current Density

To describe the motion of charges, we introduce the vector of current density $\mathbf{j}(q)$. Like any vector field, the current density has magnitude and direction, and can vary from point to point. Its direction gives the direction of charge movement at point q, while its magnitude is equal to the amount of charge passing in unit time through a surface of unit area centered at the point q and perpendicular to $\mathbf{j}$. Three conditions are assumed in this macroscopic description:

(1) The area has to be sufficiently small so that the moving charges are distributed uniformly across its surface.

(2) The surface is, by definition, perpendicular to the direction of charge movement.

(3) The amount of charge passing through the surface is measured per unit time.

All these conditions are essential; the definition of current density vector loses its meaning if any one fails to hold. Correspondingly, the vector $\mathbf{j}(q)$ can be written as

$$\mathbf{j}(q) = \frac{de}{dSdt}\mathbf{i}_0(q). \qquad [2.1]$$

Here, dS is the area of the surface perpendicular to the direction of charge motion and is sufficiently small so that the same amount of charge passes through every patch of area in unit time. In addition, de is the charge that crosses this area during the interval dt. Finally, $\mathbf{i}_0$ is a unit vector in the direction of charge movement (and by definition, is normal to the surface dS). From Eq. [2.1], the dimensions of current density are clearly

$$[\mathbf{j}] = \mathrm{C/m^2s},$$

where C stands for "Coulomb," the unit of charge. We also conclude from Eq. [2.1] that the direction $\mathbf{j}$ coincides with the direction of motion of the positive charge ($de > 0$) and is opposite to the direction of motion of the negative charge ($de < 0$). In general, charges of both signs can move in different directions, and correspondingly, the total vector $\mathbf{j}$ is

$$\mathbf{j}(q) = \mathbf{j}^+(q) + \mathbf{j}^-(q), \qquad [2.2]$$

where $\mathbf{j}^+$ and $\mathbf{j}^-$ are the current density vectors of positive and negative charges, respectively. In particular, these vectors may have opposite directions and equal magnitudes, in which case, the current density vanishes.

2.1.3. Current Lines and Tubes

As with the electric field, it is convenient to visualize the vector field $\mathbf{j}$ using field lines (in this case, also called "current lines"). By definition, the vector $\mathbf{j}$ is tangential to a current line. We will show later that all

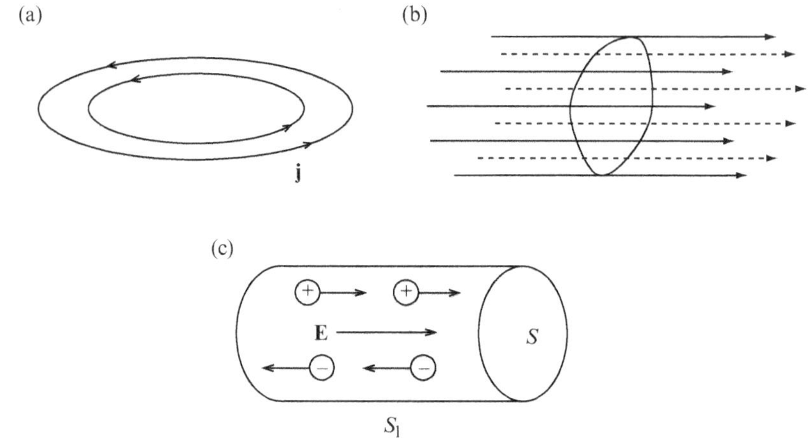

Figure 2.1 (A) Current lines, (B) current tube, and (C) illustration of Eq. [2.4].

current lines are closed (Fig. 2.1A); in addition, current lines cannot intersect, because at the point of intersection, there would be two different values of the current density vector (which would no longer be a unique vector field). It is also useful to define a current "tube" (Fig. 2.1B). Imagine an arbitrary closed contour L that does not coincide with a current line, and through each point of L, trace the line corresponding to a given current density field. This procedure defines a current "tube" with the lateral surface S_l. Because the lateral surface of the tube consists of current lines, the normal vector to this surface at each point is perpendicular to the local current density. This implies that current never leaves or enters the tube; therefore, the number of current lines within the tube does not change. A cross section through the tube is defined as a surface cutting through it that is everywhere perpendicular to the local current density vector. In studying a current density field, we can imagine that the whole conducting medium is a system of current tubes of different shapes and extensions. The cross-sectional area of an individual tube may vary from point to point, as can the magnitude of the current density at different points of its cross section. If, however, the cross section is sufficiently small, the current density vector will be constant over the cross section. Such a current tube is called an "elementary" tube.

2.1.4. Relation Between the Current Density and Charge

We now express the vector of current density $\mathbf{j}$ in terms of the densities and velocities of positive and negative charge carriers. Consider a small volume of an elementary current tube with cross section dS (Fig. 2.1C). Charges of both signs move along this tube. Let ω be the magnitude of their velocity. Then, during the time interval dt, charges located at distances less than ω dt from the surface dS, will pass through the surface (recall that the direction of current flow in the tube is perpendicular to its cross section). Therefore, the amount of charge passing through dS during dt is

$$de = \delta dV = \delta dS\omega\, dt, \qquad [2.3]$$

where δ is the volume density of charge. As follows from Eq. [2.1], the vector of the current density is

$$\mathbf{j} = \delta\boldsymbol{\omega}. \qquad [2.4]$$

The current density is, thus, directly proportional to the volume density of charge and to its velocity. This should be obvious, since the amount of charge crossing the area increases proportionally with any increase of the density or velocity of the charge. We see that the vectors $\mathbf{j}$ and $\boldsymbol{\omega}$ have the same direction if $\delta > 0$ but are opposite to each other if $\delta < 0$. (This is actually a convention that defines "positive" electric current as the flow of positive charge through the surface in the direction of its normal vector.) In general, charges of both signs move in a current tube, and we can write

$$\mathbf{j}^+ = \boldsymbol{\omega}^+\delta^+, \quad \mathbf{j}^- = \boldsymbol{\omega}^-\delta^-, \quad \mathbf{j} = \boldsymbol{\omega}^+\delta^+ + \boldsymbol{\omega}^-\delta^-,$$

or [2.5]

$$\mathbf{j}^+ = \boldsymbol{\omega}^+\delta^+, \quad \mathbf{j}^- = -\boldsymbol{\omega}^-|\delta^-|, \quad \mathbf{j} = \boldsymbol{\omega}^+\delta^+ - \boldsymbol{\omega}^-|\delta^-|,$$

where δ^+, $\boldsymbol{\omega}^+$ and δ^-, $\boldsymbol{\omega}^-$ are densities and velocities of the positive and negative charges, respectively, and $|\delta^-|$ is the absolute value of the negative charge density. The set of equations [2.5] merits several comments:

1. The vector $\mathbf{j}$ does not change under simultaneous change of the sign of the charge density and the direction of the velocity. For example, replacing the motion of negative charge by that of positive charge with the same speed (magnitude of velocity) but opposite direction gives the same field $\mathbf{j}$. This shows that electric current can conceptually always be reduced to the motion of only one type of charge (by convention, usually taken to be positive charge).
2. In metals, electrons form the current and $\mathbf{j} = \boldsymbol{\omega}^-\delta^-$. In electrolytes, however, both positive and negative charges move and often in opposite directions. Therefore, in such a case,

$$\mathbf{j}(q) = (\boldsymbol{\omega}^{+}\delta^{+} + \boldsymbol{\omega}^{-}|\delta^{-}|)\mathbf{i}_0, \qquad [2.6]$$

where $\mathbf{i}_0$ is the unit vector showing the direction of movement of positive charge. There are also cases where ions of both signs move in the same direction or where ions of only one sign move. The latter happens, for example, when positive ions move through a porous medium under the action of a pressure difference.

3. According to Eqs. [2.5], the current density vector is nonzero provided that

$$\boldsymbol{\omega}^{+}\delta^{+} \neq \boldsymbol{\omega}^{-}|\delta^{-}|. \qquad [2.7]$$

This also means that the current density $\mathbf{j}$ can vanish even when there is motion of charge, provided that charges of opposite sign move in the same direction: that is, when $\boldsymbol{\omega}^{+}\delta^{+} = \boldsymbol{\omega}^{-}|\delta^{-}|$. Vanishing of the current density does not require that the charge densities, δ^{+} and $|\delta^{-}|$, be equal.

4. The total density of charge inside of any elementary volume of a medium is

$$\delta = \delta^{+} + \delta^{-}.$$

It is helpful to point out that, often in the presence of current flow, the total charge inside the volume where the current flows is equal to zero. For example, in a uniform metal, the density of moving electrons and stationary positive ions (in the crystal structure) is zero. This is also true for homogeneous electrolytes and rocks in the absence of non-Coulomb forces. It is clear that in such cases, the volume elements of the conductor do not create an electric field.

2.2. OHM'S LAW IN DIFFERENTIAL FORM

We now consider the relationship between the vector of current density and the electric field. We will first assume that the field is a Coulomb field caused by charges; later, this assumption will be removed. Under the electric force, charge in a medium begins to move with some acceleration (Newton's law). But, the medium itself "resists" this motion, decreasing the acceleration. The mechanism producing this resistance varies from material to material. In metals, for instance, resistance to charge acceleration occurs because of the "collision" of moving electrons with ions of the crystal lattice. In electrolytes, the resistance is caused by viscous forces in the fluid. We consider such "frictional" force resisting charge motion from the macroscopic point of view. Regardless of its origin, it generally increases as the charge velocity increases and eventually becomes equal and opposite to the force caused by the electric field. At that moment, the resultant force acting on a charge vanishes and its acceleration also vanishes. In other words, the velocity of charge $\boldsymbol{\omega}$ reaches a maximum value and then remains constant, giving rise to a steady current density $\mathbf{j}$. If the forcing electric field is increased, the charge accelerates to a higher velocity before reaching again a balance of forces. Thus, it seems reasonable to assume that the velocity of charge in steady current flow is directly proportional to the Coulomb electric field $\mathbf{E}^c$, that is,

$$\boldsymbol{\omega}^+ = u^+\mathbf{E}^c \quad \text{and} \quad \boldsymbol{\omega}^- = -u^-\mathbf{E}^c, \qquad [2.8]$$

where u^+ and u^- are positive coefficients called the mobilities of the positive and negative charges, respectively. Several further comments are appropriate:

1. The process of establishment of a constant current described above uses a classical model of a charged particle and Newton's Second Law. In reality, this process involves the interaction of electromagnetic fields with the atomic structure of a conductive medium, and a detailed microscopic model requires a quantum mechanical description.
2. The values of $\boldsymbol{\omega}^+$ and $\boldsymbol{\omega}^-$ in Eqs. [2.8] correspond to the state where the resultant force is equal to zero.
3. The negative sign in the second equation of this set emphasizes the fact that negative charge moves against the electric field: that is, in the same direction as the Coulomb force $e\mathbf{E}^c$, acting on the charge.
4. The name "mobility" is well justified, since velocity increases as u^+ and u^- increase.
5. It is natural to expect that, as friction decreases, mobility of the charges increases. Substitution of Eqs. [2.8] into Eq. [2.5] gives

$$\mathbf{j} = (u^+\delta^+ + u^-|\delta^-|)\mathbf{E}^c \quad \text{or} \quad \mathbf{j} = \gamma\mathbf{E}^c, \qquad [2.9]$$

where

$$\gamma = \delta^+u^+ + |\delta^-|u^-. \qquad [2.10]$$

Equation [2.9] is called Ohm's law in differential form and the coefficient γ is the conductivity of the medium. This important physical law is based on the assumption that the velocity of charge motion in a material is directly proportional to the applied electric field $\mathbf{E}^c$. In this light, it is proper to note the following:

1. Ohm's law establishes a relationship between two different physical fields, namely, the electric field $\mathbf{E}^c$ caused by the charge and the vector of current density $\mathbf{j}$ describing the motion of the charge.
2. It follows from the macroscopic description leading to Eq. [2.9] that an electric field causes current, but not *vice versa*. In other words, a constant current does not create an electric field. A different way of seeing this is to note that there can be an electric field in a nonconducting medium where the current density is equal to zero. The relation between these fields is illustrated below, which shows that charge generates the electric field $\mathbf{E}^c$, and in a conducting medium, this field creates a movement of the charges, the field $\mathbf{j}$.

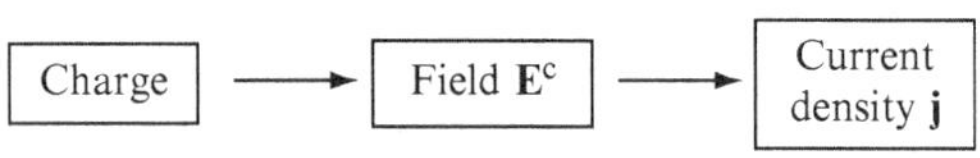

3. The sources of the electric field $\mathbf{E}^c$ need not necessarily have any direct relation to the charges that constitute the current density $\mathbf{j}$.
4. An important generalization of Ohm's law is the following: It is obvious that the process of establishment of a constant current is independent of the origin of the force acting on the charges. This force can come not only from other charges (Coulomb force) but also from other generators of force. Correspondingly, we represent the total force acting on moving charge, except for the frictional force, as a sum:

$$\mathbf{E} = \mathbf{E}^c + \mathbf{E}^{nc}. \qquad [2.11]$$

Therefore,

$$\boldsymbol{\omega}^+ = u^+\mathbf{E}, \quad \boldsymbol{\omega}^- = -u^-\mathbf{E}, \qquad [2.12]$$

and Ohm's law becomes

$$\mathbf{j} = \gamma\mathbf{E}, \quad \text{or} \quad \mathbf{j} = \gamma(\mathbf{E}^c + \mathbf{E}^{nc}). \qquad [2.13]$$

Here, $\mathbf{E}^c$ and $\mathbf{E}^{nc}$ are the Coulomb and non-Coulomb fields, respectively. The superscript "nc" emphasizes that the non-Coulomb field is not caused by charge and therefore, does not have to obey Coulomb's law. The possible sources of this force are numerous: $\mathbf{E}^{nc}$ can be electrochemical, inductive, diffusive, hydrodynamic, and other effects. In parts of a medium where the non-Coulomb forces are absent, Ohm's law is given by Eq. [2.9].

5. Ohm's law in differential form also applies to time-varying electromagnetic fields.
6. Conductivity is usually a scalar, but can be a tensor in "anisotropic" media. Anisotropic conductivity means that the direction of current flow is not necessarily the direction of the local applied electric field.
7. It follows from Eq. [2.10] that conductivity is directly proportional to the product of charge density and mobility, where the latter characterizes charge velocity. For instance, the mobility of electrons in copper is approximately equal to

$$u^{-} \approx 4.4 \times 10^{-3}\,\mathrm{m^2/Vs}.$$

This is a surprisingly small number. The electric field in a metal is generally quite small, on the order of

$$E = 1\,\mathrm{mV/m} = 10^{-3}\,\mathrm{V/m}.$$

The average velocity of electrons is, therefore,

$$\omega^{-} \approx 4.4 \times 10^{-6}\,\mathrm{m/s}.$$

This is very slow motion, requiring almost 70 h for an electron to travel 1 m. Certainly, when energy is transferred from power stations to homes, it does not travel with the electrons through wires (power lines). That would take thousands of years! In reality, there is practically an instantaneous propagation of electromagnetic energy through the space around the wires, which play a role in guiding the energy flow (wave guides).
8. The exceptionally high conductivity of metals is mainly caused by a very high density of electrons that compensates for their low mobility. Ions in electrolytes have lower mobility ($\approx 10^{-9}\,\mathrm{m^2/V\,s}$) than electrons in metals, and a lower charge density, which results in a much smaller conductivity for electrolytes than for metals.
9. Ordered movement of charges under the action of the electric and extraneous fields is always accompanied by a random motion with a relatively high velocity:

$$10^{2}\,\mathrm{m/s} < \omega < 10^{6}\,\mathrm{m/s},$$

which exceeds by many orders the ordered velocity of charges forming the current **j**. Because of this random motion, we observe random currents in materials, called fluctuating current, even in the absence of an electric field. The existence and size of the fluctuating current limits the sensitivity of measuring devices.

10. It is useful to introduce a second quantity ρ, called the "specific electrical resistivity" (or simply "resistivity"), which is numerically the inverse of conductivity:

$$\rho = \frac{1}{\gamma}. \qquad [2.14]$$

The dimensions of this quantity are

$$[\gamma] = \mathrm{C/V\,ms} \quad \text{and} \quad [\rho] = \mathrm{V\,ms/C}$$

or

$$[\gamma] = (\mathrm{ohm\,m})^{-1} \quad \text{and} \quad [\rho] = \mathrm{ohm\,m},$$

where "ohm" is a derived unit with dimensions of V s/C (the usual unit of resistance from circuit theory). One remarkable feature of resistivity is the extremely wide range of its change.

2.3. PRINCIPLE OF CHARGE CONSERVATION

2.3.1. Flux of Current Density

We now study different physical properties of the current density field $\mathbf{j}$. First of all, it is not difficult to show that the amount of charge passing per unit time through an elementary surface $\mathrm{d}S$ (Fig. 2.2A) is given by

$$\mathrm{d}I = \mathbf{j}\cdot \mathrm{d}\mathbf{S} = j\,\mathrm{d}S\cos(\mathbf{j}, \mathrm{d}\mathbf{S}), \qquad [2.15]$$

where $\mathrm{d}\mathbf{S} = \mathrm{d}S\mathbf{n}$, and $\mathbf{n}$ is the unit vector directed from the back to front side of the surface $\mathrm{d}S$. It is clear that this amount of charges is positive ($\mathrm{d}I > 0$) if they move from the back to the front face of $\mathrm{d}S$, and is negative if they pass the surface in the opposite direction. If the current density vector is tangent to the surface, the amount of charge $\mathrm{d}I$ crossing the surface vanishes. It follows from Eq. [2.15] that the flux of the vector $\mathbf{j}$ gives the net charge moving through an arbitrary surface S per unit time is (Fig. 2.2B)

$$I = \int_S \mathbf{j}\cdot \mathrm{d}\mathbf{S} = \int_S j_n \mathrm{d}S, \qquad [2.16]$$

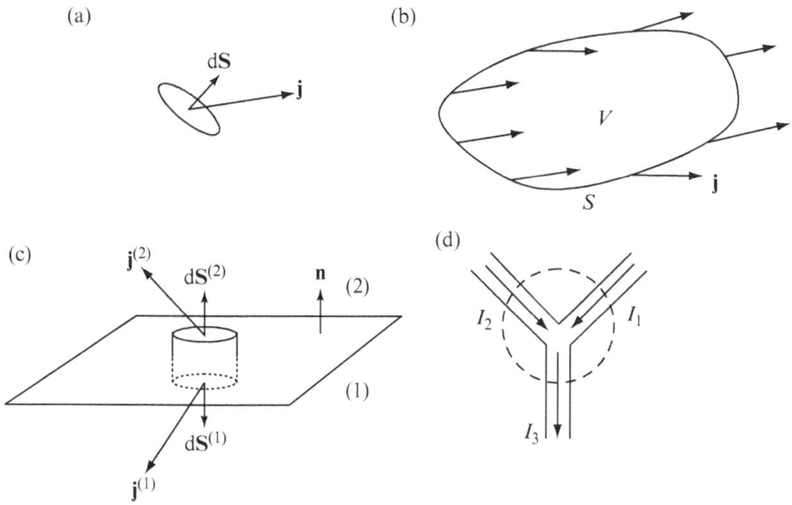

Figure 2.2 (A) Flux of **j** through an elementary surface, (B) flux of **j** through an arbitrary surface, (C) illustration of Eq. [2.21], and (D) Kirchhoff's law for circuit junctions.

where j_n is the normal component of the current density. The flux I is the algebraic sum of elementary fluxes through different parts of the surface and is called the current. The unit of current is the "ampere" or "amp" (with symbol A). By definition, its dimensions are Coulombs per second, [ampere] = A = C/s.

We already introduced the concept of a current tube and its lateral surface as a way of visualizing the movement of charges in a current density field. Since the vector **j** is tangential to the tube's lateral surface, the flux through this surface is zero and no current (or charge) enters or leaves the tube. In general, the cross-sectional area of a current tube can vary from point to point, but the total current through each cross section stays the same.

2.3.2. Principle of Charge Conservation and its Four Forms

We can now formulate the basic physical law that characterizes the behavior of the current density, namely, the principle of charge conservation. For a steady (time-invariant) field, it has the integral form

$$\oint_S \mathbf{j} \cdot d\mathbf{S} = 0. \qquad [2.17]$$

This law holds that the flux of a steady current density **j** through any closed surface is identically zero. Since the current density accounts for the movement of charge, Eq. [2.17] implies that the amount of charge entering a volume V in any time interval always equals the amount of charge leaving the volume in the same time interval. It is worth noting that this statement depends entirely on the fact that a steady state is assumed. For instance, if more net charge enters than leaves a given volume in a given time interval, then charge is accumulating in the volume over time. This, in turn, would mean that the electric field would also be changing with time. Equation [2.17] does not depend of the size and shape of the surface S surrounding an arbitrary volume V, and holds whether a medium is uniform or not. In particular, the surface S can intersect media with different conductivities. Moreover, Eq. [2.17] holds to a high degree of accuracy over a wide range of frequencies, even when the electromagnetic field and current density vary with time. Over any closed surface, there can be places where the normal component j_n is directed outward, as well as places where this component is oriented inward, and this provides the flux in and out of the volume. Indeed, if there is net charge inside some volume V, then according to Eq. [2.17], this net charge remains the same at all times as long the electric field is time invariant. In particular, the net charge can be equal to zero in spite of the presence of a flux of the current density. Two final comments are appropriate:

(1) It is possible to preserve a steady electric field and current density in a conducting and polarizable medium, only because the flux of the vector **j** through any closed surface equals zero.

(2) If the flux of current vanishes, it does not mean that there is no net charge in a conducting medium. Net charge is absent in a given volume only if the flux of the electric field is zero:

$$\oint_S \mathbf{E} \cdot d\mathbf{S} = 0,$$

and this equation alone is independent of the principle of charge conservation.

2.3.3. Different Forms of the Principle of Charge Conservation

We now derive several equations that describe the principle of charge conservation in differential form. First, consider regular points of a medium where the first derivatives of the function $\mathbf{j}(q)$ exist. Applying Gauss's theorem for such points, we have

$$\oint_S \mathbf{j} \cdot d\mathbf{S} = \int_V \operatorname{div} \mathbf{j} \, dV = 0.$$

Since this equality holds for a volume of any size, we obtain

$$\operatorname{div}\mathbf{j} = 0. \qquad [2.18]$$

Equation [2.18] is the second form of the principle of charge conservation and holds at regular points. By definition, it can be expressed in terms of derivatives of $\mathbf{j}$ and, in particular, in the Cartesian system, we have

$$\frac{\partial j_x}{\partial x} + \frac{\partial j_y}{\partial y} + \frac{\partial j_z}{\partial z} = 0, \qquad [2.19]$$

where j_x, j_y, and j_z are the Cartesian components of the vector $\mathbf{j}$. The field lines of any vector field can be either open or closed. In the case of open lines, there are terminal points where sources of the field are located. Since $\operatorname{div}\mathbf{j}=0$, we conclude that current lines as well as current tubes are always closed on themselves. In limiting cases, the lines can be closed "at infinity." In this light, rewrite Eq. [2.18] using the general form of Ohm's law:

$$\operatorname{div}\gamma(\mathbf{E}^{c} + \mathbf{E}^{nc}) = 0 \quad \text{or} \quad \operatorname{div}\gamma\mathbf{E}^{c} = -\operatorname{div}\gamma\mathbf{E}^{nc}. \qquad [2.20]$$

The latter equation shows that field lines of current density driven by a Coulomb field can terminate at points where there is a non-Coulomb field. To derive a surface analogue of Eq. [2.18], we consider the flux of the current density through a closed cylinder that intersects the interface between two media with different conductivities (Fig. 2.2C). This flux consists of three parts whose sum is equal to zero: the fluxes through the end caps and through the lateral surface,

$$\int_{S_1}\mathbf{j}\cdot d\mathbf{S} + \int_{S_2}\mathbf{j}\cdot d\mathbf{S} + \int_{S_l}\mathbf{j}\cdot d\mathbf{S} = 0.$$

Since the normal $\mathbf{n}$ is directed from the first medium to the second one, we have $d\mathbf{S}_1 = -\mathbf{n}dS$ and $d\mathbf{S}_2 = \mathbf{n}dS$ in the limit as the height of the cylinder tends to zero. The flux through the lateral surface disappears in this limit. Assuming that the end cap surfaces dS are small so that the current density is constant over each surface gives

$$\mathbf{j}^{(2)}\cdot d\mathbf{S}_2 + \mathbf{j}^{(1)}\cdot d\mathbf{S}_1 = 0 \quad \text{or} \quad j_n^{(2)} - j_n^{(1)} = 0. \qquad [2.21]$$

Thus, the normal component of the current density is continuous across an interface between media with different conductivities. It is worth re-emphasizing that this condition is needed to preserve a time-invariant electric field in a conducting medium. If the normal components of current density were not equal at both sides of the interface, then surface charge would be increasing or decreasing and the electric field would vary in time. Equation [2.21] implies, moreover, that surface charge must be present when current flows across an interface where the conductivity changes. In fact, this charge is needed to preserve continuity of the normal component of current density. To see this, imagine there is no surface charge in the vicinity of some point q at the interface between media with different conductivities, γ_1 and γ_2. Also, for simplicity, assume that there is no non-Coulomb force $\mathbf{E}^{\mathrm{nc}}$ at this point. Then, in accordance with Ohm's law, the normal components of the current density are

$$j_{\mathrm{n}}^{(1)}(q) = \gamma_1 E_n^{(1)}(q), \quad j_{\mathrm{n}}^{(2)}(q) = \gamma_2 E_{\mathrm{n}}^{(2)}(q),$$

where $E_{\mathrm{n}}^{(1)}(q)$ and $E_{\mathrm{n}}^{(2)}(q)$ are normal components of the electric field caused by all charges, which can be located anywhere. But we have assumed that there is no charge in the vicinity of q, so all charges are located at finite distances from this point. The electric field is, therefore, continuous across the interface at q, and so is its normal component $E_{\mathrm{n}}^{(1)}(q) = E_{\mathrm{n}}^{(2)}(q)$. Obviously, if the normal component of the electric field is continuous but the conductivity changes ($\gamma_1 \neq \gamma_2$), then the normal component of the current density is discontinuous, and the amount of charge at the interface must be changing in time. This is not possible with a steady electric field. The above arguments show that surface charges must arise and distribute themselves in such a way as to maintain the law of charge conservation for steady currents. We, therefore, have three forms of this principle:

$$\oint_S \mathbf{j} \cdot d\mathbf{S} = 0, \quad \operatorname{div} \mathbf{j} = 0, \quad j_{\mathrm{n}}^{(2)} = j_{\mathrm{n}}^{(1)}. \qquad [2.22]$$

It is useful to consider a special case of this principle when current flows through conductors that come together at a junction (Fig. 2.2D). This is what happens, for example, in electric circuits. Applying Eq. [2.17] to a surface S surrounding the junction and keeping in mind that there is no current flowing outside the conductors, we have

$$\oint_S \mathbf{j} \cdot d\mathbf{S} = \sum \oint_{S_i} \mathbf{j} \cdot d\mathbf{S} = \sum I_i = 0, \qquad [2.23]$$

where S_i is the cross section of conductor i and I_i is its current. Equation [2.23] expresses what is known in circuit theory as "Kirchhoff's First Law," which holds that the current arriving at a junction of wires is equal to the current that leaves the junction. This law follows from the principle of charge conservation.

2.4. SYSTEM OF EQUATIONS OF THE FIELD j

The principle of charge conservation embodied in Eq. [2.18] is actually the second equation for the current density **j**. It is instructive to derive the first equation from Ohm's law and the first equation for the electric field. We have

$$\operatorname{curl} \mathbf{j} = \operatorname{curl} \gamma \mathbf{E} = \gamma \operatorname{curl} \mathbf{E} + \nabla \gamma \times \mathbf{E}.$$

At regular points, where only a Coulomb field is present, $\operatorname{curl} \mathbf{E} = 0$, and therefore,

$$\operatorname{curl} \mathbf{j} = \nabla \gamma \times \mathbf{E}. \quad [2.24]$$

This equation shows that vortices of the field **j** are located at points where the electric field has a component in the direction perpendicular to $\nabla \gamma$. Taking into account that the tangential component of the electric field is a continuous function, the surface analogy of Eq. [2.24] is

$$j_t^{(2)} - j_t^{(1)} = (\gamma_2 - \gamma_1) E_t. \quad [2.25]$$

The right-hand side of this equation characterizes a distribution of surface vortices. Such vortices of the current density **j** do not have any direct physical meaning; they are fictitious (mathematical) generators of the field of current density, such as the vortices of the auxiliary field **D** introduced to characterize polarization of charge in a medium. To summarize, we have the following system of equations for the current density at points where non-Coulomb forces are absent.

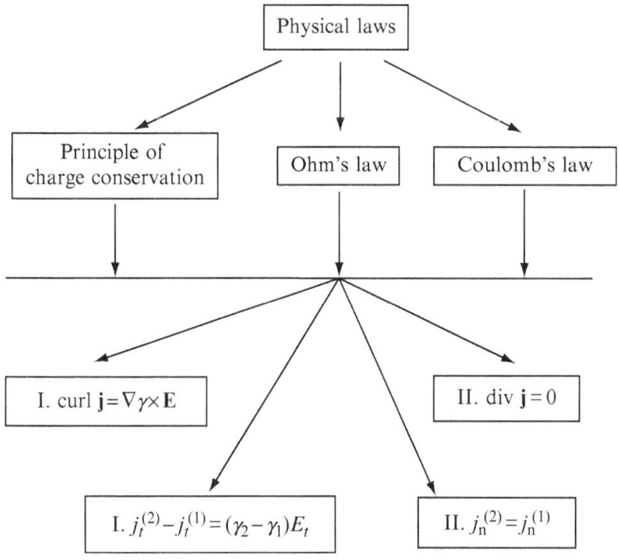

At points where a non-Coulomb force is absent, it follows from Ohm's law that $\mathbf{j} = -\gamma \nabla U$, where U is the electric field potential derived from Coulomb's law. The current density vector field is a special case of a vortex field, called a "quasipotential" field. Thus, in a conducting and polarizable medium, there are two distinct vector fields, namely, the electric field generated only by charge (if $\mathbf{E}^{\mathrm{nc}} = 0$) and the current density field driven by the electric field according to Ohm's law. Both fields are of great practical interest in electrical methods of geophysics, so we will investigate them in detail.

2.5. NON-COULOMB FORCE, OHM'S LAW, AND ELECTROMOTIVE FORCE

Coulomb's law and the principle of charge conservation imply that the electric field and the current density obey the following equations at regular points in a conducting and polarizable medium, where non-Coulomb forces are present:

$$\operatorname{curl} \mathbf{E}^{\mathrm{c}} = 0, \quad \operatorname{div} \mathbf{j} = 0. \qquad [2.26]$$

There is a linear relationship between current density and electric field (Ohm's law):

$$\mathbf{j} = \gamma \mathbf{E}^{\mathrm{c}}. \qquad [2.27]$$

In deriving these equations, we assumed that the electric field is caused by charges only, and to emphasize this, we use notation $\mathbf{E}^{\mathrm{c}}$. The first equation of the set [2.26] was derived directly from Coulomb's law. We have also mentioned that to sustain a steady current flow in a conducting medium, there must be a non-Coulomb field of a completely different origin. We now explore this subject in detail.

2.5.1. Vector Lines of the Electric Field $\mathbf{E}^{\mathrm{c}}$ and Current Density j

Assume first that the electric field is caused only by charges. Then, Ohm's law can be written as

$$\mathbf{j} = -\gamma \nabla U \quad \text{or} \quad j_l = -\gamma \frac{\partial U}{\partial l}, \qquad [2.28]$$

where U is the electric potential, l is some curve in space, and j_l is the projection of the current density vector onto this curve at each point (strictly, the projection of $\mathbf{j}$ onto the unit tangent vector to the curve). Suppose now that l coincides with a current line. By the principle of charge conservation, such a line always closes on itself, and by definition, the current density along a current line is positive. In addition, it is a fundamental fact of physics that electrical conductivity (such as mass) is never negative. Therefore, the ratio j_l/γ is positive everywhere along the curve, and we have to conclude from Eq. [2.28] that the potential decreases everywhere along the current line. But since a current line closes on itself, it follows that, after one complete passage around this line, the potential at the initial point acquires a different value. A second trip around the circuit would give a different value again. In other words, $U(q)$ would be a many-valued function. This contradicts the definition of potential derived from Coulomb's law and clearly indicates that Eq. [2.28] cannot be complete, since it is hardly prudent to abandon the fundamental principle of charge conservation ($\text{div}\,\mathbf{j}=0$) in steady current flow. We, therefore, find that a nonzero electric field cannot simultaneously satisfy the set of equations [2.26] and Ohm's law (Eq. [2.27]), $\mathbf{j}=\gamma\mathbf{E}^c$, for steady currents. We can see this in a different way working directly with these equations. Assume for convenience that the conductivity is a scalar quantity and is nonzero everywhere. On the one hand, it follows from the equation $\text{curl}\,\mathbf{E}^c=0$ that the electric field has no closed field lines. Then, in accordance with Eq. [2.27], the current density field $\mathbf{j}$ also has no closed field lines. On the other hand, it follows from the equation $\text{div}\,\mathbf{j}=0$ that the field $\mathbf{j}$ has no open field lines either. Again, making use of Eq. [2.27], this should be true for the field lines of $\mathbf{E}$. The situation can then be summarized in the following table:

Field	Open lines	Closed lines
$\mathbf{E}$	No	No
$\mathbf{j}$	No	No

It is possible to show mathematically that a vector field with neither open nor closed field lines (whose divergence and curl are zero everywhere) must be zero. We, therefore, come to the conclusion that only a field that is zero everywhere can satisfy Eqs. [2.26] and [2.27]; in other words, the absence of both open and closed field lines would mean that both $\mathbf{E}$ and $\mathbf{j}$ vanish. Since the equations in the set [2.26] were derived from known physical laws (Coulomb's law and the law of conservation of charge), we are left with only one choice: it is necessary to modify Ohm's law. These arguments also show the electric field satisfying Coulomb's law cannot, by itself, generate steady current in a conducting medium. We came to this conclusion earlier when studying the phenomenon of electrostatic induction.

2.5.2. The External and Internal Parts of a Current Tube

We can restore order to the set of equations [2.26] and [2.27] by formally introducing non-Coulomb forces to drive the current flow. To motivate the introduction of these forces, imagine the following experiment. Consider a tank made of nonconducting material with two parallel metallic plates P^+ and P^- placed at some distance from each other and bearing charges e^+ and e^- (Fig. 2.3A). These charges create an electric field directed from plate P^+ to plate P^-. If no conducting material is present, the electric field remains constant. We now partially fill the tank with an ionic conductor, leaving a portion of the plates exposed to air. Assume that the conductivity γ_e of the solution is much smaller than that of the metallic plates. The electric field $\mathbf{E}^c$ causes the current to flow in the solution, with positive ions moving toward the plate P^-, with negative charge, and negative ions moving toward the plate P^+ with positive charge. This movement of ions will cause the charge on the two plates to decrease until the electric field and current density vanish. Assuming that this decay is sufficiently slow, it can be described in a very simple manner. The electric field between the plates at some instant t is

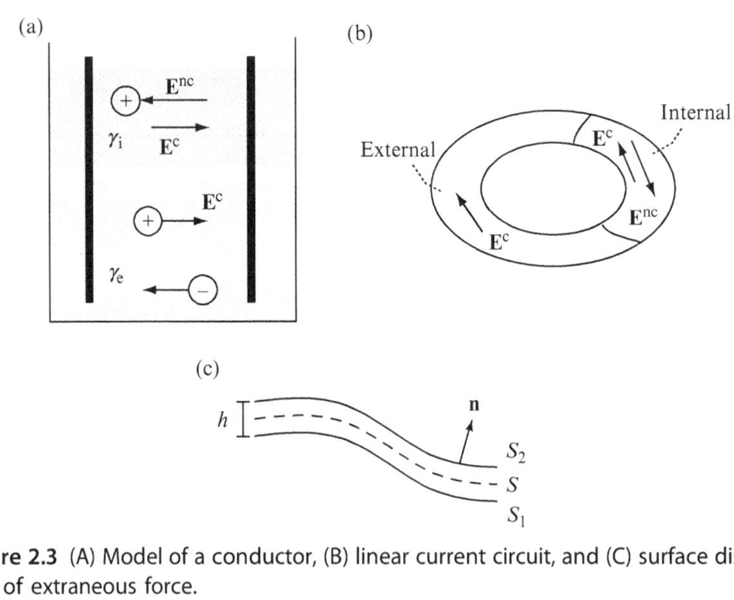

Figure 2.3 (A) Model of a conductor, (B) linear current circuit, and (C) surface distribution of extraneous force.

$$E(t) = \frac{\sigma(t)}{\varepsilon_0}, \qquad [2.29]$$

where $\sigma(t)$ is the magnitude of the charge density on the plates (i.e., the total charge on the positive plate divided by its area). Ignoring edge effects, the electric field will be perpendicular to the plates. The field $\mathbf{E}(t)$ creates the current with a density $\mathbf{j}(t)$, also directed perpendicular to the plates with magnitude:

$$j(t) = \gamma_e \frac{\sigma(t)}{\varepsilon_0}. \qquad [2.30]$$

By Ohm's law, the total current in the solution is, therefore,

$$I_e(t) = \gamma_e S_e \frac{\sigma(t)}{\varepsilon_0}, \qquad [2.31]$$

where S_e is the area of each plate that is immersed into a solution. I_e is the amount of positive and negative charges arriving every second at plates P^- and P^+, respectively. Obviously, this process discharges both plates. To maintain a constant current, we now introduce a connection between the two plates above the solution through a conductor C_i with conductivity γ_i. Thus, the full model includes two plates: a conducting solution C_e and the conductor C_i. It is clear that in both media the Coulomb electric field $\mathbf{E}^c$ has the same direction from the plate P^+ to P^-. Finally, we assume that within the conductor C_i (above the solution) there is another electric field $\mathbf{E}^{nc}$ directed opposite to $\mathbf{E}^c$. The source of this "non-Coulomb" field will be described below. Both fields drive current density in the same manner. The purpose of the field $\mathbf{E}^{nc}$ is to maintain the current flow in the circuit. In fact, in order to have positive charge moving to the plate P^+ through the conductor C_i, the magnitude of the field $\mathbf{E}^{nc}$ must be greater than that of the Coulomb field (which is driving positive charge away from the positively charged plate), that is,

$$|\mathbf{E}^{nc}| > |\mathbf{E}^c|. \qquad [2.32]$$

Therefore, we can imagine two current density fields within the conductor C_i that are oppositely directed:

$$\mathbf{j}^{nc} = \gamma_i \mathbf{E}^{nc} \quad \text{and} \quad \mathbf{j}^c = \gamma_i \mathbf{E}^c.$$

Using Eq. [2.29], the total current density in C_i is

$$\mathbf{j}(t) = \gamma_i[\mathbf{E}^c(t) + \mathbf{E}^{nc}], \quad \text{or} \quad \mathbf{j}(t) = \gamma_i\left[\mathbf{E}^{nc} - \frac{\sigma(t)}{\varepsilon_0}\right]. \qquad [2.33]$$

The non-Coulomb field $\mathbf{E}^{nc}$ is sometimes called an "extraneous field" to emphasize that it must be generated by physical processes other than simple charge accumulation. These processes can, for example, be mechanical or electrochemical in nature, or arise from the effects of filtration, adsorption, induction, and diffusion. The non-Coulomb field drives positive charge through the conductor C_i against the electric field $\mathbf{E}^c$, which arises from the accumulation of charge on the plates themselves, and compensates for discharge of the plate P^+ through the solution. (At the same time, of course, negative charge moves in the opposite direction inside C_i, increasing the charge e^- at the plate P^-.) The current inside the conductor C_i is

$$I_i(t) = \gamma_i S_i\left[\mathbf{E}^{nc} - \frac{\sigma(t)}{\varepsilon_0}\right], \qquad [2.34]$$

where S_i is the area of each plate above the solution. In this model, it is natural to distinguish two parts of the system. The first, or "external," part corresponds to the solution C_e where only a Coulomb field exists. This field is directed from the plate P^+ to P^- and drives the current I_e (Eq. [2.31]), which tends to discharge the plates. The second, or "internal," part corresponds to the C_i where both the Coulomb and extraneous forces act and where the current $I_i(t)$ restores charge on the plates. If the currents I_i and I_e are equal

$$I_i = I_e \quad \text{or} \quad \gamma_i S_i\left(\mathbf{E}^{nc} - \frac{\sigma}{\varepsilon_0}\right) = \gamma_e S_e \frac{\sigma}{\varepsilon_0}, \qquad [2.35]$$

the charges on the plates will eventually stop changing with time, and the electric and current density fields will stay constant. We can imagine that current flux in the above model consists of a system of closed current tubes with a piece located within the conductor C_i, a piece inside the metal plates, and finally, a piece passing through the solution. This shows that each complete current tube, no matter how small its cross section, must include the external and internal elements of the model to support a steady current. In some cases, a continuous distribution of the extraneous force is observed, and it can be difficult to distinguish these two parts of the current tube. To elaborate one time more on a role of the non-Coulomb (extraneous) force, assume that it appears within the internal part C_i at some instant t with value $\mathbf{E}^{nc}$. Under its action, positive and negative charges start to move in opposite directions to charge the plates; that is, positive charge moves toward P^{+} and negative charge moves toward P^{-}. These charges create their own Coulomb field, which increases with time as charge accumulates on the plates. Clearly, the two fields $\mathbf{E}^{nc}$ and $\mathbf{E}^{c}$ oppose each other. One other force acts on charges in the conductor. It is a frictional force that arises when charges move through medium C_i and is (usually) proportional to their velocity. This frictional force $\mathbf{E}_i^f$ opposes the net motion of charge and therefore, has the same direction as the field $\mathbf{E}^{c}$ (recall that Eq. [2.32] must hold initially to charge the plates). As a result of the accumulation of charge and the increase of the friction force, the acceleration decreases and eventually vanishes. Charges then move with the constant velocity, ω^{+} or ω^{-}, forming the charging current I_i. This happens because the total force acting on elementary charge de disappears:

$$\mathrm{d}e\left(\mathbf{E}^{nc} - \mathbf{E}^{c} - \mathbf{E}_i^f\right) = 0. \qquad [2.36]$$

This implies that

$$\mathbf{E}^{nc} > \mathbf{E}^{c},$$

but if the medium C_i exerts only a very small frictional force, the magnitudes of the fields $\mathbf{E}^{nc}$ and $\mathbf{E}^{c}$ nearly coincide. In the external part, ions initially accelerate and then are slowed by the friction force $\mathbf{E}_e^f$; their velocity becomes constant when

$$\mathbf{E}^{nc} = \mathbf{E}^{c} + \mathbf{E}_i^f, \quad \mathbf{E}^{c} = \mathbf{E}_e^f, \quad \text{and} \quad I_i = I_e. \qquad [2.37]$$

There is a relationship between the extraneous field $\mathbf{E}^{nc}$ and the current. Suppose the extraneous field is increased. Then, the current I_i also increases, and this results in an increase of positive charges arriving at the plate P^{+} (the same is true for negative charges on the plate P^{-}). Correspondingly, the electric field and current density also increase in the solution until currents I_e and I_i become equal. A similar process is observed if the field $\mathbf{E}^{nc}$ decreases. We conclude this section with some observations about the extraneous fields needed to drive a steady current:

1. Unlike the Coulomb force, extraneous forces create a voltage, which, in general, is path dependent; that is, the integral

$$\int_a^b \mathbf{E}^{nc} \cdot \mathbf{dl}$$

is a function of the contour l along which integration is performed.

2. The extraneous field cannot be caused by charges and, therefore, does not obey Coulomb's law. As pointed out, this field has various origins.
3. The extraneous field is usually equal to zero in places where its sources are absent, for example, in the external part of a current tube (there are some exceptions to this rule).
4. It is the combination of the extraneous field and the Coulomb field that drives a steady current in a conducting medium in presence of friction force. Of course, the process of establishing a steady current pictured in our model is only approximate, since in reality the propagation of electromagnetic fields outside and inside a conductor must be taken into account.

2.5.3. Generalization of Ohm's Law

Coulomb and extraneous fields have one common feature: in conducting media, they are related to the current density vector in the same manner. We, therefore, arrive at the general form of Ohm's law:

$$\mathbf{j} = \gamma(\mathbf{E}^c + \mathbf{E}^{nc}). \qquad [2.38]$$

This equation is valid within both internal and external pieces of any current tube. We have already postulated Eq. [2.38]. Later, after consideration of the uniqueness theorem, we will prove that each complete (closed) current tube must include a piece where an extraneous field is present. For the external piece, where $\mathbf{E}^{ext} = 0$, we again have

$$\mathbf{j} = \gamma_e \mathbf{E}^c = \gamma_e \mathbf{E}. \qquad [2.39]$$

In most practical cases of geophysics, we study the field behavior in places where extraneous fields are absent and, correspondingly, mainly use Eq. [2.39]. There are several important exceptions that require Eq. [2.38]. First, note that the voltage of the extraneous field between the end points of the internal part of a current tube is called the electromotive force:

$$\Xi = \int_a^b \mathbf{E}^{nc} d\mathbf{l}, \qquad [2.40]$$

and it characterizes the work performed by the extraneous field in moving a unit charge against Coulomb and frictional forces. For illustration, consider an example of the current circuit shown in Fig. 2.3B and assume (for convenience) that the current is just the motion of positive charges. We see that in an external part the electric field is caused by charge alone. In contrast, inside the internal part positive charge is moved by an extraneous force against the action of Coulomb's force.

2.6. DISTRIBUTION OF ELECTRIC CHARGES IN A CONDUCTING AND POLARIZABLE MEDIUM

It is useful to study the distribution of electrical charge that arises when steady current flows in a conducting and polarizable medium. We will proceed from the principle of charge conservation, Coulomb's law, and Ohm's law and first consider regular points where the extraneous field is absent, but later will pay close attention to what happens at interfaces and also take into account the effect of field $\mathbf{E}^{ext}$. In other words, we focus first on the external part of current tubes.

2.6.1. Charges at Regular Points

It follows from the principle of charge conservation and Ohm's law that

$$\operatorname{div}\mathbf{j} = \operatorname{div}(\gamma\mathbf{E}) = \gamma\operatorname{div}\mathbf{E} + \mathbf{E}\cdot\nabla\gamma = 0, \qquad [2.41]$$

where $\mathbf{E}=\mathbf{E}^c$ is the Coulomb field. Taking into account that

$$\operatorname{div}\mathbf{E} = \frac{\delta}{\varepsilon_0},$$

and substituting it into Eq. [2.41], we obtain an expression for the volume density of charges at any regular point q,

$$\delta(q) = -\varepsilon_0\frac{\mathbf{E}\cdot\nabla\gamma}{\gamma}. \qquad [2.42]$$

Since

$$\nabla\gamma = \nabla\frac{1}{\rho} = -\frac{\nabla\rho}{\rho^2},$$

Eq. [2.42] can be also written as

$$\delta(q) = \varepsilon_0\frac{\mathbf{E}\cdot\nabla\rho}{\rho}. \qquad [2.43]$$

Thus, volume charges arise in places where the conductivity (resistivity) of a medium varies and its gradient is not perpendicular to the electric field.

The sign of charges and their magnitude depend on the angle between the electric field and $\nabla\gamma$. For instance, if this angle is smaller than $\pi/2$, negative charge appears. It is clear that the amount of charge is directly proportional to the magnitudes of the electric field and $\nabla\gamma$. Note that charge conservation requires that in any volume where $\delta(q)\neq 0$ (i.e., where charge arises according to Eq. [2.43]) and where there is steady current, the amount of charge arriving into volume must balance the charge leaving it. By definition, the left-hand side of Eq. [2.43] is the sum of the densities of free and bound charges and each of them depends on the dielectric constant ε and the resistivity ρ of the medium. A typical model of the upper part of the Earth, often used in applied geophysics, consists of uniform formations having different conductivities. In this case, the volume density of charges inside each uniform region is equal to zero.

2.6.2. Surface Charges

We now study what happens at interfaces where the conductivity changes and surface distributions of charge can appear. Consider first points where $\mathbf{E}^{\text{nc}}=0$. Proceeding from Ohm's law and the surface analogy of the second equation for $\mathbf{j}$, which holds that the normal component of current density is continuous at interface, we obtain

$$\begin{aligned} j_{\text{n}}^{(2)} - j_{\text{n}}^{(1)} &= \gamma_2 E_{\text{n}}^{(2)} - \gamma_1 E_{\text{n}}^{(1)} \\ &= [(\gamma_1+\gamma_2)(E_{\text{n}}^{(2)} - E_{\text{n}}^{(1)}) + (\gamma_2-\gamma_1)(E_{\text{n}}^{(2)} + E_{\text{n}}^{(1)})]/2 = 0. \end{aligned} \quad [2.44]$$

Since

$$E_{\text{n}}^{(2)} - E_{\text{n}}^{(1)} = \frac{\sigma}{\varepsilon_0}, \quad [2.45]$$

Eq. [2.44] becomes

$$\sigma(q) = 2\varepsilon_0 \frac{\gamma_1-\gamma_2}{\gamma_1+\gamma_2} E_{\text{n}}^{\text{av}}(q) = 2\varepsilon_0 \frac{\rho_2-\rho_1}{\rho_2+\rho_1} E_{\text{n}}^{\text{av}}(q), \quad [2.46]$$

where

$$\sigma(q) = \sigma_0(q) + \sigma_{\text{b}}(q)$$

is the sum of surface densities of free and bound charges, and $E_{\text{n}}^{\text{av}}(q)$ is the average value of the normal component of the electric field in the vicinity of the interface at its point q:

$$E_{\mathrm{n}}^{\mathrm{av}}(q) = \frac{E_{\mathrm{n}}^{(1)}(q) + E_{\mathrm{n}}^{(2)}(q)}{2}.$$

We can further simplify Eq. [2.46] by using the fact that a charge located on the elementary surface d$S(q)$ creates at its vicinity a normal component of electric field equal to

$$E_{\mathrm{n}}(q) = \pm \frac{\sigma(q)}{2\varepsilon_0}.$$

Then,

$$E_{\mathrm{n}}^{(1)}(q) = E_{\mathrm{n}}^{S-q}(q) - \frac{\sigma(q)}{2\varepsilon_0}, \quad E_{\mathrm{n}}^{(2)}(q) = E_{\mathrm{n}}^{S-q}(q) + \frac{\sigma(q)}{2\varepsilon_0}, \qquad [2.47]$$

where $E_{\mathrm{n}}^{S-q}(q)$ is now the normal component which is caused only by those charges located outside the element d$S(q)$. We showed earlier that this part of the normal component is continuous across the surface; therefore,

$$E_{\mathrm{n}}^{\mathrm{av}}(q) = E_{\mathrm{n}}^{S-q}(q). \qquad [2.48]$$

Thus, the density of the surface charges is directly proportional to the contrast coefficient:

$$K_{12} = \frac{\rho_2 - \rho_1}{\rho_2 + \rho_1} \qquad [2.49]$$

and to the normal component $E_{\mathrm{n}}^{S-q}(q)$:

$$\sigma(q) = 2\varepsilon_0 K_{12} E_{\mathrm{n}}^{S-q}(q). \qquad [2.50]$$

These results show that the electric field **E** measured in electrical methods of geophysics is caused mainly by charges arising at interfaces between media with different resistivities. In fact, this feature is the physical foundation of electrical methods in geophysics. Before proceeding, it is appropriate to review some of the properties of electric current. In general, it is natural to distinguish the motion of positive and negative charge within a conducting medium. For instance, in an electrolyte, ions of both signs make up the current, generally by moving in the opposite directions. In metals, however, only negative charges (electrons) move appreciably; the positive ions remain fixed in the crystal structure. In both cases, the medium locally remains neutral. The sum of charges in every elementary volume vanishes and, correspondingly, the moving charge itself does not create an electric field. As we now know, there are also the volume and surface charges that can appear at points where the electric properties of a medium vary. If the electric field stays constant with time, these charges cannot be changing with time, as they are the sole sources of electric field in the external part of any current tube. Of course, the charges do not arise instantly. There is a (short) time interval during which propagation of the electromagnetic field plays a dominant role in setting up the steady charge distribution. Before moving on to study this distribution of charges, we consider two examples.

2.6.3. Example 1: Current Electrode in a Uniform Medium

Suppose that an isolated wire with current I is connected to a uniform medium with a spherical electrode (Fig. 2.4A). The resistivity of the electrode and surrounding medium is ρ_0 and ρ, respectively. Since both media are uniform, volume charges are absent but there is an interface between the electrode and surrounding medium where surface charges arise. It is clear that, because of the spherical symmetry, the current density and electric field have only a radial component that is normal to the interface. We have

$$E_r^{(2)} - E_r^{(1)} = \frac{\sigma}{\varepsilon_0}, \qquad [2.51]$$

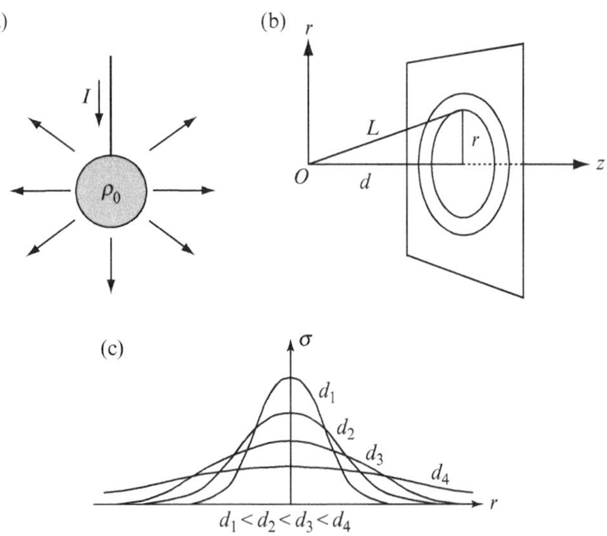

Figure 2.4 (A) Current electrode in a uniform medium, (B) current electrode in the presence of plane interface, and (C) distribution of surface charges.

where $E_r^{(2)}$ and $E_r^{(1)}$ are radial components of the electric field at the external and internal sides of the electrode surface, respectively, and σ is the density of charge on this surface. Using continuity of the normal component of the current density and Ohm's law for the external part of a current tube, Eq. [2.51] gives

$$\rho j_r^{(2)} - \rho_0 j_r^{(1)} = (\rho - \rho_0) j_r = \frac{\sigma}{\varepsilon_0}$$

or

$$\sigma = \varepsilon_0 (\rho - \rho_0) j_r. \qquad [2.52]$$

Therefore, the total charge on the electrode surface is

$$e = \oint_S \sigma \, dS = \varepsilon_0 (\rho - \rho_0) \oint_S j_r dS \quad \text{or} \quad e_A = \varepsilon_0 (\rho - \rho_0) I, \qquad [2.53]$$

where I is the current in the wire feeding the electrode from the source. (It is assumed that all the current leaves the electrode through its surface; strictly, the integral in Eq. [2.53] is over the entire electrode surface except for the point where the wire connects to the electrode.) This charge, in fact, creates the electric field that has only a radial component:

$$E_r(p) = \frac{e_A}{4\pi\varepsilon_0 L_{0p}^2} = \frac{(\rho - \rho_0) I}{4\pi L_{0p}^2},$$

where L_{0p} is the distance from the electrode center to the observation point p. It follows from Eq. [2.53] that the charge on the electrode is extremely small, since $\varepsilon_0 \approx 10^{-11}$. Notice, however, that the electric field is independent of this constant (canceled by the reciprocal of ε_0 in Coulomb's law), and is a function of only the total current, the medium resistivity, and the distance from the electrode. Because formation resistivity is usually many orders greater than that of the electrode, $\rho \ggg \rho_0$, we can write

$$e_A = \varepsilon_0 \rho I \quad \text{and} \quad E(p) = \frac{\rho I}{4\pi L_{0p}^2}. \qquad [2.54]$$

Note that if current goes from the medium to the electrode, negative charges appear on its surface. Thus, we demonstrated that the charge on the electrode surface creates an electric field as if it were located in free space, not in a conducting and polarizable medium. We will explore this further below. Under the action of this field, current flows radially outward in the conducting medium surrounding the electrode, with density

$$j_r(p) = \gamma E_r = \frac{I}{4\pi L_{0p}^2}. \qquad [2.55]$$

Of course, this result could also be derived from the spherical symmetry and the fact that at any distance from the electrode the total current I must be flowing through a spherical surface of area $4\pi L_{0p}^2$ centered on the electrode. The electrical potential is easily calculated to be

$$U(p) = \frac{\rho I}{4\pi L_{0p}}. \qquad [2.56]$$

The equipotential surfaces are spherical and include the electrode surface itself. It is interesting to calculate the potential on this surface. Letting the electrode radius be a and using Eq. [2.54], we have

$$U_{\text{el}} = \frac{\rho I}{4\pi a} = \frac{e_A}{4\pi\varepsilon_0 a},$$

where e_A is the charge on the electrode. This is a general result for the potential at the surface of a charged conducting sphere. The difference of potential, or the voltage between two equipotential surfaces, is

$$U(a) - U(b) = \frac{\rho I}{4\pi}\left(\frac{1}{L_{oa}} - \frac{1}{L_{ob}}\right). \qquad [2.57]$$

This is the simplest example of field behavior in the presence of the steady current flow, but is nevertheless important for understanding the physical basis of electric methods in geophysics. Moreover, we see from Eqs. [2.54] that the charge on the electrode surface is defined by the current and resistivity of the medium, and is independent of the medium's dielectric properties (its polarizability). This means that if the dielectric constant ε varies in the medium, the total charge e and field $\mathbf{E}$ stays the same. This sounds strange, because we know that an electric field can polarize charge in a medium, creating a secondary electric field. In fact, the charge e that arises on the electrode includes both the free or mobile charge from the current and bound charge from polarization. We will later discuss this paradox in some detail.

2.6.4. Example 2: Induced Charges on a Plane Interface

We now assume that the electrode A carrying current I is surrounded by a medium of resistivity ρ_1, and at a distance d from the electrode, there is a plane boundary across which there is a second medium of resistivity ρ_2 (Fig. 2.4B). In this case, charges arise on the electrode surface and on the plane boundary between the two media, and at every point, these charges create electric fields. We restrict ourselves to a study of charge distribution only. By Eq. [2.54], the charge on the electrode surface is

$$e_A = \varepsilon_0 \rho_1 I \quad \text{if} \quad \rho \ggg \rho_0.$$

To calculate the charge density on the plane interface, we will use Eq. [2.50], which holds that the density of surface charges is directly proportional to the continuous component of the normal electric field at the interface. In Chapter 1, we showed that the continuous normal component of the field across at a given point q on an arbitrary interface is the field caused by all other charges in space, except for charge at the point q. Because the boundary is plane, however, another simplification occurs: again, as we showed in Chapter 1, charges on a plane surface located away from the point q do not create a normal component of the field at q. (This fact should be obvious from Coulomb's law and symmetry.) This means that the field $E_n^{S-q}(q)$ in Eq. [2.50] is caused only by the charge on the electrode:

$$E_n^{S-q} = E_n^A, \qquad [2.58]$$

where $E_n^A(q)$ is the normal component of this field. To determine this component, we can use cylindrical coordinates centered at the spherical electrode A, with the z-axis perpendicular to the interface (Fig. 2.4B). Then, using Eq. [2.54] (Coulomb's law) and elementary geometry, the normal component of the field at any point of the boundary is

$$E_n(q) = \frac{e_A}{4\pi\varepsilon_0 L^2}\frac{d}{L} = \frac{e_A d}{4\pi\varepsilon_0 (r^2 + d^2)^{3/2}}, \qquad [2.59]$$

where d is the distance from the electrode to the plane and r is the distance from the point q to the z-axis (Fig. 2.4B). From Eq. [2.50], the surface density of charge is, therefore,

$$\sigma(q) = \frac{K_{12} e_A}{2\pi} \frac{d}{(r^2 + d^2)^{3/2}}. \qquad [2.60]$$

We can say that, because of the plane geometry, there is no interaction between charges at the interface, which simplifies the charge calculation. It follows from Eq. [2.60] that the density of charge is maximum at the point where the z-axis intersects the plane ($r=0$, $z=d$), and decreases as r increase. Close to the axis, where $r \ll d$, the density decreases slowly, but later when r exceeds d, this decay becomes stronger, and $\sigma(q)$ eventually decreases as r^{-3}. It also follows that the highest charge density on the interface gets larger as the electrode gets closer to the interface. At the same time, it starts to decrease more rapidly at smaller values of r (Fig. 2.4C). When the electrode is placed on the interface, the charge density tends to infinity at the point ($r=0$, $z=d$), and vanishes at all other points of the boundary since at such points $E_n^A = 0$. We now calculate the total charge induced on the plane interface. It is convenient first to find the elementary charge de_S induced inside the ring with the radius r and thickness dr (Fig. 2.4B). The area of this ring is

$$dS = 2\pi r\,dr,$$

and since the charge density is axially symmetric (around the z-axis), we have

$$de_S = \sigma\,dS = K_{12}e_A d\frac{r\,dr}{(r^2+d^2)^{3/2}}.$$

Integrating with respect to r from zero to infinity gives

$$e_S = K_{12}e_A d\int_0^\infty \frac{r\,dr}{(r^2+d^2)^{3/2}} = K_{12}e_A.$$

Thus, the total charge on the interface is

$$e_S = K_{12}e_A. \qquad [2.61]$$

This interesting result shows that the total charge induced on the interface e_S is independent of the distance of the current electrode from the interface. This distance influences only on the charge distribution over the boundary. For instance, as the electrode approaches the interface, charge becomes concentrated near the point $r=0$. In the limit, as the electrode touches the interface, all the charge resides on the electrode, and the total charge is

$$e = e_A + e_S = (1+K_{12})e_A. \qquad [2.62]$$

By definition

$$K_{12} = \frac{\rho_2 - \rho_1}{\rho_2 + \rho_1},$$

which varies as $-1 < K_{12} < 1$. Consequently, the surface charge e_S cannot exceed in magnitude the electrode charge e_A. If the medium in which the electrode is situated is more conductive, both charges have the same sign, but in the opposite case when $\rho_1 > \rho_2$, they have a different sign. In the limit, when the second medium is either an ideal conductor ($\rho_2 = 0$), or an ideal insulator ($\rho_2 = \infty$), the surface charge equals $-e_A$ and $+e_A$, respectively. Therefore, if the electrode is placed at the interface with an insulator, such as air, its charge is doubled:

$$e_A = 2\varepsilon_0 \rho_1 I \quad \text{if} \quad d = 0. \qquad [2.63]$$

2.6.5. Volume Density of Charges Related to an Extraneous Field

We next examine the distribution of charges in a conducting and polarizable medium where an extraneous field is present. We start from the general form of Ohm's law:

$$\mathbf{j} = \gamma(\mathbf{E}^c + \mathbf{E}^{nc}).$$

Repeating the transformations traced earlier, we now obtain

$$\operatorname{div} \mathbf{j} = \operatorname{div} \gamma(\mathbf{E}^c + \mathbf{E}^{nc}) = \operatorname{div} \gamma\mathbf{E}^c + \operatorname{div} \gamma\mathbf{E}^{nc} = 0,$$

or

$$\gamma \operatorname{div} \mathbf{E}^c + \mathbf{E}^c \cdot \nabla\gamma + \operatorname{div} \gamma\mathbf{E}^{nc} = 0,$$

or

$$\operatorname{div} \mathbf{E}^c = -\frac{\mathbf{E}^c \cdot \nabla\gamma}{\gamma} - \frac{\operatorname{div} \gamma\mathbf{E}^{nc}}{\gamma}. \qquad [2.64]$$

By definition, the right-hand side of this equation characterizes the density of sources of the Coulomb field. Therefore, the volume density of free and bound charges related to the extraneous field is

$$\delta^{ext} = -\varepsilon_0 \frac{\operatorname{div}(\gamma\mathbf{E}^{nc})}{\gamma}. \qquad [2.65]$$

For instance, if a medium is homogeneous, $\gamma = \text{constant}$, and we have

$$\delta^{\text{ext}}(q) = -\varepsilon_0 \operatorname{div} \mathbf{E}^{\text{nc}}. \qquad [2.66]$$

On the contrary, when conductivity varies but the field $\mathbf{E}^{\text{nc}}$ is constant,

$$\delta^{\text{ext}}(q) = -\varepsilon_0 \mathbf{E}^{\text{nc}} \cdot \nabla\gamma. \qquad [2.67]$$

It follows from Eq. [2.65] that extraneous charge disappears where $\operatorname{div} \gamma\mathbf{E}^{\text{nc}} = 0$. This consideration shows that if the extraneous force and conductivity of a medium are known, we can calculate from Eq. [2.67] the distribution of charges associated with the field $\mathbf{E}^{\text{ext}}$, and therefore, the electric field $\mathbf{E}^{\text{c}}$ caused by these charges. This can be done by either applying Coulomb's law or the differential equations for the field.

2.6.6. Surface Density of Charges Related to an Extraneous Field

We apply again the principle of charge conservation at points of the interface, that is, the continuity of the normal component of the current density, obtaining

$$j_{\text{n}}^{(2)} - j_{\text{n}}^{(1)} = \gamma_2 E_{\text{n}}^{(2)} - \gamma_1 E_{\text{n}}^{(1)} = 0,$$

or

$$\gamma_2 E_{\text{n}}^{(2)\text{c}} - \gamma_1 E_{\text{n}}^{(1)\text{c}} + \operatorname{Div} \gamma\mathbf{E}^{\text{nc}} = 0, \qquad [2.68]$$

where

$$\operatorname{Div} \gamma\mathbf{E}^{\text{nc}} = \gamma_2 E_{\text{n}}^{(2)\text{nc}} - \gamma_1 E_{\text{n}}^{(1)\text{nc}}. \qquad [2.69]$$

Here, $E_{\text{n}}^{(2)\text{nc}}$ and $E_{\text{n}}^{(1)\text{nc}}$ are normal components of the extraneous field at the front and back sides of the surface, respectively. As shown before, the difference of the first two terms in Eq. [2.68] can be written in the form

$$\frac{1}{2}\left[(\gamma_2 - \gamma_1)\left(E_{\text{n}}^{(2)\text{c}} + E_{\text{n}}^{(1)\text{c}}\right) + (\gamma_2 + \gamma_1)\left(E_{\text{n}}^{(2)\text{c}} - E_{\text{n}}^{(1)\text{c}}\right)\right],$$

or

$$(\gamma_2 - \gamma_1) E_{\text{n}}^{\text{av,c}} + \gamma^{\text{av}} \frac{\sigma}{\varepsilon_0},$$

since

$$E_{\text{n}}^{(2)\text{c}} - E_{\text{n}}^{(1)\text{c}} = \frac{\sigma}{\varepsilon_0}, \quad E_{\text{n}}^{\text{av,c}} = \frac{E_{\text{n}}^{(2)\text{c}} + E_{\text{n}}^{(1)\text{c}}}{2}, \quad \gamma^{\text{av}} = \frac{\gamma_1 + \gamma_2}{2}.$$

Thus, in place of Eq. [2.68], we have

$$(\gamma_2 - \gamma_1)E_n^{av,c} + \frac{\gamma_{av}}{\varepsilon_0}\sigma(q) + \mathrm{Div}\,\gamma\mathbf{E}^{nc} = 0,$$

or

$$\sigma(q) = 2\varepsilon_0 K_{12} E_n^{av,c} - \varepsilon_0 \frac{\mathrm{Div}\,\gamma\mathbf{E}^{nc}}{\gamma^{av}}. \qquad [2.70]$$

The charge caused by the extraneous field is clearly

$$\sigma^{nc}(q) = -\varepsilon_0 \frac{\mathrm{Div}\,\gamma\mathbf{E}^{nc}}{\gamma^{av}}. \qquad [2.71]$$

In particular, near the end points of the internal part of the elementary current tube, we have

$$\sigma_+ = \varepsilon_0 \frac{\gamma_i}{\gamma^{av}} E_{n+}^{nc}, \quad \sigma_- = -\varepsilon_0 \frac{\gamma_i}{\gamma^{av}} E_{n-}^{nc}, \qquad [2.72]$$

since the field $\mathbf{E}^{nc}$ vanishes at the external part of the tube (by definition). Of course, these charges create an electric field at any point of a current tube. In the case when $E_{n+}^{nc} = E_{n-}^{nc} = E_n$ and $\gamma_i = \gamma_e$, we have

$$\sigma_+ = \varepsilon_0 E_n^{nc} \quad \text{and} \quad \sigma_- = -\varepsilon_0 E_n^{nc}. \qquad [2.73]$$

2.6.7. The Contact Electromotive Force

Now, we assume that the extraneous force $\mathbf{E}^{nc}$ is distributed within a thin surface layer S of thickness h (Fig. 2.3C). The boundary surfaces of this layer are S_1 and S_2, and the normal $\mathbf{n}$ (to the average surface S) is directed from S_1 to S_2. In general, the conductivities of the media inside and outside the layer are different and the thickness h may vary. As is usual, it is convenient to represent the current density as a system of elementary current tubes. Then, every element of the layer plays the role of the internal part of a corresponding tube. It is obvious that charges arising on surfaces (Eqs. [2.71]) are maintained by the extraneous force, while the Coulomb field tends to discharge them. Assuming that the layer is very thin, it is natural to assume that the field $\mathbf{E}^{ext}$ is constant over its thickness:

$$\Xi = \int_1^2 \mathbf{E}^{nc} \cdot d\mathbf{l} = h\mathbf{E}^{nc}, \qquad [2.74]$$

where $d\mathbf{l}$ is a line element locally perpendicular to elementary boundary surfaces, dS_1 and dS_2, which carry charges of the same magnitude but opposite sign. The charge situated on surfaces S_1 and S_2 creates the electric field in a conducting medium. If an observation point is located at distances greatly exceeding h, the system of charges can be treated as the double layer. As shown in Chapter 1, it is a simple matter to find a relation between the electromotive force and the density of dipole moments $\eta(q)$ on the surface:

$$\eta(q) = \varepsilon_0 \Xi = \varepsilon_0 h E_n^{nc}, \qquad [2.75]$$

Since η is a function of the product h and $\mathbf{E}^{nc}$, consider the limiting case where

$$h \to 0 \quad \text{and} \quad E_n^{nc} \to \infty,$$

but the dipole moment $\eta(q)$ remains constant. This configuration is the same as the model of the double layer considered in Chapter 1. Then, using the results derived there, we have near the double layer:

$$U_+ - U_- = \Xi_c, \quad \frac{\partial U_+}{\partial s} - \frac{\partial U_-}{\partial s} = \frac{\partial \Xi_c}{\partial s}, \quad \text{and} \quad \frac{\partial U_+}{\partial n} = \frac{\partial U_-}{\partial n}, \qquad [2.76]$$

where s characterizes a direction tangential to a double layer, and

$$\Xi = \Xi_c = \frac{\eta(q)}{\varepsilon_0} \qquad [2.77]$$

is called the contact electromotive force. As shown in Chapter 1, using the concept of solid angle, it is relatively simple to find the field caused by the double layer.

2.6.8. Total Charge is Independent of the Dielectric Constant in Steady Flow

Earlier, we saw that the charge on an electrode surface does not depend on the dielectric constants of the electrode and the surrounding medium. Let us examine this fundamental fact in more detail. We have shown that the electric field of Coulomb's law satisfies the following equations at regular points in a conducting medium:

$$\text{curl}\, \mathbf{E}^c = 0, \quad \text{div}\, \gamma \mathbf{E}^c = -\,\text{div}\, \gamma \mathbf{E}^{nc}.$$

Clearly, to determine the field from these equations requires knowing the distribution of electromotive force and conductivity but not the dielectric constant. At the same time, we know that polarization causes the appearance of bounded charges, which, in turn, give rise to the field $\mathbf{E}$. It is natural to ask, how is it possible to ignore the influence of polarization (i.e., the dielectric constant) on the behavior of electric field in a conducting and polarizable medium? In order to resolve this paradox, let us, for simplicity, consider regular points of the external part of the current tube. Taking Eqs. (1.190) and [2.42],

$$\delta_b = \frac{1-\varepsilon_r}{\varepsilon_r}\delta_0 - \frac{\mathbf{E}\cdot\nabla\varepsilon}{\varepsilon_r} \quad \text{and} \quad \delta = -\varepsilon_0\frac{\mathbf{E}\cdot\nabla\gamma}{\gamma}, \qquad [2.78]$$

and writing

$$\delta = \delta_0 + \delta_b,$$

we have separately for the density of free and bound charges:

$$\delta_0 = -\varepsilon_0\frac{\mathbf{E}\cdot\nabla\gamma}{\gamma} - \frac{1-\varepsilon_r}{\varepsilon_r}\delta_0 + \frac{\mathbf{E}\cdot\nabla\varepsilon}{\varepsilon_r}$$

or

$$\delta_0 = -\varepsilon\frac{\mathbf{E}\cdot\nabla\gamma}{\gamma} + \mathbf{E}\cdot\nabla\varepsilon \quad \text{and} \quad \delta_b = \frac{1-\varepsilon_r}{\varepsilon_r}\delta_0 - \frac{\mathbf{E}\cdot\nabla\varepsilon}{\varepsilon_r}. \qquad [2.79]$$

For illustration, consider three special cases.

Case 1. In a region where the medium is homogeneous, $\nabla\gamma=0$, $\nabla\varepsilon=0$, and there is neither free nor bound charges, the total charge is zero.

Case 2. In a region where the polarizability α varies ($\nabla\varepsilon\neq 0$), but conductivity is constant ($\nabla\gamma=0$), both free and bound charges appear and we have for the total charge:

$$\delta = \mathbf{E}\cdot\nabla\varepsilon + \frac{1-\varepsilon_r}{\varepsilon_r}\mathbf{E}\cdot\nabla\varepsilon - \frac{\mathbf{E}\cdot\nabla\varepsilon}{\varepsilon_r} = 0.$$

The two types of charges compensate each other. Certainly, it was not obvious to expect this effect.

Case 3. In a region where conductivity varies, but not the polarizability, we have

$$\delta_0 = -\varepsilon\frac{\mathbf{E}\cdot\nabla\gamma}{\gamma}, \quad \delta_b = \frac{1-\varepsilon_r}{\varepsilon_r}\delta_0 = \left(1-\frac{1}{\varepsilon_r}\right)\varepsilon\frac{\mathbf{E}\cdot\nabla\gamma}{\gamma}.$$

Again, bound and free charges appear, and they depend on both the conductivity and dielectric constant, but the total charge is a function of resistivity only. Next, consider points at interfaces where, as follows from Eqs. (1.193) and [2.50], we have

$$\sigma_b = -\frac{\varepsilon_2 - \varepsilon_1}{\varepsilon_r^{av}} E_n^{av} + \frac{1 - \varepsilon_r^{av}}{\varepsilon_r^{av}} \sigma_0 \quad \text{and} \quad \sigma = 2\varepsilon_0 K_{12} E_n^{av}, \qquad [2.80]$$

whence

$$\sigma_0 = \left[\frac{\varepsilon^{av}}{\rho^{av}} (\rho_2 - \rho_1) + (\varepsilon_2 - \varepsilon_1) \right] E_n^{av}, \qquad [2.81]$$

and again the free and bound surface charges are functions of both electric parameters of a medium. For instance, it is simple matter to see that, if $\rho_1 = \rho_2$ but $\varepsilon_1 \neq \varepsilon_2$, the free and bound charges compensate each other. The same conclusion follows in the general case when both polarizability and conductivity vary. Summarizing, we can say that, in general, free and bound charges arise when steady current flows in a conducting and dielectric medium. But their sum, the total charge, is always independent of the dielectric constant as well as its electric field.

2.7. SYSTEM OF EQUATIONS OF THE ELECTRIC FIELD AND POTENTIAL

2.7.1. Introduction

We have described the general features of the electric field and current density in a conducting and polarizable medium, including (a) the behavior of these fields at regular points and near interfaces between media with different resistivities, (b) the distribution of charges in a conducting medium, (c) the current density field and its relation to the electric field **E** through Ohm's law, and (d) the role of extraneous fields in providing the ultimate driving force for steady current flow. We now begin to examine the basic question of how to determine the electric field in a conducting medium with specified extraneous sources. This is often called "solution of the forward problem" or simply "forward modeling."

2.7.2. Solution of Forward Problems for Steady Current in Conducting Media

Calculation of the electric field for different configurations is clearly the cornerstone of quantitative interpretation of electrical measurements in geophysics. In this light, it is instructive to recall that modeling of the gravity field for a given mass distribution is based on the direct use of the equation

$$\mathbf{g}(p) = -\gamma \int_V \frac{\delta(q)\mathbf{L}_{qp}}{L_{qp}^3} \mathrm{d}V,$$

and requires (at most) a numerical integration. Here, γ is the gravitational constant and L_{qp} is the distance between an elementary mass and an observation point. The density of mass $\delta(q)$ is a physical parameter of the model, and is usually independent of the field $\mathbf{g}$. In other words, when a model of a medium is fully specified, the sources of the field are known. This fact allows us to make use of Newton's law and determine the field by direct integration. In contrast, the equation for the electric field that follows from Coulomb's law,

$$\mathbf{E}(p) = \frac{1}{4\pi\varepsilon_0} \int_V \frac{\delta(q)\mathbf{L}_{qp}}{L_{qp}^3} \mathrm{d}V + \frac{1}{4\pi\varepsilon_0} \int_S \frac{\sigma(q)\mathbf{L}_{qp}}{L_{qp}^3} \mathrm{d}S,$$

cannot usually be used to calculate the electric field in a conducting medium because the density of charges, the source of the field, is not just a physical parameter of the medium, but instead depends both on its physical properties (the distribution of resistivity) and on the electric field, the quantity to be computed. In particular, as shown earlier, the total charge density (sum of free and bound charge) is given by

$$\delta(q) = -\varepsilon_0 \rho \mathbf{E} \cdot \nabla \frac{1}{\rho}, \quad \sigma(q) = 2\varepsilon_0 \frac{\rho_2 - \rho_1}{\rho_2 + \rho_1} E_{\mathrm{n}}^{\mathrm{av}}(q).$$

This means, in effect, that any given model of a medium can support an infinite number of charge distributions, depending on the location of extraneous sources and electric field that drive the current flow. This is the same "closed-circle problem" described in Chapter 1 and is the reason that we need to formulate boundary-value problems to solve for the electric field. It also accounts for the fundamental difference between the theory of gravity and electric methods. Modeling of electric methods in geophysics would be much simpler if we knew the charge distribution in advance and could simply apply Coulomb's law.

2.7.3. System of Equations of the Electric Field in a Conducting and Polarizable Medium

Before describing boundary-value problems, we first derive the full system of equations for the electric field in a conducting medium. As follows from vector analysis, such a system should include equations for the curl and divergence of the field. From Chapter 1, we know that the electric field satisfies the following equations at regular points:

$$\text{curl}\,\mathbf{E} = 0 \quad \text{and} \quad \text{div}\,\varepsilon\mathbf{E} = \delta_0. \qquad [2.82]$$

The second equation, however, has two serious shortcomings:

1. It contains the density of free charges δ_0, which is usually unknown in advance.
2. It contains the dielectric constant, which we have just shown is not required for calculation of the electric field in a conducting medium. To overcome these problems, we will proceed from Coulomb's law, Ohm's law, and the principle of charge conservation. This gives at regular points

$$\text{curl}\,\mathbf{E} = 0 \quad \text{and} \quad \text{div}\,\gamma\mathbf{E} = -\,\text{div}\,\gamma\mathbf{E}^{\text{nc}}. \qquad [2.83]$$

Also, at the boundary between media with different conductivities, we have the following interface conditions easily derived from Eqs. [2.83]:

$$E_{2t} - E_{1t} = 0, \quad \gamma_2 E_{2n} - \gamma_1 E_{1n} = -\,\text{Div}\,\gamma\mathbf{E}^{\text{nc}}, \qquad [2.84]$$

where

$$\text{Div}\,\gamma\mathbf{E}^{\text{nc}} = \gamma_2 E_{2n}^{\text{nc}} - \gamma_1 E_{1n}^{\text{nc}}.$$

To highlight the importance of Eqs. [2.83] and [2.84], we lay out below the full system of equations for the electric field in steady current flow (Eqs. [2.85]). To illuminate the meaning of these equations, we note the following:

1. This complete system of equations no longer contains the dielectric constant, and shows that the electric field in steady current flow through a conducting medium does not depend on the polarizability.
2. Since the conductivity and extraneous force can be specified, there is only one unknown in this system, namely, the electric field **E**. Terms involving the density of free or bound charges do not appear explicitly, but can be recovered from the second equation. For instance, at regular points we have

$$\text{div}\,\gamma\mathbf{E} = \gamma\,\text{div}\,\mathbf{E} + \mathbf{E}\cdot\nabla\gamma = -\,\text{div}\,\gamma\mathbf{E}^{\text{nc}}$$

or

$$\delta = -\varepsilon_0\frac{\mathbf{E}\cdot\nabla\gamma}{\gamma} - \varepsilon_0\frac{\nabla\cdot\gamma\mathbf{E}^{\text{nc}}}{\gamma}.$$

Similarly, the presence of the surface charge is implicit in the continuity of the normal component of the current density at the interface.

3. The field equations were obtained from three physical laws: Coulomb's law, Ohm's law, and the principle of charge conservation. The set (2.85), therefore, contains the same information as these laws. If we suppose, for example, that the first equation does not hold,

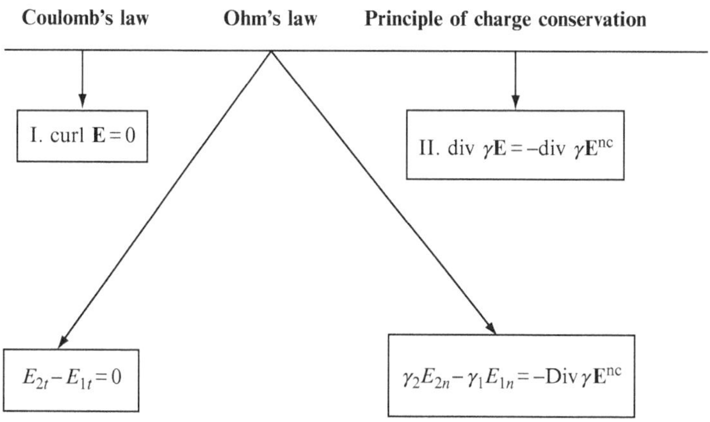

[2.85]

$$\text{curl}\,\mathbf{E} \neq 0 \quad \text{or} \quad E_{2t} \neq E_{1t},$$

it would imply that the field **E** no longer obeys Coulomb's law and is caused by vortices as well as charges. Besides, the voltage of such a field would be path dependent. If the second equation did not hold,

$$\text{div}\,\gamma\mathbf{E} = \text{div}\,\mathbf{j} \neq 0 \quad \text{or} \quad j_{2n} \neq j_{1n},$$

it would imply that charge varies with time, and therefore, that the field itself is not time invariant (steady).

4. The system (2.85) does not describe field behavior caused by a double layer of charge, a special case considered in detail in Chapter 1.

5. Any electric field created by constant charges is a solution of the set (2.85) regardless of the distribution of resistivity and the position of extraneous sources creating a primary field. This happens because every Coulomb field and current density must obey all three physical laws. Alternatively, we can say that this system has an infinite number of solutions, which occurs, of course, because the partial differential equations require additional boundary conditions for a unique solution. We now specify these conditions by reintroducing the scalar potential for the electric field.

2.7.4. System of Equations for the Potential *U*

We introduce again, as in Chapter 1, the scalar potential of the electric field, and consider the system of equations describing its behavior in the case of steady current flow. The first equation of the set (2.85) is satisfied identically if

$$\mathbf{E} = -\nabla U.$$

Using Ohm's law and substituting into the second equation, $\operatorname{div}\mathbf{j}=0$, gives at regular points

$$\nabla\cdot(\gamma\nabla U) = \nabla\cdot(\gamma\mathbf{E}^{\text{ext}}). \qquad [2.86]$$

The conditions on the electric field at interfaces between media of different conductivities become

$$\frac{\partial U_1}{\partial s} = \frac{\partial U_2}{\partial s} \quad \text{and} \quad \gamma_2\frac{\partial U_2}{\partial n} - \gamma_1\frac{\partial U_1}{\partial n} = \operatorname{Div}(\gamma\mathbf{E}^{\text{nc}}), \qquad [2.87]$$

where U_1 and U_2 are values of the potential on different sides of the interface and s and n are directional tangent and normal, respectively, to the interface. Since components of the electric field have finite values everywhere, the potential U must be continuous at the interface. This also follows from the definition of potential,

$$U(p_2) - U(p_1) = \int_{p_2}^{p_1} \mathbf{E}\cdot d\mathbf{l},$$

because when points p_1 and p_2 located on opposite sides of the interface approach each other, the integral vanishes (again assuming that the field remains finite everywhere). The continuity of the potential across an interface

$$U_1 = U_2 \qquad [2.88]$$

applies at each point and ensures that the tangential derivatives $\partial U/\partial s$ is also continuous. That is, the conditions

$$U_1 = U_2 \quad \text{and} \quad \frac{\partial U_1}{\partial s} = \frac{\partial U_2}{\partial s}$$

are equivalent. Correspondingly, in place of Eqs. [2.87], we obtain

$$U_1 = U_2 \quad \text{and} \quad \gamma_2 \frac{\partial U_2}{\partial n} - \gamma_1 \frac{\partial U_1}{\partial n} = \text{Div}\, \gamma \mathbf{E}^{\text{nc}}. \qquad [2.89]$$

Summarizing these results, we arrive at the system of equations describing the behavior of the potential in a conducting medium (Eqs. [2.94]). As usual, we have several comments:

1. It is more convenient to solve the forward problem for the scalar potential $U(p)$ than the vector equations [2.85] for the electric field.
2. Equation [2.86] is essentially Poisson's equation. In fact, expanding the left-hand side gives

$$\gamma \nabla^2 U + \nabla\gamma \cdot \nabla U = \text{div}\, \gamma \mathbf{E}^{\text{nc}} \quad \text{or} \quad \nabla^2 U = -\frac{\nabla\gamma \cdot \nabla U}{\gamma} + \frac{\nabla \cdot \gamma \mathbf{E}^{\text{nc}}}{\gamma}, \qquad [2.90]$$

where, as shown above, the right-hand side characterizes the density of volume charges.
In particular, at points where $\mathbf{E}^{\text{nc}} = 0$, we have

$$\nabla \cdot (\gamma \nabla U) = 0. \qquad [2.91]$$

3. In many practical cases of electrical methods in geophysics, we can assume that the medium is piecewise uniform; that is, it is composed of different regions of constant resistivity separated by interfaces. Under these conditions, the equation for the potential is drastically simplified and we have

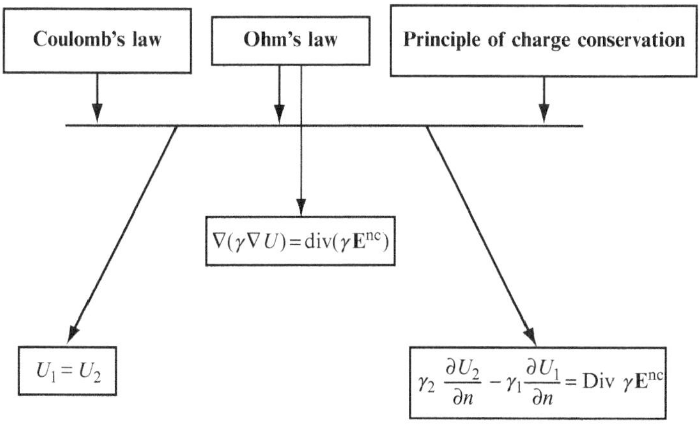

[2.92]

$$\nabla^2 U = \nabla \cdot \mathbf{E}^{nc} \qquad [2.93]$$

separately in each region, plus the conditions [2.89] at each interface. In particular, the potential obeys Laplace's equation in regions outside extraneous sources:

$$\nabla^2 U = 0. \qquad [2.94]$$

It is difficult to overestimate the importance of this equation in electrical methods.

4. The case of an anisotropic medium requires a special treatment, and will be studied later.
5. The system of equations for the electric field (Eq. [2.94]) does not apply to situations where there is a double layer.
6. The potential $U(p)$ of any time-invariant electric field is a solution of the system [2.94], which does not define the potential uniquely. Moreover, for each electric field, the equation $\mathbf{E} = -\nabla U$ determines the potential only up to a constant. Thus, the set [2.94] has an infinite number of solutions.

2.8. UNIQUENESS THEOREM

In Chapter 1, we studied a set of equations for the potential similar to the system [2.94] and derived specific conditions that provide unique solutions. With some small modifications taking into account the variation of conductivity, we again develop the uniqueness theorem and formulate boundary-value problems for steady current flow. For simplicity, we consider the external part of the current tubes where $\mathbf{E}^{nc} = 0$, but the proof requires only small modifications in the presence of the extraneous field.

2.8.1. Difference of Two Arbitrary Solutions of Eq. [2.91] and Gauss Theorem

Suppose that the potential $U(p)$ is considered in a volume V surrounded by surfaces S_1 and S_2. Also assume that there is an internal interface S_{12} in V across which the conductivity changes abruptly (Fig. 2.5A). The potential U is a solution of Eq. [2.91] at regular points of this volume. It may be appropriate to notice that surfaces S_1 and S_2 may have different meanings.

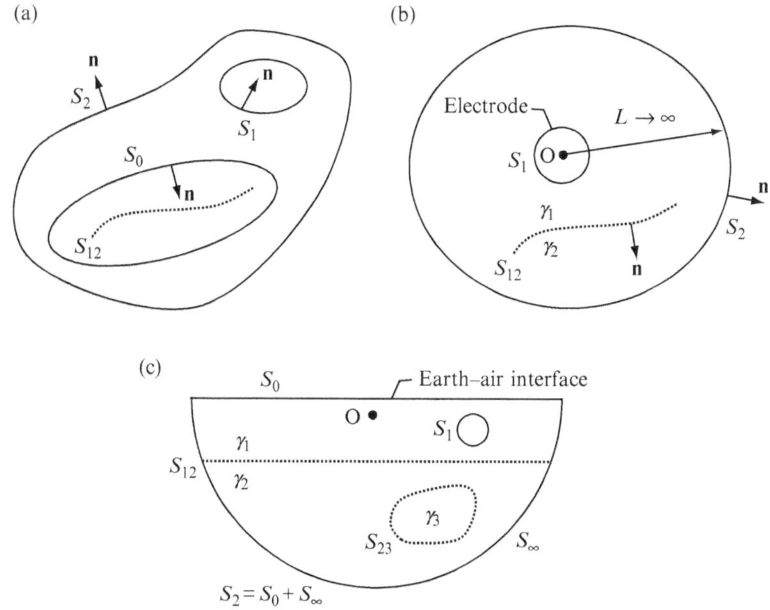

Figure 2.5 (A) Illustration of Eq. [2.99], (B) Model 1, and (C) Model 2.

They can be surfaces of the current electrode or the earth's surface or a spherical surface of infinitely large radius, etc. To determine conditions at the boundary surfaces that uniquely define the field, let us assume that there are two different solutions of the system [2.91], $U^{(1)}$ and $U^{(2)}$,

$$\nabla \cdot (\gamma \nabla U^{(1)}) = 0, \quad \nabla \cdot (\gamma \nabla U^{(2)}) = 0$$

and

$$U_1^{(1)} = U_2^{(1)}, \quad \gamma_1 \frac{\partial U_1^{(1)}}{\partial n} = \gamma_2 \frac{\partial U_2^{(1)}}{\partial n}, \tag{2.95}$$

$$U_1^{(2)} = U_2^{(2)}, \quad \gamma_1 \frac{\partial U_1^{(2)}}{\partial n} = \gamma_2 \frac{\partial U_2^{(2)}}{\partial n},$$

where γ_1, U_1 and γ_2, U_2 are the conductivity and potential at the back and front sides of the interface S_{12}, respectively. Next, consider the difference of these solutions:

$$U^{(3)} = U^{(2)} - U^{(1)},$$

and in accordance with Eqs. [2.95], it satisfies the following conditions:

$$\nabla \cdot (\gamma \nabla U^{(3)}) = 0 \quad \text{and} \quad U_1^{(3)} = U_2^{(3)}, \quad \gamma_1 \frac{\partial U_1^{(3)}}{\partial n} = \gamma_2 \frac{\partial U_2^{(3)}}{\partial n}, \quad \text{at } S_{12}. \tag{2.96}$$

Next, introduce the vector **X**

$$\mathbf{X} = U^{(3)} \gamma \nabla U^{(3)}, \tag{2.97}$$

and make use of Gauss's theorem

$$\int_V \operatorname{div} \mathbf{X} \mathrm{d}V = \oint_{S_1} \mathbf{X} \cdot \mathrm{d}\mathbf{S} + \oint_{S_2} \mathbf{X} \cdot \mathrm{d}\mathbf{S} + \oint_{S_0} \mathbf{X} \cdot \mathrm{d}\mathbf{S}. \tag{2.98}$$

Here, S_0 is a surface surrounding the interface S_{12} where the vector **X** is a discontinuous function because conductivities γ_1 and γ_2 are different. Correspondingly, Gauss's theorem is applied to the volume V^*, which is confined by the surfaces S_1, S_2, and S_0. Substituting Eq. [2.97] into Eq. [2.98] and differentiating, we obtain

$$\int_{V^*} U^{(3)} \nabla (\gamma \nabla U^{(3)}) \mathrm{d}V + \int_{V^*} \gamma (\nabla U^{(3)})^2$$
$$= \oint_{S_1} U^{(3)} \gamma \frac{\partial U^{(3)}}{\partial n} \mathrm{d}\mathbf{S} + \oint_{S_2} U^{(3)} \gamma \frac{\partial U^{(3)}}{\partial n} \mathrm{d}\mathbf{S} + \oint_{S_0} U^{(3)} \gamma \frac{\partial U^{(3)}}{\partial n} \mathrm{d}\mathbf{S}. \tag{2.99}$$

As S_0 approaches the interface S_{12}, the integral over S_0 reduces to integrals over opposite sides of the interface. Taking into account the fact that at the back and front sides of S_{12}

$$\mathbf{n}_1 = \mathbf{n}, \quad \mathbf{n}_2 = -\mathbf{n},$$

we have

$$\oint_{S_0} U^{(3)} \gamma \frac{\partial U^{(3)}}{\partial n} \mathrm{d}\mathbf{S} = \oint_{S_{12}} \left[U_1^{(3)} \gamma_1 \frac{\partial U_1^{(3)}}{\partial n} - U_2^{(3)} \gamma_2 \frac{\partial U_2^{(3)}}{\partial n} \right] \mathrm{d}\mathbf{S}, \text{ as } V^* \to V. \tag{2.100}$$

Equations [2.96] and [2.99] are thus, greatly simplified:

$$\int_V \gamma (\nabla U^{(3)})^2 \mathrm{d}V = \oint_{S_1} \gamma U^{(3)} \frac{\partial U^{(3)}}{\partial n} \mathrm{d}\mathbf{S} + \oint_{S_2} \gamma U^{(3)} \frac{\partial U^{(3)}}{\partial n} \mathrm{d}\mathbf{S}. \tag{2.101}$$

This last equation relates values of the potential and the field on the boundary surfaces S_1 and S_2 to its values inside the volume.

2.8.2. Boundary-Value Problems

Suppose that the surface integrals in Eq. [2.101] vanish. Then, the left-hand side is also zero:

$$\int_V \gamma(\nabla U^{(3)})^2 \mathrm{d}V = 0.$$

Since the integrand cannot be negative,

$$\gamma(\nabla U^{(3)})^2 \geq 0,$$

and we conclude that

$$\nabla U^{(3)} = 0. \qquad [2.102]$$

Since $U^{(3)}$ is the difference of two solutions $U^{(1)}$ and $U^{(2)}$ of the system [2.94], we see that two arbitrary solutions of the system [2.94] differ by a constant if the surface integrals of Eq. [2.101] vanishes. The field $\mathbf{E}(p)$ is defined uniquely, of course, since

$$\mathbf{E} = -\nabla U = -\nabla(U + C).$$

We now formulate boundary conditions such that the surface integrals in Eq. [2.101] vanish. Using the results from Chapter 1, we obtain three forms of the boundary conditions on these surfaces:

$$\text{I.}\, U = \phi(q), \quad \text{II.}\, \frac{\partial U}{\partial n} = \psi(q) \quad \text{III.} \oint_S \gamma \frac{\partial U}{\partial n} \mathrm{d}S = Q. \qquad [2.103]$$

Here, q is any point of the surfaces, Q is a given value, and $\phi(q)$ and $\psi(q)$ are specified functions. In the last case, S is an equipotential surface. For example, assume that there are two potentials $U^{(1)}$ and $U^{(2)}$ with the same values on the boundary surfaces S_1 and S_2:

$$U^{(1)}(S) = U^{(2)}(S).$$

Then, their difference $U^{(3)}$ is zero at these surfaces, and according to Eq. [2.101], the volume integral vanishes. As follows from Eq. [2.102], at each point of the volume V

$$U^{(2)}(p) = U^{(1)}(p) + C,$$

where C is a constant. Since the two functions coincide on the surfaces S_1 and S_2, we conclude that the constant equals zero. In other words, the boundary condition $U(S)=\phi(q)$ uniquely defines the potential in the volume V. In fact, any one of the conditions [2.103] in combination with system [2.94] uniquely defines the electric field. We now formulate three boundary-value problems that are most useful in the theory of electric methods.

I. Dirichlet's boundary-value problem

(a) At regular points,

$$\nabla\cdot(\gamma\nabla U)=0.$$

(b) At the interface S_{12},

$$U_1=U_2,\quad \gamma_1\frac{\partial U_1}{\partial n}=\gamma_2\frac{\partial U_2}{\partial n}.$$

(c) At boundaries S_1 and S_2,

$$U(S_1)=\phi_1(q),\quad U(S_2)=\phi_2(q).$$

II. Neumann's boundary-value problem

(a) At regular points,

$$\nabla\cdot(\gamma\nabla U)=0.$$

(b) At the interface S_{12},

$$U_1=U_2,\quad \gamma_1\frac{\partial U_1}{\partial n}=\gamma_2\frac{\partial U_2}{\partial n}.$$

(c) At boundaries S_1 and S_2,

$$\frac{\partial U}{\partial n}=\psi(S_1),\quad \frac{\partial U}{\partial n}=\psi(S_2).$$

III. The third boundary-value problem

(a) At regular points,

$$\nabla\cdot(\gamma\nabla U)=0.$$

(b) At the interface S_{12},

$$U_1=U_2,\quad \gamma_1\frac{\partial U_1}{\partial n}=\gamma_2\frac{\partial U_2}{\partial n}.$$

(c) At equipotential boundaries S_1 and S_2,

$$\oint_{S_1} \gamma \frac{\partial U}{\partial n} d\mathbf{S} = -I_1, \quad \oint_{S_1} \gamma \frac{\partial U}{\partial n} d\mathbf{S} = -I_2,$$

where I_1 and I_2 are given currents. A few comments perhaps will help to clarify the role of the uniqueness theorem:

1. Every boundary problem consists of two parts, namely, (a) the system of equations for the potential (Eqs. [2.94]) and (b) the boundary conditions.
2. All three boundary conditions always have a clear physical meaning and it is usually very simple to formulate them.
3. The type of boundary conditions can vary from point to point on a given surface, and they can be different at different boundary surfaces. At the same time, it is necessary to note that at all points of these surfaces, one of the boundary conditions must be specified.
4. In deriving the system of equations [2.94] and formulating the uniqueness theorem, we considered the model of a medium with one interface S_{12}. However, it should be obvious that these results remain valid when there are several interfaces between media with different resistivities. From Eqs. [2.94], at every interface, both the potential and normal component of the current density have to be continuous.
5. The uniqueness theorem does not provide an algorithm for calculating the field, but does outline the steps needed to find the electric field. Moreover, once a solution is found that satisfies all the conditions, we know that it is unique. To illustrate typical boundary-value problems for steady current flow, we consider two models often used in electrical methods of geophysics.

2.8.2.1 Model 1

Suppose that current I enters a medium through an electrode A located in the vicinity of the point 0 (Fig. 2.5B). The medium is everywhere uniform except at the interface S_{12}. Let us extract information about field behavior essential to the formulation of a unique boundary-value problem. First, at regular points, the potential U satisfies Laplace's equation:

$$\nabla^2 U = 0,$$

since γ is constant. Then, it is obvious from Coulomb's law that the potential caused by all surface charges tends to zero as the distance L_{0p} from the electrode increases without limit:

$$U \to 0 \quad \text{as} \quad L_{0p} \to \infty.$$

In other words, we can say that on the spherical surface S_2 with infinitely large radius the potential is equal to zero, and we have a boundary condition "at infinity." At the interface S_{12}, both the potential and the normal component of the current density must be continuous:

$$U_1 = U_2, \quad \gamma_1 \frac{\partial U_1}{\partial n} = \gamma_2 \frac{\partial U_2}{\partial n}.$$

We, thus, have equations for potential everywhere except at the electrode surface S_1, which along with S_2 confines the volume V. Since the electrode conductivity is many orders of magnitude greater than that of the surrounding medium, it is possible to treat S_1 as an equipotential surface:

$$U(S_1) = \text{constant}.$$

Also, it is natural to assume that the total current I through the electrode surface is known, and we have from Ohm's law,

$$\oint_{S_1} \gamma \frac{\partial U}{\partial n} d\mathbf{S} = I,$$

where n is the normal directed into the electrode. In this case, the boundary-value problem can be posed as follows:

1. At regular points of a medium, the potential is a solution of Laplace's equation

$$\nabla^2 U = 0.$$

2. At the equipotential boundary surface of the electrode,

$$\oint_{S_1} \gamma_1 \frac{\partial U_1}{\partial n} d\mathbf{S} = I.$$

3. The potential and normal component of the current density are continuous functions across an interface:

$$U_1 = U_2, \quad \gamma_1 \frac{\partial U_1}{\partial n} = \gamma_2 \frac{\partial U_2}{\partial n}.$$

4. At the boundary surface S_2 with an infinitely large radius, the potential tends to zero:

$$U \to 0, \quad \text{as} \quad L_{0p} \to \infty.$$

It follows from the uniqueness theorem that these four conditions uniquely define the potential, and therefore, the electric field. It is useful to notice that the boundary condition near the source A can often be simplified. Suppose that the current electrode A is a small sphere with radius a. Then, the potential at any point p can be written as a sum

$$U(p) = U_0(p) + U_s(p),$$

where U_0 is the potential due to the charge on the electrode surface, while U_s is the potential caused by charges that appear at the interface. Then, taking into account that

$$e_A = \varepsilon_0 \rho_1 I,$$

we have

$$U(p) = \frac{\rho_1 I}{4\pi L_{0p}} + U_s,$$

where ρ_1 is the resistivity of a medium that directly surrounds this electrode. If we assume that the electrode radius is sufficiently small, then, in approaching its surface, the potential U_s caused by relatively remote charges tends to some finite value, whereas the potential U_0 becomes very large. Therefore, the boundary condition near the source can be written approximately as

$$U \to U_0 = \frac{\rho_1 I}{4\pi L_{0p}} \quad \text{if} \quad L_{0p} \to a.$$

In the limit, we can replace the surface charge by a point charge, and this replacement affects the field only in the vicinity of the electrode. From the physical point of view, it should be obvious that this boundary condition is valid when the electrode is small but has an arbitrary shape, as long as the surface S_1 is located at some distance from the electrode. Thus, we have the following boundary-value problem widely used in the theory of electric methods:

$$\begin{aligned}
&1.\quad U \to U_0 \quad \text{as} \quad L_{0p} \to 0,\\
&2.\quad \nabla^2 U = 0,\\
&3.\quad U_i = U_{i+1}, \quad \gamma_i \frac{\partial U_i}{\partial n} = \gamma_{i+1} \frac{\partial U_{i+1}}{\partial n},\\
&4.\quad U \to 0 \quad \text{if} \quad L_{0p} \to \infty.
\end{aligned} \qquad [2.104]$$

Here, we have one obvious generalization by assuming that the medium contains several interfaces. Certainly, the boundary condition $U \to U_0$ is much simpler than the equality

$$\oint_{S_1} \gamma \frac{\partial U}{\partial n} d\mathbf{S} = I$$

but the latter is more general.

2.8.2.2 Model 2

This model consists of an upper nonconducting half-space (air) and a piecewise uniform conducting medium, which represents the earth model (Fig. 2.5C). We will formulate the boundary-value problem for the potential in a conducting half-space surrounded by the electrode surface S_1 and the boundary surface S_2. The latter includes the earth–air interface S_0 and a half-spherical surface S_∞ of infinitely large radius:

$$S_2 = S_0 + S_\infty.$$

Taking into account that the normal component of the current density equals zero at the boundary with a nonconducting medium, it is simple to formulate the boundary condition at the earth's surface. The normal component of the electric field must vanish at the conducting side of this surface. (Note that on the other side, this component may have any value.) Thus, the boundary-value problem is formulated in the following way:

1. At regular points,

$$\nabla^2 U = 0.$$

2. At the interface S_i,

$$U_i = U_{i+1}, \quad \gamma_i \frac{\partial U_i}{\partial n} = \gamma_{i+1} \frac{\partial U_{i+1}}{\partial n}. \qquad [2.105]$$

3. At the surface S_1 near the source of the primary field,

$$U \to U_0 = \frac{\rho_1 I}{4\pi L_{0p}}.$$

4. At the earth's surface S_0,

$$\frac{\partial U}{\partial n} = 0.$$

5. At the surface S_∞ (condition at infinity),

$$U \to 0 \quad \text{if} \quad L_{0p} \to \infty.$$

Notice that, because the normal component of the electric field is known at all points of the earth's surface ($E_n = 0$), we are able to formulate the boundary-value problem in the conducting medium (earth) without having to worry about its value in the upper half-space (air). This illustrates how the uniqueness theorem is important for developing the theory of electric methods, and we will see many other examples of its application. Before moving on to develop general methods for solving boundary-value problems, we look at one further application.

2.8.2.3 The Need for an Extraneous Field to Drive Steady Current

Consider an arbitrary current tube in a conducting medium and suppose that an extraneous field is absent. Then, taking into account that the complete tube must be closed (by the conservation of current), the potential has to satisfy the following conditions:

1. At regular points in the tube's interior,

$$\nabla \cdot (\gamma \nabla U) = 0.$$

2. At the lateral surface,

$$\frac{\partial U}{\partial n} = 0,$$

since the normal component of the current density is zero. Since the lateral surface of the tube encloses the volume of space occupied by the tube, the above conditions constitute a Neumann boundary-value problem and, from the uniqueness theorem, define the field. It is clear that a constant potential $U(p)$ satisfies both conditions, in which case, there is no electric field. This argument works for any current tube and means that an electric field cannot exist in steady current flow without an extraneous field $\mathbf{E}^{nc}$.

Having formulated the main boundary-value problems for the electric field in a conducting and polarizable medium, we will next discuss methods of solving these problems to calculate the field **E**. There are three general such methods: (1) the method of separation of variables, (2) the method of finite differences or finite elements, and (3) the method of integral equations. In the first method, a solution of the boundary-value problem is usually found in analytical form as an infinite series or integral involving known functions. The other methods, generally called numerical methods, usually solve a set of linear algebraic equations that approximate the exact solution. The study of all three methods is the subject of applied mathematics and is beyond the scope of this monograph. In subsequent chapters, we will develop the method of separation of variables to solve for the field in models with relatively simple configurations involving media of different resistivities. Such models are of a great practical interest in developing the basic theory of the most electric methods, but more complex models requiring numerical methods are often needed in analyzing actual measurements at the earth's surface and especially in boreholes. Before moving on to consider some of the models commonly used, we conclude this chapter by studying some general features of the electric field and steady current flow in conducting media, including the concepts of resistance, electrical work, and dissipation.

2.9. RESISTANCE

2.9.1. Ohm's Law and the Concept of Resistance

The electrical resistance of wires and circuit elements is one of the fundamental concepts of electrical circuit theory. In geophysics, we deal mainly with current flow in volume conductors, and the concept of resistance is also the leading concept, which is important for practical understanding of electric methods. With this in mind, consider an arbitrary element C of a current tube, confined by its lateral surface S_l and two cross sections S_1 and S_2 forming end caps (Fig. 2.6A). Since, by definition, the current

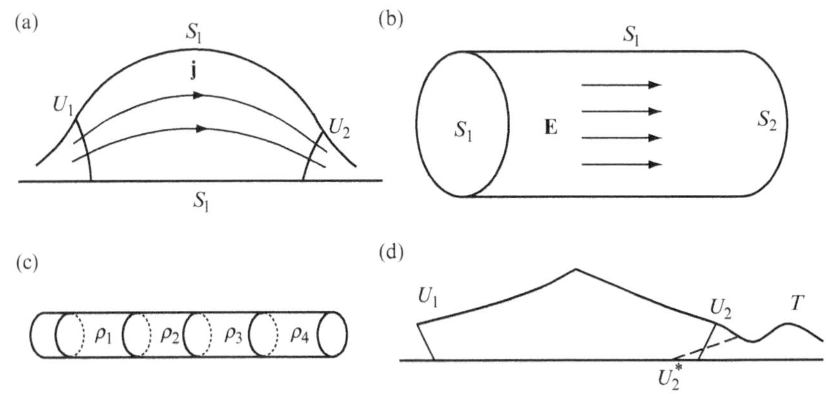

Figure 2.6 (A) Voltage in a current tube, (B) cylindrical conductor, (C) quasilinear conductor, and (D) influence of different parts of a current tube on resistance.

density is perpendicular to a cross section through a current tube, each cross section is an equipotential surface. If a piece of the current tube contains no extraneous sources (i.e., $\mathbf{E}^{nc}=0$ in the tube volume), then by definition, the difference of potential or voltage between two equipotential surfaces S_1 and S_2 is simply

$$V_{12} = U_1 - U_2 = \int_1^2 \mathbf{E}^c \cdot \mathbf{dl} = \int_1^2 \rho\mathbf{j} \cdot \mathbf{dl}, \qquad [2.106]$$

where the last equality comes from Ohm's law, written in the form

$$\rho\mathbf{j} = \mathbf{E}^c.$$

The integral at the right-hand side of Eq. [2.106] is path independent; integration can be performed along any path l having terminal points on the surfaces S_1 and S_2. For instance, if the path coincides with a current line, along which $\cos(\mathbf{j}, \mathbf{dl}) = 1$, Eq. [2.106] becomes

$$V_{12} = U_1 - U_2 = \int_1^2 \rho\mathbf{j}\,\mathbf{dl}. \qquad [2.107]$$

It is clear that, within a given tube, current lines l that are longer and pass through more resistive parts of a medium must carry a lower current density than shorter lines passing through more conductive regions (Fig. 2.6A). This shows that the current density, while everywhere perpendicular to a cross section, does not have a uniform magnitude. (The same is true of the electric field.) Let us rewrite Eq. [2.106] as

$$U_1 - U_2 = IR_{12}, \qquad [2.108]$$

where

$$R_{12} = \int_1^2 \frac{\rho\mathbf{j} \cdot \mathbf{dl}}{I}. \qquad [2.109]$$

Here,

$$I = \int_{S_1} \mathbf{j} \cdot \mathbf{dS} = \int_{S_2} \mathbf{j} \cdot \mathbf{dS}$$

is the current in the tube C, and R_{12} is called the resistance of the element of the current tube between the surfaces S_1 and S_2. This quantity has the usual dimension of resistance from circuit theory, V s/m, a unit called the "ohm." A few comments will help to elucidate the concept of resistance in steady current flow through volume conductors:

1. Resistance is always positive, since by definition, the current through the tube is positive ($I > 0$) when the path of integration is along a current line from S_1 to S_2 (current lines point in the direction of positive current density). When integrating along the opposite direction, the voltage changes sign and so does the current.
2. The resistance R_{12} defined by Eq. [2.109] has physical meaning only for current tubes, that is, when the surfaces S_1 and S_2 are equipotential ones and S_l is the lateral surface of the tube. Indeed, suppose that surface S_1 is not an equipotential surface. Then, changing the position of the starting point of the integral would change the voltage and therefore, the resistance of the conductor C. The value of R_{12} would then be ambiguous. This also happens if the bounding surface S_l is not the lateral surface of a current tube, since then, the current I would vary along the tube and would have no well-defined value. We see that, for a given distribution of the electric field, the concept of resistance has meaning only for certain regions of a conducting medium.
3. It follows from Eq. [2.108]

$$I = \frac{U^{(1)} - U^{(2)}}{R_{12}}, \qquad [2.110]$$

which is called Ohm's law in integral form. Unlike the differential form $\mathbf{j} = \gamma \mathbf{E}$, the integral form is valid only for time-invariant fields or, more precisely, in cases when the effects of electromagnetic induction can be neglected. One can say that the function R_{12} characterizes the ability of a conductor C to resist the flow of current. When the current magnitude has unit value, the resistance is numerically equal to the difference of potentials $U^{(1)} - U^{(2)}$. In particular, locations along the path where the potential drop is greater, have larger resistance.

2.9.2. Dependence of R_{12} on Parameters of the Medium

We will use the uniqueness theorem to determine configurations that define the resistance R_{12} uniquely. It is clear that the potential of the electric field within the external element of a current tube C obeys the following conditions (Fig. 2.6A):

(a) At regular points, the potential is a solution of the equation

$$\nabla \cdot (\gamma \nabla U) = 0.$$

(b) At points of the lateral surface S_l, the normal component of the electric field is zero:

$$\frac{\partial U}{\partial n} = 0.$$

(c) At the equipotential surfaces S_1 and S_2, we have a given value of current

$$\int_{S_1} \gamma \frac{\partial U}{\partial n} \mathrm{d}\mathbf{S} = I \quad \text{and} \quad \int_{S_2} \gamma \frac{\partial U}{\partial n} \mathrm{d}\mathbf{S} = -I,$$

where the normal n is directed to the outside of the current tube volume and I is the current in the tube. It follows from the uniqueness theorem that these three conditions uniquely define the electric field $\mathbf{E}$ inside the conductor C. In other words, there can be only one distribution of the fields $\mathbf{E}$ and $\mathbf{j}$ if the current I is specified. Now, suppose that the current equals I_1 and correspondingly, the electric field is equal to $\mathbf{E}_1(q)$, which of course satisfies all the boundary conditions. Let us increase the current by a factor of m so that $I_2 = mI_1$. This change increases only the magnitude of the electric field at each point by the same factor m, but does not change its direction:

$$\mathbf{E}_2(q) = m\mathbf{E}_1(q).$$

The potential created by the new field $\mathbf{E}_2(q)$ is, therefore, equal to

$$U_2(q) = mU_1(q),$$

where q is a point of the current tube. It is obvious that if $U_1(q)$ is a solution of the equation $\nabla \cdot (\gamma \nabla U) = 0$, the function $U_2(q)$ obeys the same equation. This new field also satisfies the boundary conditions at surfaces S_1 and S_2, as well as that at the lateral surface because

$$\frac{\partial U_2}{\partial n} = m \frac{\partial U_1}{\partial n} = 0.$$

Thus, in accordance with the uniqueness theorem, we can say that our assumption is correct, and the function $\mathbf{E}_2(q) = m\mathbf{E}_1(q)$ describes the electric field in the conductor C if the current through its cross sections equals I_2. This allows us to conclude that the ratio

$$\frac{\mathbf{E}}{I} \quad \text{or} \quad \frac{\rho \mathbf{j}}{I}$$

is independent of the current magnitude, and implies that the resistance given by Eq. [2.109] is completely defined by the dimensions and shape of the material of the conductor occupying the current tube and by the distribution of its resistivity, but it does not depend on what happens outside this volume as long as the surfaces S_1 and S_2 remain equipotentials and the normal component of current density vanishes on the lateral surface S_l.

2.9.3. The Calculation of Resistance

As follows from Eq. [2.109], in order to determine the resistance of some part of a medium, which is an element of the current tube, it is necessary to know the current density field for a given configuration of extraneous sources, which is rather complicated in general and is equivalent to solving the boundary-value problem. Nevertheless, there are cases where calculation of resistance is relatively simple. First, suppose that the magnitude of the current density **j** is constant over any cross section of a conductor *C*. Then,

$$\frac{\mathbf{j}\cdot d\mathbf{l}}{I}=\frac{j\,dl}{I}=\frac{dl}{S}$$

and

$$R_{12}=\int_1^2 \rho\frac{dl}{S}. \qquad [2.111]$$

In particular, if the conductor is uniform ($\rho=\text{constant}$),

$$R_{12}=\rho\int_1^2 \frac{dl}{S}. \qquad [2.112]$$

2.9.4. Resistance of a Cylindrical Conductor

Suppose that a conductor *C* is a uniform cylinder confined by a lateral surface S_l and two equipotential surfaces S_1 and S_2 of arbitrary shape (Fig. 2.6B). In this case, the electric field is remarkably simple: namely, it is constant within a conductor and tangent to its lateral surface. In fact, the potential of such a field is a linear function, and therefore, satisfies Laplace equation and the boundary conditions

$$\frac{\partial U}{\partial n}=0 \text{ at } S_l$$

and

$$\gamma\int_{S_1}\frac{\partial U}{\partial n}d\mathbf{S}=-\gamma E_n S=I, \quad \gamma\int_{S_2}\frac{\partial U}{\partial n}d\mathbf{S}=-I.$$

Thus, from the uniqueness theorem, we see that the field **E** is uniform. Correspondingly, the current density field is also uniform and Eq. [2.109] for the resistance simplifies greatly:

$$R_{12} = \rho \frac{l}{S}, \qquad [2.113]$$

where l and S are the length and cross section of the cylinder, respectively. This is the expression for the resistance of a cylindrical wire widely used in circuit theory. As we know, the electric field **E** within a uniform cylinder is caused by charges located on its lateral surfaces. As was just proved, these charges are distributed in such a way that at each point the electric and current density fields are parallel to the lateral surface. A conductor often consists of several parts connected in series, each of which is a uniform cylinder with a length that greatly exceeds dimensions of its cross section (Fig. 2.6C). These elements of a current tube are called "quasilinear" and their resistivity is defined by Eq. [2.113]. It is obvious that if the resistivity of the cylinder ρ does not change within each cross section but does vary along the conductor, then the current density field **j** still remains constant and tangential to the lateral surface S_l. At the same time, the electric field changes along the cylinder, but is constant over a cross section and directly proportional to the resistivity:

$$E = \rho j = \rho \frac{I}{S}.$$

Therefore, in place of Eq. [2.111], we have

$$R_{12} = \frac{1}{S} \int_1^2 \rho \, \mathrm{d}l. \qquad [2.114]$$

2.9.5. Influence of a Medium Beyond a Conductor on Its Resistance R_{12}

Earlier, we showed that the resistance of the conductor C has meaning if surfaces S_1 and S_2 are equipotential surfaces and S_l is the lateral surface of a current tube. It is easy to imagine changes outside a conductor C that cause its potential to vary along these surfaces so that they cease to be equipotential. For instance, let us introduce an inhomogeneity T outside the conductor (Fig. 2.6D). Electrical charges arise at its surface, and they can create a field that is not perpendicular to the surface S_2, because the equipotential surface with T present now occupies a new position S_2^*. This example illustrates that resistance R_{12} can depend not only on parameters of the conductor C but also on the surrounding distribution of resistivity that determines the current density field over the conductor. There are cases, however, where the surfaces S_1 and S_2 will always remain equipotentials. For instance, this occurs if they are the surfaces of ideal (perfect) conductors or cross sections of the quasilinear parts of a current tube (Fig. 2.7A and B). In fact, these conditions hold in practice when all the surfaces involved are interfaces between a conductor and a medium that has much higher conductivity. We see that in such cases the value of resistance R_{12} does not change when the electric field in other parts of a medium changes. In particular, a change of sources in these places does not affect the resistance of a conductor as along as its lateral surface S_l remains the lateral surface of the current tube. To further illustrate the concept of the resistance, we consider two examples from theory of electrical circuits and geophysics.

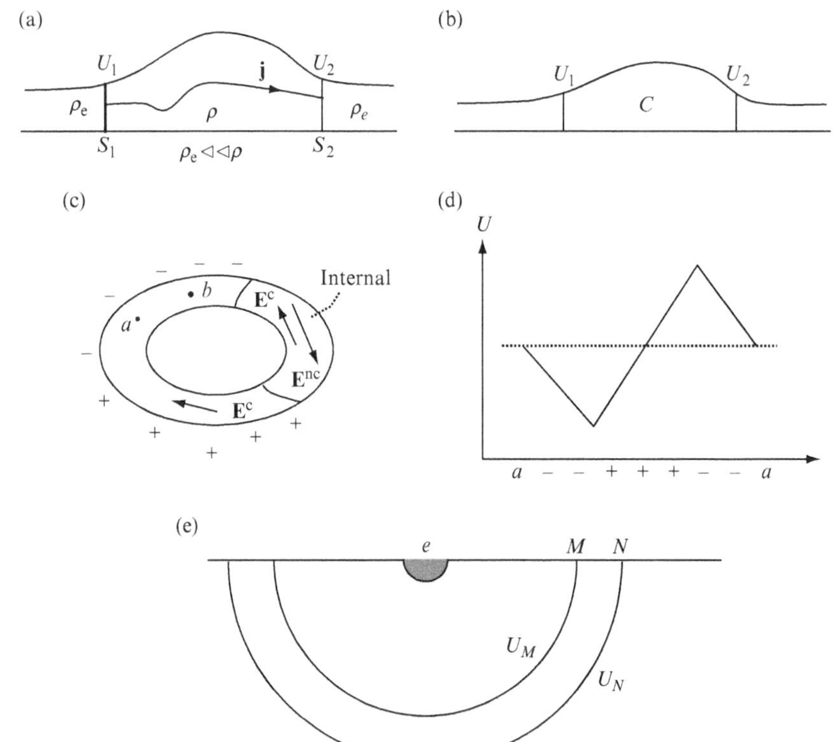

Figure 2.7 (A) Case of very low resistivity, (B) quasilinear conductors, (C) linear circuit, (D) distribution of potential, and (E) three-electrode array.

2.9.6. Behavior of the Electric Field and Potential Along a Quasilinear Circuit

Suppose that the external part of the circuit is a uniform conductor with a constant cross section S and the extraneous field $\mathbf{E}^{ext}$ acts within the internal part of this system (Fig. 2.7C). First, consider the external part where the current density field is uniform, because both the current I and the cross section S are constant. From Ohm's law,

$$j = \gamma E \quad \text{or} \quad \mathbf{E} = \rho \mathbf{j},$$

we conclude that the electric field inside this part of the conductor (the external part) is also constant. This electric field in the external part is caused only by charge, according to Coulomb's law, and it is natural to ask, where is this charge located? Certainly, the charge is not inside the conductor in the current, because the conductor is uniform (γ is constant), and from the equation

$$\delta = -\varepsilon_0 \frac{\mathbf{E} \cdot \nabla \gamma}{\gamma},$$

it follows that the volume density is equal to zero. Also charge cannot be concentrated only near the ends of the internal part of the current tube, because the field caused by two such charge distributions of opposite sign will not be uniform along the circuit. Finally, the charge cannot be located outside the circuit, because such charge will not have any influence on the field inside the conductor (by the principle of electrostatic induction). Thus, there is only one place where the charge that is the source of the uniform electric field can be located: namely, on conductor surface. As soon as a current I is established, charge appears on this surface and distributes itself in such a way that the electric field becomes uniform within the external part of the current tube. The distribution of charge is schematically shown in Fig. 2.7C. In addition, we note the following features that hold for the quasilinear circuit above:

1. If the position of the conductor and its shape change in the external part of the circuit (where the extraneous force is absent), the current I remains the same. This happens because the distribution of charges on the surface of the conductor changes in such a manner as to maintain the same electric field at all points inside the conductor. (As was pointed out earlier, the appearance of these charges is actually induced by the propagation of an electromagnetic field from an internal part of the circuit through a surrounding medium.)

2. If the cross section or conductivity of the conductor changes in the external part of the circuit, the electric field inside the conductor becomes nonuniform but its behavior is again governed by charges. For instance, if the conductivity varies, both surface and volume charges arise, but they are distributed in such a way that their electric field provides the same current through every cross section of the conductor.
3. In general, there are two types of charges on the conductor surface. One type creates the electric field inside, while the other plays a different but equally important role. The second type of charge maintains the condition that the current depends only on the electromotive force and the circuit resistance. In fact, assume that there is a charge somewhere outside the circuit (Fig. 2.7C) that would, by Coulomb's law, create a strong electric field **E** inside the conductor. We know, however, that the current I will remain the same inside the conductor because charges that appear on the conductor surface cancel the field of external charges at any point inside the circuit (by the phenomenon of electrostatic induction).
4. In the internal part of the circuit, where the extraneous field is present, the current density is defined by both the extraneous and Coulomb fields. The Coulomb field is caused by charges that arise on the surface of a conductor if it is uniform, and within the conductor if it is not.

2.9.7. Potential Along a Circuit

We now examine the behavior of the potential along a circuit. First, consider the voltage between two arbitrary points a and b of a circuit (Fig. 2.7C). If there is no extraneous field within this interval (a, b), we have

$$U(a) - U(b) = \int_a^b \mathbf{E}^c \cdot d\mathbf{l} = \int_a^b \rho \mathbf{j} \cdot d\mathbf{l},$$

or

$$U(a) - U(b) = I \int_a^b \frac{\rho j \cdot d\mathbf{l}}{I} = IR_{ab}, \qquad [2.115]$$

where R_{ab} is the resistance of the circuit between the points a and b. In particular, when these points coincide with end points of the internal part of the circuit, we obtain

$$U_+ - U_- = IR_e. \qquad [2.116]$$

Here, R_e is the resistance of the external part of the circuit. Next, suppose that the electromotive force Ξ is located somewhere between points a and b in the internal part of the circuit. Then, applying Ohm's law

$$\mathbf{j} = \gamma(\mathbf{E}^c + \mathbf{E}^{nc}) \quad \text{or} \quad \mathbf{E}^c = \rho\mathbf{j} - \mathbf{E}^{nc},$$

we have

$$U(a) - U(b) = \int_a^b \mathbf{E}^c \cdot \mathbf{dl} = I\int_a^b \frac{\rho\mathbf{j}\cdot\mathbf{dl}}{I} - \int_-^+ \mathbf{E}^{nc}\cdot\mathbf{dl},$$

since the extraneous force is present only between the end points of the internal part (by definition). The last equality can be written as

$$U(a) - U(b) = IR_{ab} - \Xi, \qquad [2.117]$$

where

$$\Xi = \int_-^+ \mathbf{E}^{nc}\cdot\mathbf{dl}$$

is the electromotive force, and the integral along the extraneous field, that is, along the current lines. R_{ab} is the sum of the resistance of the internal part R_i and the resistance of that section of the external part located between a and b. Considering the whole circle when $a=b$ and $U(a)=U(b)$, we obtain

$$\Xi = I(R_i + R_e) = IR, \qquad [2.118]$$

where R is the total resistance of the entire circuit. From Eqs. [2.116] and [2.118], we find the relationship between the potential difference at terminal points and the electromotive force:

$$U_+ - U_- = \frac{\Xi R_e}{R_i + R_e}. \qquad [2.119]$$

Thus, this difference of potentials is practically equal to the electromotive force if the resistance of the external part is much greater than that of the internal one:

$$U_+ - U_- \approx \Xi \quad \text{if} \quad R_e \gg R_i.$$

It follows from Eq. [2.115] that the potential decreases along the current line within the external part, while it increases inside the internal part. In fact, we have

$$U_{-} - U(b) = \int_{-}^{b} \mathbf{E}^{c} \cdot \mathbf{dl} \quad \text{or} \quad U(b) = U_{-} - \int_{-}^{b} \mathbf{E}^{c} \cdot \mathbf{dl}.$$

Since the Coulomb's field $\mathbf{E}^c$ and the element of the current line $\mathbf{dl}$ have opposite directions, the integral is negative and $U(b) > U_{-}$. Here, the point b belongs to the internal part of the circuit. The behavior of the potential is shown in Fig. 2.7D.

2.9.8. Three-Electrode Array

Next, consider a conducting half-space with an array of three electrodes located at its surface (Fig. 2.7E). The array consists of one current electrode A and two receiver electrodes M and N arranged along a line. Current I_A through the electrode A (located at the point 0), induces a charge $e_A = 2\varepsilon_0 \rho I_A$ on its surface and creates a primary electric field in a conducting medium

$$\mathbf{E}_0 = \frac{\rho I_A}{2\pi L_{0p}^3} \mathbf{L}_{0p},$$

where ρ is the resistivity of the medium and L_{0p} is the distance between this electrode and an observation point p. If there are interfaces with different resistivities within the medium, surface charges appear at these interfaces and create a secondary electric field $\mathbf{E}_s$. The total field at any point p is a sum of the primary and secondary fields:

$$\mathbf{E}(p) = \mathbf{E}_0(p) + \mathbf{E}_s(p).$$

This vector field is described by a potential $U(p)$, and the points M and N are located on two equipotential surfaces, and the voltage can be measured with a voltmeter connected to the electrodes:

$$V_{mn} = \int_{M}^{N} \mathbf{E}^{c} \cdot \mathbf{dl} = U_M - U_N.$$

To eliminate the influence of a change of the current I_A, the ratio

$$\frac{V_{MN}}{I_A} = \frac{U_M - U_N}{I_A}$$

is calculated. By Ohm's law, this quantity is equal to the resistance R_{MN} of that part of the medium which is confined by the top interface (with the nonconducting half-space) and the two equipotential surfaces passing through points M and N, respectively:

$$R_{mn} = \frac{U_M - U_N}{I_{AB}}. \quad [2.120]$$

This result is characteristic of most electrical methods in geophysics: an array of electrodes measures the resistance of the medium between equipotential surfaces. For illustration, consider two special cases of the three-electrode array. First, suppose that the electrode N is placed far away from the current electrode A so that we can let $U_N = 0$, and in effect obtain a two-electrode array AM. In borehole geophysics, where the electrodes are lowered into a hole drilled into the earth, such an array is usually called a normal or potential probe. As follows from Eq. [2.120], this array measures the resistance of the medium confined by the equipotential surface $U = U_M$ passing through electrode M and a spherical surface of an infinitely large radius. It is natural to expect that those elements of the medium located relatively far from the electrode A make a very small contribution to this resistance. Next, consider the opposite case where the distance between the receiver electrodes is so small that the electric field within this interval is practically constant. This gives

$$R_{mn} = \frac{V_{mn}}{I_{AB}} = \frac{MN}{I_A} E. \quad [2.121]$$

Here, E is the component of the electric field at points of the line MN. In borehole geophysics, such an array is called a gradient or lateral probe. It is worth pointing out that in a uniform medium a three-electrode array measures the resistance of the spherical layer with thickness MN and the center at point A, but in general when the medium is nonuniform, the a shape of the equipotential surface can be rather complicated.

2.10. GROUNDING RESISTANCE

We consider several further examples to illustrate the concept of resistance of a medium and its measurement.

2.10.1. Grounding Resistance of a Spherical Electrode

Earlier, we have shown that the electric field generated by a charge on a small (spherical) electrode has spherical symmetry and its potential at any point p is

$$U(p) = \frac{\rho I}{4\pi L_{0p}}. \quad [2.122]$$

Take two spherical and equipotential surfaces with radii L_{0a} and L_{0b}, respectively (Fig. 2.8A). Then, the difference of potentials is

$$U(a) - U(b) = \frac{\rho I}{4\pi}\left(\frac{1}{L_{0a}} - \frac{1}{L_{0b}}\right),$$

and the resistance of this spherical layer is, therefore,

$$R_{ab} = \frac{U(a) - U(b)}{I} = \frac{\rho}{4\pi}\left(\frac{1}{L_{0a}} - \frac{1}{L_{0b}}\right). \qquad [2.123]$$

The resistance depends on the layer thickness and the distance from the electrode, as well as on the resistivity of the medium. It is instructive to represent the whole medium around the electrode as a system of spherical layers with their center at the point 0 (Fig. 2.8A). Each layer has its own resistance. Taking into account that the current I goes through all the layers, we can say that they are all connected in series. Correspondingly, the resistance of the entire medium to the current that flows from the electrode can be written as follows:

$$R_g = \frac{\rho}{4\pi}\left(\frac{1}{L_{0a}} - \frac{1}{L_{01}} + \frac{1}{L_{01}} - \frac{1}{L_{02}} + \frac{1}{L_{02}} - \frac{1}{L_{03}} + \cdots\right), \text{ or } R_g = \frac{\rho}{4\pi a}, \qquad [2.124]$$

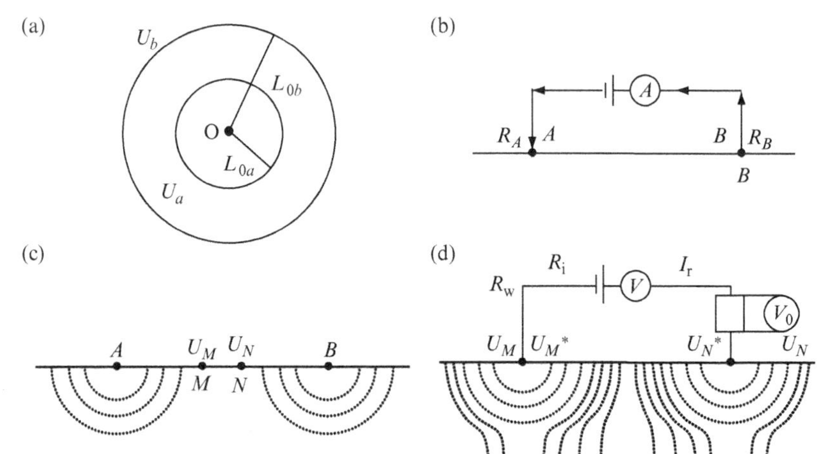

Figure 2.8 (A) Resistance of spherical layer, (B) current line, (C) distribution of potential in the absence of receiver line, and (D) influence of grounding resistance of receiver line.

where a is radius of the electrode. R_g is usually called the electrode or grounding resistance. It does not characterize the resistance of the electrode material or that of any particular surface between the electrode and the medium since their influence is negligible. In contrast, this quantity describes the resistance of the whole medium to the current leaving the electrode. Correspondingly, Eq. [2.124] follows from Eq. [2.123], if we let $L_{0b}=\infty$. Grounding resistance is an important concept in electrical methods of geophysics. According to Eq. [2.124], it strongly increases as the radius of the electrode decreases. Such behavior is expected: as the electrode's surface area decreases, the current density increases around the electrode (if the current stays constant), and the higher current density flowing through a medium causes the potential to drop more rapidly near the electrode. Replacing L_{0a} by a in Eq. [2.123], we represent the resistance of the spherical layer bounded by the spherical electrode surface and the sphere with radius L_{0b} as

$$R_{0b} = R_g\left(1 - \frac{a}{L_{0b}}\right). \quad [2.125]$$

This equations shows that if $L_{0b} \gg a$, the resistance R_{0b} of this part of a medium differs only slightly from the grounding resistance. For instance, the spherical layer of radius $L_{0b}=10a$ around the electrode contributes 90% of R_g. Since the electrodes are generally small, the thickness of such a layer is usually about 0.5–1 m. In other words, a uniform medium near an electrode gives the main contribution to the grounding resistance. With this in mind, consider the second example.

2.10.2. The Current Line of an Array

Any practical system for making electrical measurements with steady currents includes the following elements: (a) an internal part (current source) with an electromotive force Ξ, (b) a linear conductor (wire) with resistance R_w, (c) at least two electrodes with grounding resistance R_A and R_B that are in contact with a volume conductor (e.g., the earth), (d) an ammeter, and (e) the volume conductor itself, a medium with resistivity ρ that can vary from point to point. According to Ohm's law, the current in this measurement circuit is (Fig. 2.8B)

$$I = \frac{\Xi}{R_A + R_B + R_w + R_i}, \quad [2.126]$$

where R_i is the sum of the resistances of the current source and ammeter. As we know, the grounding resistances R_A and R_B depend on the distribution of the medium's resistivity everywhere regardless of distances from electrodes. In other words, a change of resistivity anywhere beneath the earth's surface, in principle, changes the current I (Eq. [2.126]). Therefore, it appears that measurements of the current with this simple two-electrode array will give information about the electric properties of a medium. In practice, however, the grounding resistance is mainly defined by the resistivity of the medium in regions very close to the electrode, and according to Eq. [2.125], the dimension of this region seldom exceeds 1 m. Correspondingly, the current I is practically insensitive to a change of the resistivity beyond this range. Because of this, the two-electrode array is not used much except for some applications in borehole geophysics. The influence of grounding resistance is the reason why most conventional electrode arrays for measuring steady currents consists of two different parts, namely:

(a) A current or "source" line AB that includes the electromotive force and ammeter. The sole function of this part is to create the electric field in a conducting medium.

(b) A measurement or "receiver" line MN that contains a voltmeter to measure the voltage between electrodes located off the current line, caused by charges arising in the medium. The layout and separation of the source and receiver lines are the distinguishing features of arrays in most electric methods.

2.10.3. Influence of Grounding Resistance on Measurement of Voltage

To complete the measurement picture, we now investigate the role of grounding resistance in measuring a voltage. Suppose that we wish to measure the difference of potentials between points M and N, located, for example, on the earth's surface:

$$V_{MN} = U_M - U_N = \int_M^N \mathbf{E} \cdot d\mathbf{l}.$$

The distribution of equipotential surfaces in a conducting medium is shown in Fig. 2.8C. To measure a voltage, we insert two electrodes in the vicinity of these points connected to each other through a voltmeter. Since the integral is path independent, the latter can be also written as

$$V_{MN} = \int_M^N E\,\mathrm{d}l, \qquad [2.127]$$

where E is a component of the field directed along the straight line between these electrodes. Very often, the distance MN is so short that the field component E practically does not change within this interval. Then, Eq. [2.127] is simplified:

$$V_{MN} = E \cdot MN,$$

and the electric field is easily calculated. Now, we will describe the distribution of the potential in a conducting medium in the presence of receiver electrodes M and N. Near these electrodes, a part of the current goes into the circuit while a part moves through the conducting medium; one can say that there are two circuits connected in parallel (Fig. 2.8D). Earlier, we showed that surface charges arise when the current density intersects the boundary between regions of different resistivities. Since current enters the measurement circuit M and leaves it through N, negative and positive charges must appear on the surfaces of electrodes M and N, respectively. The electric field of these charges decreases the potential in the vicinity of the electrode M, while it becomes greater at the electrode N. In other words, the presence of the receiver electrodes, connected together, causes a distortion of equipotential surfaces near measuring electrodes (Fig. 2.8D); that is,

$$V_{MN}^* = U_M^* - U_N^* \neq V_{MN} = U_M - U_N,$$

where U_M^* and U_N^* are potentials of electrodes M and N, respectively, when some current goes through the receiver line. Thus, near the electrodes, when part of the current moves from the volume to the linear wire conductor (or *vice versa*), we observe a relatively strong drop of the potential:

$$U_M - U_M^* \quad \text{and} \quad U_N^* - U_N.$$

By the definition of resistance, these differences can be expressed as

$$U_M - U_M^* = I_r R_M, \quad U_N^* - U_N = I_r R_N, \qquad [2.128]$$

where I_r is the current in the measuring circuit, while R_M and R_N are the grounding resistances of electrodes M and N (usually unknown). Putting these relations together gives

$$(U_M - U_N) - (U_M^* - U_N^*) = I_r(R_M + R_N) \quad \text{or}$$
$$V_{MN} - V_{MN}^* = I_r(R_M + R_N). \qquad [2.129]$$

We, therefore, see that the measured voltage V_{MN}^* and the "ideal" voltage V_{MN} (the voltage present when the measuring electrodes are absent) differ from each other and the difference is directly proportional to the sum of the grounding resistances. For this reason, it is necessary to take into account, or reduce, the influence of grounding resistance in practical measurement systems. Let us outline three such methods for accomplishing this.

Method 1. Suppose that at the instant when the input V_{MN}^* is measured, the current I_r through receiver electrodes equals zero. Then, in accordance with Eq. [2.129], the influence of the grounding resistance vanishes and

$$V_{MN}^* = V_{MN}. \qquad [2.130]$$

This can be achieved by outfitting the receiver circuit with a source of "compensation" voltage V_0 which can be varied to enforce Eq. [2.130] at the time of measurement. In other words, the compensation voltage in the measurement circuit is adjusted to have the same magnitude as V_{MN}^* but opposite sign. Correspondingly, at such instant the current I_r equals zero, and the influence of the grounding resistance vanishes. This is called the "compensation method" and was widely used for a long time, because the sensitivity of voltmeters was, until recently, relatively low.

Method 2. It follows from Ohm's law that the voltage V_{MN}^* between the receiver electrodes is

$$V_{MN}^* = IR,$$

where R is the total resistance of the receiver circuit, which includes the wire resistance R_w, the internal resistance of the voltmeter R_i, and the two grounding resistances of the electrodes M and N:

$$V_{MN} = I_r(R_i + R_w + R_M + R_N). \qquad [2.131]$$

The product I_rR_i is the voltage between the terminal points of the voltmeter, which is the quantity that is actually measured. If the resistance R_i is chosen so that the inequality

$$R_i \ggg R_M + R_N + R_w$$

holds within a range of possible change of the ground resistance, we have simply

$$V_{MN} = I_r R_i. \qquad [2.132]$$

In other words, the voltage measured by the voltmeter is practically equal to the difference of potentials between points M and N when the influence of the receiver electrodes is absent. Thus, the influence of grounding resistance can be minimized by using voltmeters with a high internal resistance that can measure tiny currents accurately. Such sensors are now widely available and this approach has become conventional in electric methods of geophysics.

Method 3. Finally, the goal of eliminating the influence of grounding resistance can be achieved by making two measurements. In the first, a known voltage V_0 is introduced into the receiver circuit, and the output of the voltmeter N_0 allows a calibration of this circuit in the presence of unknown grounding resistance. Next, the voltage V_{MN} caused by the current line is measured and the voltmeter output is N. This value is corrected using the calibration measurement to give

$$V_{MN} = \frac{V_0}{N_0} N.$$

2.11. WORK DONE BY COULOMB AND EXTRANEOUS FORCES, JOULE'S LAW

As is well known, the motion of charge in a conducting medium (current) is always accompanied by the generation of heat. From the macroscopic point of view, this can be explained by some kind of friction. We now describe the relationship between the current density and electric field caused by Coulomb and extraneous forces and the amount of heat produced by a steady current flow. We will first assume that a conductor is an element of the current tube C_{12}. Earlier, we demonstrated that a constant current can exist in a conducting medium only if there is an extraneous electric force, which must have an origin different from Coulomb's law. Electromagnetic induction, for instance, is an important type of the extraneous force widely used to generate constant currents. Mechanical forces can also provide transportation of charges, for example, with electrostatic generators. Forces of electrochemical origin and of filtration also play an important role. These and similar devices are usually called current "sources." Now, we consider the work done by a current source and the distribution of its energy along a current tube. In general, a moving charge is subjected to the action of the Coulomb and extraneous forces (Ohm's law), which perform work that can be easily calculated. In fact, during the time interval dt, the amount of a charge passing through every cross section of the element C_{12} is the same, and it is equal to

$$\mathrm{d}e = I\,\mathrm{d}t, \qquad [2.133]$$

where I is the current. By definition, the amount of charge that enters and leaves the element C_{12} is the same. The elementary work of the field

$$\mathbf{E} = \mathbf{E}^{\mathrm{c}} + \mathbf{E}^{\mathrm{nc}}$$

along the displacement **dl** is

$$\mathrm{d}A = \mathbf{F}\cdot\mathbf{dl} = I\,\mathrm{d}t(\mathbf{E}^{\mathrm{c}} + \mathbf{E}^{\mathrm{nc}})\cdot\mathbf{dl}, \qquad [2.134]$$

since the force acting on charge de is

$$\mathbf{F} = I\,\mathrm{d}t(\mathbf{E}^{\mathrm{c}} + \mathbf{E}^{\mathrm{nc}}).$$

Thus, the work involved in translating the charge between two sections S_1 and S_2 during 1 s is

$$A = I\int_1^2 (\mathbf{E}^{\mathrm{c}} + \mathbf{E}^{\mathrm{nc}})\cdot\mathbf{dl}. \qquad [2.135]$$

This quantity has the dimension of power (joules per second) and in S.I. units is measured in watts (W). As experiments show, this energy produced by the action of the Coulomb and extraneous forces is transformed into heat, Q,

$$Q = I\int_1^2 (\mathbf{E}^{\mathrm{c}} + \mathbf{E}^{\mathrm{nc}})\cdot\mathbf{dl}, \qquad [2.136]$$

within some element C_{12} of the current tube. Equation [2.136] is, in effect, a version of Joule's law and is valid provided that chemical reactions and other factors do not consume energy from the current source. It follows from Eq. [2.136] that the amount of heat appearing in whole current tube during 1 s is equal to

$$Q = I\oint \mathbf{E}^{\mathrm{c}}\cdot\mathbf{dl} + \oint \mathbf{E}^{\mathrm{nc}}\cdot\mathbf{dl} = P. \qquad [2.137]$$

Since the circulation of the Coulomb electric field (the first term) vanishes,

$$P = I \oint \mathbf{E}^{\mathrm{nc}} \cdot \mathbf{dl} \qquad [2.138]$$

is the work performed by the current source per unit time. Thus, all the energy of the current source is transformed into heat. As usual, several comments are appropriate:

1. The equality $\oint \mathbf{E}^{\mathrm{c}} \cdot \mathbf{dl} = 0$ means that the work done by Coulomb's field along the current tube is zero.
2. If the external part of the field $\mathbf{E}^{\mathrm{nc}}$ is absent in a given piece of current tube, it follows from Eq. [2.136] that

$$Q = \int_{+}^{-} \mathbf{E}^{\mathrm{c}} \cdot \mathbf{dl}, \qquad [2.139]$$

where "+" and "−" are the end points of the current source. This equality shows that the work of the electric field $\mathbf{E}^{\mathrm{c}}$ is converted into heat continuously, and illustrates once again that an extraneous force, or source of energy, is needed to maintain the field. Within the internal part of the circuit, the work of the total field, $\mathbf{E}^{\mathrm{c}} + \mathbf{E}^{\mathrm{nc}}$, is transformed into heat.

3. Joule's law (Eq. [2.136]) can be expressed in terms of the current, resistance, voltage, and electromotive force of the extraneous field. In fact, using Ohm's law, we have

$$Q = I(U_1 - U_2 + \Xi^{\mathrm{nc}}) = I(IR_{12} + \Xi^{\mathrm{nc}}), \qquad [2.140]$$

where R_{12} is the resistance of the element C_{12} of current tube, while U_1 and U_2 are potentials at points of the cross sections S_1 and S_2, respectively. Then, in every element of the external part, we have

$$Q = I^2 R_{12}, \qquad [2.141]$$

that is, the heat that appears per second (the "heating") is directly proportional to the resistance and square of the current. In particular, for the whole external part, Eq. [2.140] gives

$$Q_{\mathrm{e}} = I(U_+ - U_-) = I^2 R_{\mathrm{e}}. \qquad [2.142]$$

In the internal part of the current tube, the heating is

$$Q_{\mathrm{i}} = I \int_{-}^{+} (\mathbf{E}^{\mathrm{c}} + \mathbf{E}^{\mathrm{nc}}) \cdot \mathbf{dl} \;\text{ or }\; Q_{\mathrm{i}} = I\Xi + I(U_- - U_+). \qquad [2.143]$$

Note that vectors **dl** and **E** have opposite directions and $U_+ > U_-$; that is, charges move in the direction of increase in potential. Equation [2.143] can be written as

$$P = I\Xi = I(U_+ - U_-) + Q_i = Q_e + Q_i = Q. \qquad [2.144]$$

The first term on the right-hand side of this equation describes the work of the Coulomb field $\mathbf{E}^c$, and is related to a movement of charge $I\,dt$ per unit time along the external part of the circuit:

$$A_e = I(U_+ - U_-) = I\int_+^- \mathbf{E}^c \cdot d\mathbf{l}.$$

In other words, this term characterizes the amount of energy drawn from the current source in order to maintain the constant electric field. Certainly, this source must perform at least this work A_e, but in reality, it also produces additional work A_i. For instance, in an electrostatic voltage generator, some part of the work is converted to frictional heat in the mechanical parts of the device. This additional work A_i can be accounted for by introducing an internal resistance R_i of the current source, which absorbs heat energy, so that

$$Q_i = A_i.$$

We can also easily imagine that a current source has an internal resistance, and, then, according to Joule's law, hearting that appears within the source is

$$Q_i = I^2 R_i.$$

4. As shown earlier, some of the energy generated at the source travels from the internal to external parts of the current tube. This happens because energy that originates near the source moves through the surrounding medium as electromagnetic energy and is converted into energy of the electric field. Correspondingly, transformation of the source energy into heat can be represented in the form shown below.
5. Strictly speaking, the total amount of heat associated with the constant current does not exactly coincide with the Joule heating Q given by Eq. [2.144]. This happens because of a phenomenon called "Peltier heating" that occurs near the contact between different conductors. In addition, Thomson heating occurs if a temperature gradient arises in the conductor. The influence of these factors is usually negligible.

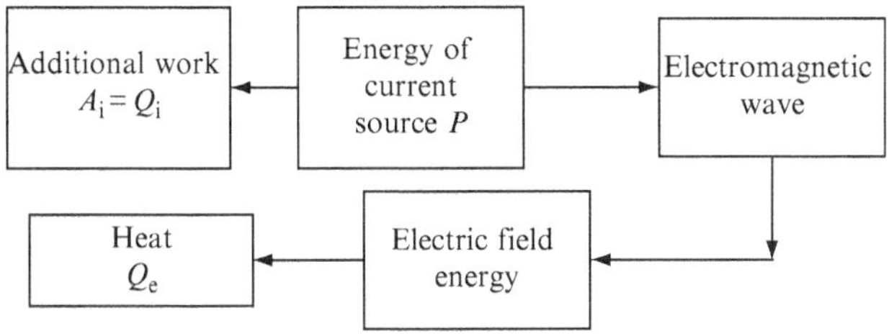

To conclude this chapter, let us generalize the formulas derived for a current tube to arbitrary points of the conductor. In an arbitrary elementary volume $dV = d\mathbf{S} \cdot d\mathbf{l}$, the amount of charge passing through the surface dS during time dt is

$$de = (\mathbf{j} \cdot d\mathbf{S})dt,$$

and the work performed by the Coulomb and extraneous forces is

$$(\mathbf{j} \cdot d\mathbf{S})[(\mathbf{E}^c + \mathbf{E}^{nc}) \cdot d\mathbf{l}]dt.$$

Therefore, the amount of heat appearing in this volume per second is

$$dQ = [\mathbf{j} \cdot (\mathbf{E}^c + \mathbf{E}^{nc})(d\mathbf{S} \cdot d\mathbf{l})] = \mathbf{j} \cdot (\mathbf{E}^c + \mathbf{E}^{nc})dV.$$

At the same time, the work done by the extraneous force is

$$dP = \mathbf{j} \cdot \mathbf{E}^{nc} dV.$$

Thus, in an arbitrary unit volume of the conductor, two processes occur continuously during steady current flow: (a) the extraneous force performs work and (b) heat is generated. The rate of each process per unit volume is given by

$$\frac{dP}{dV} = \mathbf{j} \cdot \mathbf{E}^{nc}, \quad \frac{dQ}{dV} = \mathbf{j} \cdot (\mathbf{E}^c + \mathbf{E}^{nc}) = \rho \mathbf{j}^2, \qquad [2.145]$$

where ρ is the resistivity of the medium. In particular, in an elementary volume with no extraneous force, we have

$$\frac{dP}{dV} = 0, \quad \frac{dQ}{dV} = \mathbf{j} \cdot \mathbf{E}^c = \frac{\mathbf{E}^2}{\rho} = \rho \mathbf{j}^2. \qquad [2.146]$$

CHAPTER THREE

The Electric Field, Current Density, and Potential in a Conducting Medium

In this chapter, we begin to study in detail the behavior of the electric field, current density, and the potential in conducting and nonconducting media. The models are chosen to facilitate understanding of the theory and applications of electric methods in geophysics.

3.1. INFLUENCE OF INHOMOGENEITY ON THE ELECTRIC FIELD AND CURRENT DENSITY

3.1.1. Distribution of the Electric Field, Potential, and Current Density

Suppose that a region with conductivity γ_i (the "inhomogeneity") is surrounded by a uniform medium with conductivity γ_e (Fig. 3.1A). Let the field in the absence of the inhomogeneity be $\mathbf{E}_0$, which is usually called the primary field. As we know, the primary field causes charges to appear at the surface of the inhomogeneity, and this induced charge creates its own electric field $\mathbf{E}_s$, called the secondary field. Eventually, equilibrium (steady state) is reached in which the density of charge on the surface of the inhomogeneity is given by Eq. [2.50]:

$$\sigma(q) = 2\varepsilon_0 \frac{\rho_i - \rho_e}{\rho_i + \rho_e} E_n^{\text{av}}, \qquad [3.1]$$

where ρ_i and ρ_e are the resistivities of the inhomogeneity and surrounding medium, respectively, and $E_n^{\text{av}}(q)$ is the average value of the total electric field along a unit vector $\mathbf{n}$ directed normal to the surface at q and pointing into the inhomogeneity. We can assume that, in the absence of the field $\mathbf{E}_0$, the net charge on the confined body is zero. It follows from the principle of charge conservation that the total induced surface charge $\sigma(q)$ in the presence of the primary field will also be zero when the steady state

Methods in Geochemistry and Geophysics, Volume 44
ISSN 0076-6895, DOI: 10.1016/S0076-6895(10)44003-2
© 2010 Elsevier B.V.
All rights reserved.

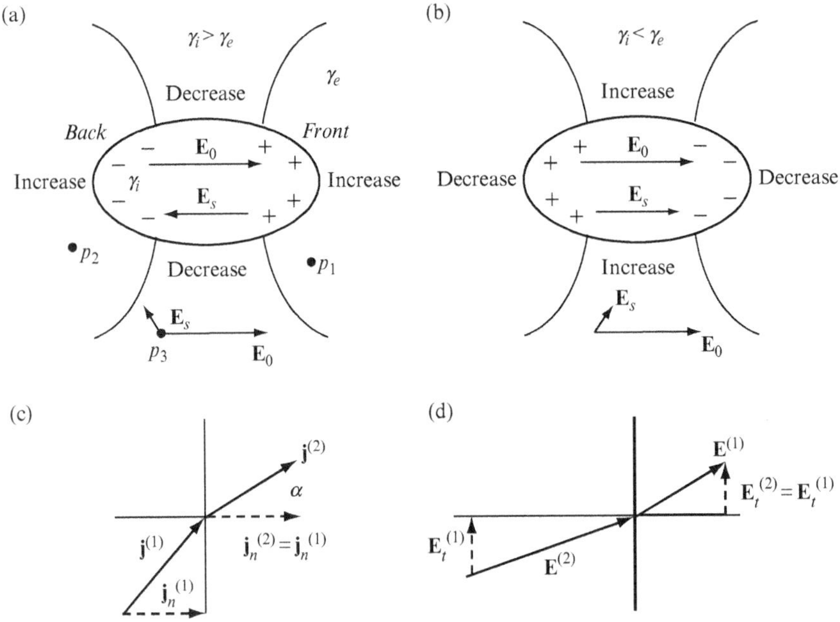

Figure 3.1 The field near a confined inhomogeneity when (A) $\gamma_i > \gamma_e$ and (B) $\gamma_i < \gamma_e$; (C) current density at an interface; (D) field lines at an interface.

is reached. In most practical cases, this means that positive charge appears on one side of the inhomogeneity while negative charge appears on the opposite side. As a result, there is often a single, closed curved on the surface along which the density $\sigma(q)$ is zero. In general, the distribution of charge can be more complicated. This charge is the source of the secondary field $\mathbf{E}_s$. The total electric field outside and inside of the inhomogeneity is therefore written as the sum,

$$\mathbf{E} = \mathbf{E}_0 + \mathbf{E}_s. \qquad [3.2]$$

Before analyzing the field in detail, several comments are helpful. (1) According to Coulomb's law, the secondary field $\mathbf{E}_s$ is related to the charge density $\sigma(q)$ by

$$\mathbf{E}_s(p) = \frac{1}{4\pi\varepsilon_0} \oint_S \frac{\sigma(q)\mathbf{L}_{qp}}{L_{qp}^3} \, dS,$$

where $\sigma(q)$ is usually unknown. (2) The fields $\mathbf{E}_0$ and $\mathbf{E}_s$ generally have different magnitudes and directions. (3) Clearly, only the secondary field can contain information about the resistivity, shape, dimensions, and location of the inhomogeneity. In geophysical applications, these quantities, along with ρ_e, are called the model's "geoelectric" parameters. (4) The potential and current density can be also represented as the sum of two fields,

$$U = U_0 + U_s,$$

and

$$\mathbf{j}_i = \gamma_i(\mathbf{E}_0 + \mathbf{E}_s), \ \mathbf{j}_e = \gamma_e(\mathbf{E}_0 + \mathbf{E}_s)$$

at points inside and outside the inhomogeneity, respectively. In examining the field behavior, it is natural to distinguish two cases.

Case 1: $\gamma_i > \gamma_e$

First, assume that the inhomogeneity is a body (of relatively simple shape) that is more conductive than the surrounding medium ($\gamma_i > \gamma_e$ or $\rho_i < \rho_e$). This means that the "contrast coefficient" in Eq. [3.1],

$$\frac{\rho_i - \rho_e}{\rho_i + \rho_e},$$

is negative. The primary electric field is directed from what we will call the "back" to the "front" side of the inhomogeneity (Fig. 3.1). Unlike in the case of electrostatic induction (when the inhomogeneity is surrounded by a nonconducting medium), the secondary field in this configuration will not completely cancel the primary field inside the conductor. The average normal component of the total field $E_n^{\text{av}}(q)$ will then be usually positive at the back side, directed along the normal vector pointing into the body, and negative at the front side, directed opposite the normal vector (in fact, it is the normal vector that changes its absolute direction in moving from the back to front side of the body; the total field maintains the same direction; see Fig. 3.1). Since the contrast coefficient is negative, then according to Eq. [3.1], negative charge appears on the back side while positive charge appears at the front side. Consider now the component of the secondary field directed along the primary field. It is convenient to represent the secondary field as the sum of fields from the positive and negative induced charges (with the obvious notation):

$$\mathbf{E}_s = \mathbf{E}_s^+ + \mathbf{E}_s^-.$$

At an observation point p_1 located outside and in front of the inhomogeneity, the field $\mathbf{E}_s^+$ has a positive projection along the direction of the primary field $\mathbf{E}_0$, while the field $\mathbf{E}_s^-$ has a negative projection. Since the amount of positive and negative charge is the same, but the negative charge is located at greater distance from the observation point, we can say that at p_1

$$|\mathbf{E}_{s0}^+| > |\mathbf{E}_{s0}^-|,$$

and therefore that the primary field is enhanced at p_1 because of the inhomogeneity; that is,

$$|\mathbf{E}_0 + \mathbf{E}_{s0}| > |\mathbf{E}_0|, \qquad [3.3]$$

where $\mathbf{E}_{s0}$ is the vector component of the secondary field in the direction of the primary field. Next, consider a point p_2 located outside, but behind the body. In this case, the fields of the positive and negative charges reverse roles: $\mathbf{E}_s^-$ is directed toward the body and $\mathbf{E}_s^+$ is directed away from the body. Because the negative charges are closer to p_2, we again conclude that in this region, the field is enhanced in the direction of the primary field. Finally, take a point p_3 in the surrounding medium, located somewhere around an equatorial plane. As is seen from Fig. 3.1A, both $\mathbf{E}_s^+$ and $\mathbf{E}_s^-$ have vector components opposite to the primary field, that is,

$$|\mathbf{E}_0 + \mathbf{E}_{s0}| < \mathbf{E}_0. \qquad [3.4]$$

Summarizing the results for the three regions, we see that, in front of and behind the conductive inhomogeneity, the field is enhanced in the direction of the primary field, and in the region around the equatorial plane, it decreases. At the boundaries of these regions, this component of the secondary field must be zero. Consider next the behavior of the field inside the inhomogeneity when $\gamma_i > \gamma_e$ Since the positive and negative charges are located on the front and back sides of the inhomogeneity, their field produces a component opposed to the primary field. Correspondingly, the total field in this direction is less than the primary field. As the ratio of conductivities γ_i/γ_e increases, the secondary field inside the inhomogeneity also increases, approaching the magnitude of the primary field. In the limit, we have the case of an ideal conductor where the total electric field vanishes inside the body. It follows from this analysis that the voltage between two arbitrary points of the inhomogeneity decreases as its conductivity γ_i increases. In other words, the better the conductor, the smaller the electric field inside. This feature can be seen as characteristic of a conductor. Next, we study the secondary potential U_s. From the equation

$$U_s(p) = \frac{1}{4\pi\varepsilon_0} \oint_S \frac{\sigma(q)}{L_{qp}} \, dS,$$

it follows that the potential is positive if the observation point p is closer to the front than to the back side of the inhomogeneity, and negative if observation points are closer to the back side. Correspondingly, there is a surface passing through the inhomogeneity where the potential U_s equals zero. In the case of the ideal conductor, the potential of the total field remains constant at all points of this body even if the potential of the primary field changes. Let us also briefly discuss the current density. Outside the inhomogeneity, the behavior of the current density is similar to that of the electric field since $\mathbf{j}_e = \gamma_e \mathbf{E}$. Inside the body, the current density increases as its conductivity increases despite the decrease in the total electric field. Since the amount of charge passing through any elementary surface cannot be infinitely large, the magnitude of the current density must tend to some finite value that depends on the shape and dimensions of the inhomogeneity. We will see that for certain simple shapes, the direction of $\mathbf{j}$ inside the body coincides with that of the primary field.

Case 2: $\gamma_i < \gamma_e$

Suppose now that body is less conductive (more resistive) than its surroundings, $\gamma_i < \gamma_e$. In the presence of a uniform primary field, positive charge now appears on its back side and negative charge on its front side (Fig. 3.1B). The distribution of the field is opposite to that described in the previous case. For example, the total field is enhanced inside the inhomogeneity as well as in the region around the equatorial plane, while it is diminished in front and behind the body. Now, as the resistivity ρ_i of the body increases (i.e., its conductivity decreases), the electric field inside gradually increases and, in the limiting case of an insulator, it reaches some finite value that depends on the shape and size of the inhomogeneity. As can be expected, the positions of zones with positive and negative values of the potential U_s are reversed from the previous case. Finally, it is obvious that the behavior of the current density outside the inhomogeneity follows that of the electric field, whereas inside, it must vanish as the resistivity increases without limit.

3.1.2. Electric Field and Current Density at the Inhomogeneity Surface

Let us examine the behavior of the field lines of **E** and **j** near the interface. Recall the interface conditions that follow from the field equations,

$$E_t^i = E_t^e \quad \text{and} \quad j_n^i = j_n^e, \qquad [3.5]$$

where the indices "*i*" and "*e*" denote the fields at different sides of the interface, in the inhomogeneity and surrounding medium, respectively. We also have from Ohm's law,

$$\frac{j_t^i}{\gamma_i} = \frac{j_t^e}{\gamma_e} \quad \text{and} \quad \gamma_i E_n^i = \gamma_e E_n^e. \qquad [3.6]$$

Using both sets of equations, we obtain

$$\frac{j_t^i}{\gamma_i j_n^i} = \frac{j_t^e}{\gamma_e j_n^e} \quad \text{and} \quad \frac{E_n^i \gamma_i}{E_t^i} = \frac{E_n^e \gamma_e}{E_t^e}. \qquad [3.7]$$

The direction of the electric field **E** near the interface can be characterized by the angle α between the field vector and the normal **n** (Fig. 3.1C):

$$\tan \alpha = \frac{E_t}{E_n}.$$

The second equation of the set [3.7] can be written as

$$\frac{\gamma_i}{\tan \alpha_i} = \frac{\gamma_e}{\tan \alpha_e} \quad \text{or} \quad \frac{\tan \alpha_i}{\tan \alpha_e} = \frac{\gamma_i}{\gamma_e}. \qquad [3.8]$$

Of course, the same result is obtained if we proceed from the first equation of Eq. [3.7]. According to Eq. [3.8], the field lines of **E** and **j** refract near the boundary in such a way that the value of tan α is proportional to the conductivity of each medium. The field lines approach the normal in the medium with higher resistivity as if trying to reduce their path length in this medium. If one of the angles equals 0 or $\pi/2$, the other angle has the same value (provided that both conductivities are finite). In these two cases, the field lines do not refract, even though one of the field components has a discontinuity: the normal component of the electric field and the tangential component of the current density are discontinuous across the interface. This example shows that the field lines of the current density **j** concentrate inside the inhomogeneity, as well as in front and behind it, when the body is more conductive than its surroundings, $\gamma_i > \gamma_e$. This is accompanied by rarefaction of the current lines near the lateral sides of the body. When the body is more resistive, the field lines of **j** are more rarefied in front of and behind the body, as well as inside it. This again shows that current tends to concentrate in the more conductive medium. If the inhomogeneity is an ideal conductor ($\rho_i = 0$), then its surface becomes an equipotential, where the tangential component of the field $\mathbf{E}^e$ vanishes. This means that the vector lines of both **E** and **j** are perpendicular to the surface of an ideal conductor. Thus, although the electric field inside an ideal conductor vanishes, the current lines pass through it continuously. In the opposite case, when the body is an insulator ($\gamma_i = 0$), the normal component of the current density vanishes near the interface ($j_n^e = 0$), and therefore E_n^e also vanishes. Consequently, the lines of both fields outside an insulator are tangent to its surface.

3.1.3. Dependence of the Secondary Field on the Distance From An Inhomogeneity

One other feature of the field caused by the inhomogeneity is worth exploring. Since the total charge induced on the body is zero, its secondary field at distances much larger than any of its linear dimensions approaches that of an electric dipole (Eq. [3.118]):

$$U_s(p) \to \frac{1}{4\pi\varepsilon_0} \frac{\mathbf{M} \cdot \mathbf{L}_{qp}}{L_{qp}^3}, \text{ if } L_{qp} \to \infty, \qquad [3.9]$$

where $\mathbf{L}_{qp}$ is a vector from any point in the body to the observation point and **M** is the dipole moment of the inhomogeneity. **M** is proportional to the total induced positive charge and also depends on the primary field $\mathbf{E}_0$, as well as on the body's shape and dimensions. The field inside the body is usually nonuniform, but there are exceptions: For example, when the inhomogeneity has a relatively simple shape (such as spherical or ellipsoidal) and the primary field is uniform in its vicinity, the dipole moment **M** is directly proportional to the field $\mathbf{E}_0$. Moreover, it has the same direction if $\gamma_i > \gamma_e$, and the opposite direction if $\gamma_i < \gamma_e$. The normal component of the electric field is discontinuous at the surface of the inhomogeneity, and its jump is defined by the principle of charge conservation:

$$j_n^e = j_n^i \text{ or } \frac{E_n^e}{E_n^i} = \frac{\gamma_i}{\gamma_e}.$$

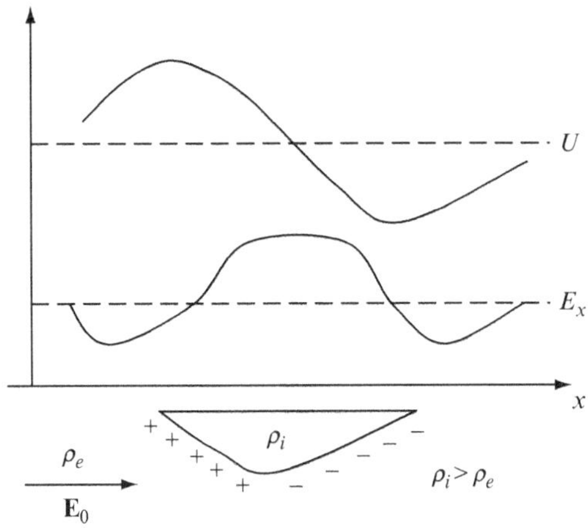

Figure 3.2 Schematic behavior of the potential and horizontal component of the electric field above an inhomogeneity.

Figure 3.2 illustrates the behavior of the potential and the horizontal component of the electric field above an inhomogeneity that is more resistive than the surrounding medium.

Such bodies can be the target of investigation when electric methods are used for profiling and geologic mapping, but in exploration, they are mainly a source of "geological noise" that obscures the response of good conductors.

3.1.4. The Case of Weak Influence of Inhomogeneity

It follows from Eq. [3.1] that charge is absent if the normal component of the primary field E_{0n} is zero at every point of an interface between two media. In such cases, continuity of the normal component of the current density is automatically satisfied without the appearance of surface charge. For instance, suppose that the primary field $\mathbf{E}_0$ is directed along the strike of a geological structure called a dike, which is a thin, highly elongated body of rock intruded into surrounding bedrock. Such bodies can often be modeled as a two-dimensional (2D) structure of infinite extent. If $\mathbf{E}_0$ is directed perpendicular to the structure's cross section, then it never crosses a conductivity boundary and the density of surface charge σ is zero everywhere. Therefore, a secondary field $\mathbf{E}_s$ is absent, and the total electric field coincides with the primary field everywhere. In such cases, it is impossible to discover the dike with only electrical measurements. Of course, in reality, such bodies are not infinitely elongated: there is always charge on their lateral surfaces, even if the amount of charge is small and located far away from observation points. At the same time, the current density can be very large within an elongated conductor since

$$j_i = \gamma_i E_0 \gg \gamma_e E_0, \quad \text{if } \gamma_i \gg \gamma_e.$$

Thus, measurements of the steady magnetic field created by the current allow the detection of such conductors. In the following sections, we study bodies of simple shapes where it is possible to obtain exact expressions for the electric field, potential, and current density in the presence of a uniform primary field. We begin with the simplest case, a conducting sphere.

3.2. A CONDUCTING SPHERE IN A UNIFORM ELECTRIC FIELD

3.2.1. Introduction

Suppose a sphere with radius a and conductivity γ_2 is situated in medium with conductivity γ_1 in the presence of a uniform electric field $\mathbf{E}_0$ (Fig. 3.3). Before setting up and solving the boundary-value problem, let us discuss several obvious features of the primary and secondary fields. First of all, note that the model is not completely realistic because a uniform primary field pervading all of space would require that its source have infinite power. Nevertheless, this approximation is useful for studying the field near the sphere when sources of the primary field are located far away from observation points. In fact, the secondary field itself is defined only by the behavior of the primary field in the volume occupied by a sphere. Since the uniform field $\mathbf{E}_0$ intersects the entire surface of the sphere, both positive and negative charges are induced on its surface, but the sphere remains neutral overall. It is natural to assume that the distribution of

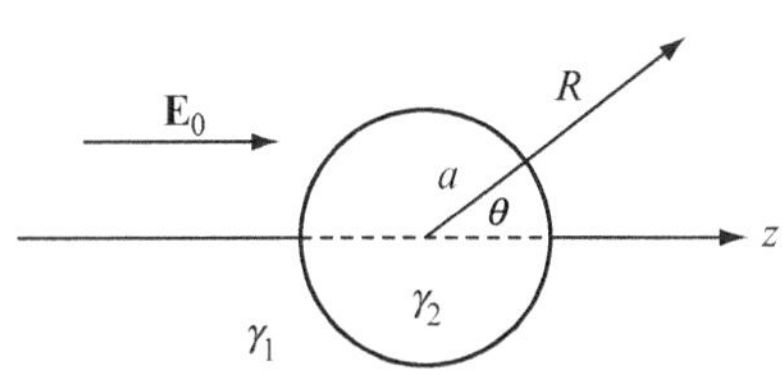

Figure 3.3 Sphere in uniform electric field.

charge has axial symmetry around a line that passes through the sphere's center in the direction of the primary field (Fig. 3.3). The potential, electric field, and current density should have the same symmetry. This is only a starting assumption, but we will quickly prove that it is correct with the help of the uniqueness theorem.

3.2.2. Formulation of the Boundary-Value Problem

We will use a spherical system of coordinates R, θ, ϕ with the origin at the center of the sphere and the z-axis directed along the primary field. In this system, axial symmetry simply means that all field quantities should be independent of the azimuthal angle ϕ. Then, by the uniqueness theorem derived in Chapter 2, the boundary-value problem is:

1. $\nabla^2 U = 0$ at regular points,

2. $$U_1 = U_2 \quad \text{and} \quad \gamma_1 \frac{\partial U_1}{\partial R} = \gamma_2 \frac{\partial U_2}{\partial R}, \quad \text{if } R = a, \qquad [3.10]$$

3. $U_1 \to U_0, \quad \text{as } R \to \infty,$

where U_0 is the potential of the primary field. The boundary condition at infinity should be obvious, since the secondary field of induced charges eventually decays to zero far away from the sphere. Given the simplicity of the problem, we construct a solution proceeding from a physical understanding of the field behavior. The potential of the primary field can be determined easily since the field $\mathbf{E}_0$ has only a component along z-axis, and therefore

$$E_0 = -\frac{\partial U_0}{\partial z}, \quad \text{or} \quad U_0 = -E_0 z + C_0, \qquad [3.11]$$

where C_0 is a constant that can be set to zero since it has no effect on the electric field. (This is equivalent to setting a reference value of zero for the potential of the primary field at the plane $z=0$.) We then have

$$U_0 = -E_0 z = E_0 R \cos\theta. \qquad [3.12]$$

Consider now the secondary field caused by surface charges. As we know, the total induced charge equals zero, which means that far away the secondary field is equivalent to that of an electric dipole. We will assume that this behavior holds everywhere outside of the sphere ($R > a$). Let $\mathbf{M}$ be the net dipole moment of the charge induced on the sphere. The vector $\mathbf{M}$

acts at the origin of the sphere and is directed along the primary field $\mathbf{E}_0$. The potential of the secondary field outside the sphere is then

$$U_s = \frac{\mathbf{M}\cdot\mathbf{R}}{4\pi\varepsilon_0 R^3} = \frac{M\cos\theta}{4\pi\varepsilon_0 R^2}, \qquad [3.13]$$

where the magnitude of the dipole moment M is a constant to be determined. Thus, the total potential outside of the sphere is

$$U_1(R,\theta) = -E_0 R\cos\theta + \frac{M\cos\theta}{4\pi\varepsilon_0 R^2}, \quad \text{if } R \geq a. \qquad [3.14]$$

We also assume that charge is distributed on the surface in such a way that the electric field inside the sphere is constant and directed along the z-axis. This is perhaps the only assumption that is not immediately suggested by the symmetry of the model. It is, however, the simplest possible field behavior inside the sphere, and with it, the total field within a body can be represented as

$$\mathbf{E}_2 = C\mathbf{E}_0,$$

and

$$U_2 = -CE_0 z = -CE_0 R\cos\theta, \quad \text{if } R \leq a, \qquad [3.15]$$

where C is a constant.

3.2.3. Solution of the Boundary-Value Problem

Arguing from physical principles, we have thus constructed the potential given by Eqs. [3.14] and [3.15]. We can now use the uniqueness theorem to check if our assumptions are correct and, in the process, determine the unknowns M and C. This means simply that we will attempt to satisfy the conditions (Eqs. [3.10]) of the boundary-value problem. First, it is clear that the potentials representing the field outside and inside the sphere (Eqs. [3.14] and [3.15], respectively) satisfy Laplace's equation everywhere in their separate domains of validity, excluding the surface of the sphere itself. That is, the potential of the uniform field (Eqs. [3.15]) satisfies Laplace's equation in the region $R < a$, whereas the uniform field plus electric dipole potential (Eqs. [3.14]) satisfies Laplace's equation in $R > a$. This fact can be checked by directly substituting these functions into Laplace's equation. The condition at infinity is also satisfied by the potential [3.14]. We now check if the unknown constants M and C are determined uniquely from the interface conditions requiring continuity of the potential and of the radial component of the current density at the sphere surface. Using Eqs. [3.14] and [3.15], we have

$$\left(-E_0 a + \frac{M}{4\pi\varepsilon_0 a^2}\right) \cos\theta = -CE_0 a \cos\theta \text{ (continuity of potential)}$$

and

$$\gamma_1\left(-E_0 - \frac{2M}{4\pi\varepsilon_0 a^3}\right) \cos\theta = -\gamma_2 CE_0 \cos\theta \text{ (continuity of radial current density).}$$

We thus have two equations for two unknowns,

$$-E_0 + \frac{M}{4\pi\varepsilon_0 a^3} = -CE_0 \quad \text{and} \quad \gamma_1\left\{E_0 + \frac{2M}{4\pi\varepsilon_0 a^3}\right\} = \gamma_2 CE_0. \qquad [3.16]$$

Such a linear system does not always have a unique solution or even any solution. If the system [3.16] could not be solved, it would mean that one or more assumptions were wrong. Equations [3.16], however, have the unique solution,

$$M = 4\pi\varepsilon_0 \frac{\gamma_2 - \gamma_1}{\gamma_2 + 2\gamma_1} a^3 E_0, \qquad C = \frac{3\gamma_1}{\gamma_2 + 2\gamma_1}. \qquad [3.17]$$

Since the conductivity can never be negative, this solution is always well-defined. Thus, the potentials outside and inside the sphere are

$$\begin{aligned} U_1 &= -E_0 R \cos\theta + \frac{\gamma_2 - \gamma_1}{\gamma_2 + 2\gamma_1} a^3 E_0 \frac{\cos\theta}{R^2}, \quad \text{if } R \geq a, \\ U_2 &= -\frac{3\gamma_1}{\gamma_2 + 2\gamma_1} E_0 R \cos\theta, \quad \text{if } R \leq a. \end{aligned} \qquad [3.18]$$

Equations [3.18] satisfy all the conditions of the boundary-value problem [3.10], so we can conclude from the uniqueness theorem that this is the unique potential of the electric field caused by charges on the surface of a sphere in a uniform primary field. The theorem ensures that there is no other solution. This vividly illustrates again the value of the theorem: whatever approach is used to find a trial solution for a given problem, the theorem provides an easy way to check whether it is the actual solution of the problem.

3.2.4. Distribution of Charges

Now, we will describe some features of the secondary field inside and outside of the sphere. First, consider the distribution of surface charge, which can be calculated from the surface analogue of the second equation for the electric field:

$$\sigma = \varepsilon_0 (E_{2n} - E_{1n}) = \varepsilon_0 (E_{1R} - E_{2R}), \qquad [3.19]$$

where the last equality holds since the unit vectors **n** and **R** have opposite directions. Taking derivatives with respect to R, we obtain

$$\frac{\partial U_1}{\partial R} = -E_0 \cos\theta - 2\frac{\gamma_2 - \gamma_1}{\gamma_2 + 2\gamma_1}\frac{a^3}{R^3}\cos\theta \quad \text{and} \quad \frac{\partial U_2}{\partial R} = -\frac{3\gamma_1}{\gamma_2 + 2\gamma_1}E_0 \cos\theta.$$

Since

$$E_R = -\frac{\partial U}{\partial R},$$

we have

$$\sigma = \varepsilon_0 \left(1 + 2\frac{\gamma_2 - \gamma_1}{\gamma_2 + 2\gamma_1} - \frac{3\gamma_1}{\gamma_2 + 2\gamma_1} \right) E_0 \cos\theta,$$

or

$$\sigma = 3\varepsilon_0 \frac{\gamma_2 - \gamma_1}{\gamma_2 + 2\gamma_1} E_0 \cos\theta. \qquad [3.20]$$

Thus, the density of charges σ is distributed symmetrically with respect to the z-axis, and varies as $\cos\theta$. The magnitude of the charge density peaks at $\theta=0$, where the primary field is perpendicular to the surface, and it is equal to zero in the equatorial plane, $\theta=\pi/2$. As follows from Eq. [3.20], the charges within the intervals $0<\theta<\pi/2$ and $\pi/2<\theta<\pi$ have opposite signs. The dependence of the charge density on conductivities is defined by the coefficient K_{12}:

$$K_{12} = \frac{(\gamma_2/\gamma_1) - 1}{(\gamma_2/\gamma_1) + 2}, \qquad [3.21]$$

which varies between $-1/2$ and $+1$. It is clear that the same relatively weak dependence on conductivity holds for the secondary electric field, too. It follows from Eq. [3.21] that coefficient K_{12} is close to unity ("saturates") even when the sphere conductivity only slightly exceeds that of the surrounding medium. This means that spheres having very different conductivities can create practically the same field. In other words, electric measurements are highly sensitive to the presence of an inhomogeneity, allowing detection of a body having only a small contrast with the surrounding medium, but do not easily permit determination of the conductivity of the body.

3.2.5. Behavior of the Secondary Electric Field and Vector of Current Density

Next, let us consider the behavior of the electric field inside and outside the sphere. From Eq. [3.18], the electric field inside the sphere $\mathbf{E}_2$ is uniform and directed along the z-axis:

$$E_2 = -\frac{\partial U_2}{\partial z},$$

or

$$E_2 = \frac{3\gamma_1}{\gamma_2 + 2\gamma_1} E_0 = \frac{3}{2 + (\gamma_2/\gamma_1)} E_0. \quad [3.22]$$

Correspondingly, the secondary field caused by charges is

$$E_{2s} = \frac{1 - (\gamma_2/\gamma_1)}{2 + (\gamma_2/\gamma_1)} E_0. \quad [3.23]$$

When the sphere is more conductive than its surroundings ($\gamma_2 > \gamma_1$), the primary and secondary fields have opposite directions inside the conductor. As the sphere conductivity γ_2 increases, the magnitude of secondary field increases and approaches that of the primary field. Therefore, the total field decreases and, as in the case of the ideal conductor ($\gamma_2 \to \infty$), tends to zero. At the same time, the total current density inside the sphere is

$$j_2 = \frac{3\gamma_2}{2 + (\gamma_2/\gamma_1)} E_0,$$

which, as γ_2 increases, approaches a limit independent of the sphere conductivity,

$$j_2 \to 3\gamma_1 E_0 = 3j_1. \quad [3.24]$$

The magnitude of the current density tends to a value that is three times larger than that in the absence of the sphere. This is different from the electrostatic case (where the conductor sits in an insulating medium of zero conductivity) in which both the electric field and current density vanish inside the sphere no matter how small its conductivity. In the case of a more resistive sphere, the electric field $\mathbf{E}_2$ increases as its resistivity increases. In the limit of an insulator, we have

$$E_2 \to 1.5E_0, \quad \text{as} \quad \rho_2 \to \infty, \tag{3.25}$$

but the current vanishes. The secondary field outside the sphere is equivalent to that of the electric dipole, and from Eqs. [3.18], we have

$$E_R = -\frac{\partial U}{\partial R}, \quad E_\theta = -\frac{1}{R}\frac{\partial U}{\partial \theta},$$

or

$$E_{2R} = 2\frac{\gamma_2 - \gamma_1}{\gamma_2 + 2\gamma_1}\frac{a^3}{R^3}E_0 \cos\theta, \quad E_{1\theta} = \frac{\gamma_2 - \gamma_1}{\gamma_2 + 2\gamma_1}\frac{a^3}{R^3} \sin\theta. \tag{3.26}$$

The behavior of this field was studied in detail in Chapter 1. It is useful to note that the moment of the dipole given by Eqs. [3.26],

$$M = 4\pi\varepsilon_0 \frac{\gamma_2 - \gamma_1}{\gamma_2 + 2\gamma_1} a^3 E_0, \tag{3.27}$$

depends on the volume of the sphere, the contrast in conductivity, and the magnitude of the primary field. Although highly stylized, the field around a spherical inhomogeneity is useful for understanding different aspects of electrical methods in mining and engineering geophysics and also for studying the influence of geological noise in electromagnetic methods such as magnetotelluric soundings. We now consider the field of a slightly more complex 2D body, an elliptical cylinder.

3.3. ELLIPTICAL CYLINDER IN A UNIFORM ELECTRIC FIELD

Consider next an elliptical cylinder located in a uniform medium where the primary electric field $\mathbf{E}_0$ is constant and perpendicular to the cylinder's axis (Fig. 3.4A). In solving the boundary-value problem, we have to consider only those solutions of Laplace's equations that provide continuity of the potential and the normal component of the current density at the cylinder surface as well as the condition at infinity. First, we focus on general solutions of Laplace's equation appropriate to the cylindrical geometry.

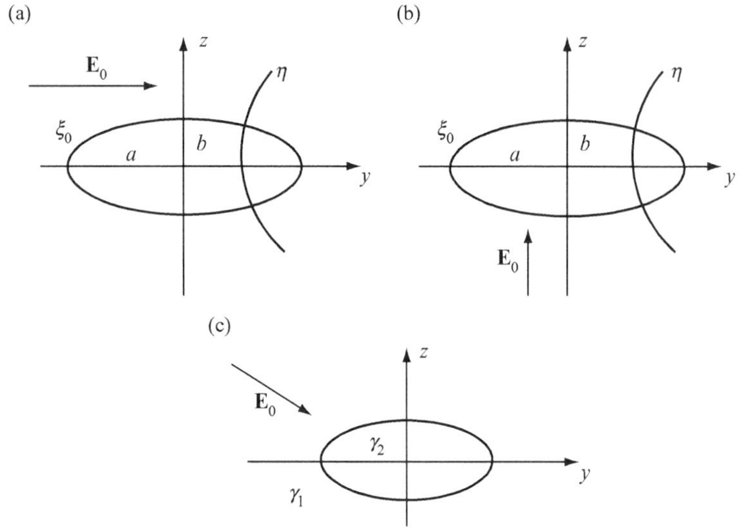

Figure 3.4 Elliptical cylinder in a uniform primary field directed along the (A) major and (B) minor axis. (C) Elliptical cylinder with arbitrary direction of the primary field.

3.3.1. Elliptical System of Coordinates and Solution of Laplace's Equation

To solve this problem, we will introduce an elliptical system of coordinates ξ, η in the (y, z) plane in the following way:

$$y = c \cosh \xi \cos \eta, \quad z = c \sinh \xi \sin \eta, \qquad [3.28]$$

where $c = (a^2 - b^2)^{1/2}$ is the eccentricity of the cylinder, and a and b are its major and minor semiaxes, respectively. The coordinates are defined by two families of elliptical and hyperbolic cylinders that are orthogonal to each other and have the same focus at the points $y = \pm c$, $z = 0$. The domains of the coordinates are

$$0 \le \xi < \infty, \quad 0 \le \eta < 2\pi,$$

where the coordinate η is measured from the y-axis. This system is natural to the problem because the cylinder surface coincides with one of the coordinate surfaces, $\xi = \xi_0$, which greatly simplifies the boundary-value problem. The metric coefficients are

$$h_1 = h_2 = c(\cosh^2 \xi - \cos^2 \eta)^{1/2},$$

and, therefore, Laplace's equation has the simple invariant form,

$$\frac{\partial^2 U}{\partial \xi^2} + \frac{\partial^2 U}{\partial \eta^2} = 0. \qquad [3.29]$$

The secondary field is caused by induced charges on the cylinder surface, and as usual, we represent the potential inside and outside the cylinder in the form

$$U_1 = U_0 + U_s, \text{ if } \xi \geq \xi_0$$

and [3.30]

$$U_2, \text{ if } \xi \leq \xi_0.$$

Consider first the case where the primary field is directed along the major axis (y-axis). Its potential U_0 is

$$U_0 = -E_0 y = E_0 c \cosh \xi \cos \eta,$$

or

$$U_0 = -\frac{E_0 c}{2}\left(e^{\xi} + e^{-\xi}\right) \cos \eta. \qquad [3.31]$$

Substituting into Eq. [3.29], it is easy to see that functions of the type

$$e^{\xi} \cos \eta \quad \text{and} \quad e^{-\xi} \cos \eta$$

satisfy Laplace's equation.

3.3.2. Expressions for the Potential and Electric Field

To simplify determination of the potential, we make two assumptions about the field behavior, namely (a) inside the cylinder, the field remains uniform and it is still directed along the y-axis; (b) outside the cylinder, the secondary field decays with distance from the cylinder as $\exp(-\xi)$ and depends on the angle η in the same manner as the primary field. Correspondingly, we write

$$\begin{aligned} U_1 &= -cE_0\left(\cosh \xi + A\, e^{-\xi}\right) \cos \eta \quad \text{and} \\ U_2 &= -cE_0 B \cosh \xi \cos \eta, \end{aligned} \qquad [3.32]$$

where A and B are unknown constants independent of the coordinates ξ and η. It is obvious that functions U_1 and U_2 are solutions of Laplace's equation at regular points and that U_1 obeys the condition at infinity:

$$U_1 \to U_0, \quad \text{as } \xi \to \infty.$$

From the two continuity conditions at the cylinder interface ($\xi = \xi_0$),

$$U_1(\xi_0, \eta) = U_2(\xi_0, \eta), \quad \gamma_1 \frac{\partial U_1}{\partial \xi} = \gamma_2 \frac{\partial U_2}{\partial \xi},$$

we obtain a system of equations for the unknowns:

$$\cosh \xi_0 + A \exp(-\xi_0) = B \cosh \xi_0$$

and [3.33]

$$\gamma_1[\sinh \xi_0 - A \exp(-\xi_0)] = \gamma_2 B \sinh \xi_0.$$

The solution of this system gives

$$A = \frac{[1 - (\gamma_2/\gamma_1)] \sinh \xi_0 \exp(\xi_0)}{1 + (\gamma_2/\gamma_1) \tanh \xi_0}$$

and [3.34]

$$B = \frac{1 + \tanh \xi_0}{1 + (\gamma_2/\gamma_1) \tanh \xi_0}.$$

Inasmuch as the system has nonzero solution, our assumptions about the field behavior are correct, and the functions U_1 and U_2 (Eqs. [3.32] and [3.34]) are the electric potential for this model. The electric field outside the cylinder is

$$E_\xi = -\frac{1}{h_1}\frac{\partial U}{\partial \xi}, \quad E_\eta = -\frac{1}{h_2}\frac{\partial U}{\partial \eta},$$

$$E_{1\xi} = \frac{E_0}{(\cosh^2\xi - \cos^2\eta)^{1/2}}[\sinh \xi - A \exp(-\xi)] \cos \eta$$

and

$$E_{1\eta} = \frac{E_0}{(\cosh^2\xi - \cos^2\eta)^{1/2}}[\cosh \xi + A \exp(-\xi)] \sin \eta, \quad \text{if } \xi > \xi_0.$$

[3.35]

Inside the cylinder, we have

$$E_{2\xi} = \frac{E_0 B}{(\cosh^2\xi - \cos^2\eta)^{1/2}} \sinh \xi \cos \eta$$

and [3.36]

$$E_{2\eta} = \frac{E_0 B}{(\cosh^2\xi - \cos^2\eta)^{1/2}} \cosh \xi \sin \eta, \text{ if } \xi < \xi_0.$$

Note that the vector

$$\mathbf{E}_2 = E_{2\xi}\mathbf{i}_\xi + E_{2\eta}\mathbf{i}_\eta$$

is directed along the y-axis. Here, $\mathbf{i}_\xi$, $\mathbf{i}_\eta$ are unit vectors along the coordinate lines in the elliptical system of coordinates.

3.3.3. The Field in the Presence of the Circular Cylinder

We now specialize this solution to the case of circular cylinder, when $a=b$. It follows from Eq. [3.28] that the distance r from the origin is

$$r = c(y^2 + z^2)^{1/2} = c(\cosh^2\xi\ \cos^2\eta + \sinh^2\xi\ \sin^2\eta)^{1/2}.$$

Performing a transition from the elliptical to cylindrical coordinates (r, ϕ), and taking into account that $c = (a^2 - b^2)^{1/2} \to 0$, we conclude that $\xi \gg 1$ and

$$\cosh\xi = \sinh\xi = \frac{\exp(\xi)}{2}.$$

Also, we have

$$\tan\phi = \frac{z}{y} = \frac{\sinh\xi\ \sin\eta}{\cosh\xi\ \cos\eta} = \tan\eta.$$

Thus, if $a \to b$,

$$r \to c\cosh\xi \quad \text{and} \quad \eta \to \phi. \qquad [3.37]$$

In particular,

$$a = b = c\frac{\exp(\xi_0)}{2},$$

and the expressions for coefficients A and B become

$$A = \frac{a}{c}\frac{1 - (\gamma_2/\gamma_1)}{1 + (\gamma_2/\gamma_1)}\exp(\xi_0) \quad \text{and} \quad B = \frac{2}{1 + (\gamma_2/\gamma_1)}.$$

Correspondingly, the potentials of the electric field are

$$U_1(r,\phi) = -E_0 r\cos\phi - \frac{1 - (\gamma_2/\gamma_1)}{1 + (\gamma_2/\gamma_1)}\frac{a^2}{r}\cos\phi$$

and [3.38]

$$U_2(r,\phi) = -E_0\frac{2}{1 + (\gamma_2/\gamma_1)}r\cos\phi = -\frac{2E_0}{1 + (\gamma_2/\gamma_1)}y.$$

Using

$$E_r = -\frac{\partial U}{\partial r}, \quad E_\phi = -\frac{1}{r}\frac{\partial U}{\partial \varphi},$$

we arrive at the known formulas for the electric field outside the cylinder

$$E_{1r} = E_0 \cos\phi + \frac{(\gamma_2/\gamma_1) - 1}{(\gamma_2/\gamma_1) + 1}\frac{a^2}{r^2} \cos\phi$$

and

$$E_{1\phi} = -E_0 \sin\phi - \frac{(\gamma_2/\gamma_1) - 1}{(\gamma_2/\gamma_1) + 1}\frac{a^2}{r^2} \sin\phi.$$

Inside the cylinder, the field is constant:

$$E_{2y} = \frac{2E_0}{1 + (\gamma_2/\gamma_2)}.$$

The equipotential surfaces of the primary field are clearly planes perpendicular to the y-axis. The presence of the cylinder distorts these surfaces, but at large distances from the cylinder, they tend asymptotically again to planes. Inside the cylinder, however, the field is constant (as in the case of the sphere) and the equipotential surfaces for the total field remain normal to the y-axis.

Let us return to the general case of the elliptical cylinder and consider the distribution of surface charges. In this light, it is useful to write down some relations:

$$\tanh\xi_0 = \frac{b}{a}, \quad \sinh\xi_0 = \frac{b/a}{[1 - (b^2/a^2)]^{1/2}}$$

and [3.39]

$$\cosh\xi_0 = \frac{1}{[1 - (b^2/a^2)]^{1/2}}, \quad \exp(\xi_0) = \left(\frac{1 + (b/a)}{1 - (b/a)}\right)^{1/2}.$$

3.3.4. Distribution of Surface Charges and Behavior of Fields E and j

The surface charge density σ is defined by the discontinuity of the normal component of the electric field

$$\sigma = \varepsilon_0(E_{1\xi} - E_{2\xi}),$$

and, from Eqs. [3.35] and [3.36], we have

$$\sigma = \varepsilon_0 \frac{[(\gamma_2/\gamma_1) - 1](1 + \tanh \xi_0) \sinh \xi_0 \cos \eta}{[1 + (\gamma_2/\gamma_1) \tanh \xi_0](\cosh^2\xi_0 - \cos^2\eta)^{1/2}} E_0. \qquad [3.40]$$

The latter shows that the charge density is everywhere finite and vanishes at points where the primary field is tangent to the cylinder ($\eta = \pi/2$). It then gradually increases toward the y-axis, $\eta = 0$.

$$\sigma(\xi_0, 0) = \varepsilon_0 E_0 \frac{[(\gamma_2/\gamma_1) - 1](1 + \tanh \xi_0)}{1 + (\gamma_2/\gamma_1) \tanh \xi_0}, \qquad [3.41]$$

since

$$\cosh^2\xi - \sinh^2\xi = 1.$$

In particular, at points $\eta = 0$ and $\eta = \pi$ we have

$$\sigma = \pm\varepsilon_0 E_0 \left(\frac{\gamma_2}{\gamma_1} - 1\right), \text{ if } \frac{\gamma_2}{\gamma_1}\frac{a}{b} \ll 1.$$

Next, we will describe the field outside the cylinder proceeding from Eqs. [3.35] and [3.36]. First, it is clear that if the cylinder is much more resistive than the surrounding medium ($\gamma_2 \ll \gamma_1$), the secondary electric field depends only weakly on the conductivity of the cylinder or medium and is mainly defined by the value of b/a When the cylinder has much higher conductivity than the surrounding medium ($\gamma_2 \gg \gamma_1$) and the ratio of axes is not small, or more precisely when $(\gamma_2/\gamma_1)(b/a) \gg 1$, the secondary field is controlled by geometric parameters only. In contrast, when the cylinder is highly elongated in the direction of the field $\mathbf{E}_0$, that is, when

$$\frac{\gamma_2}{\gamma_1}\frac{b}{a} \ll 1 \quad \text{and} \quad \frac{b}{a} \ll 1, \qquad [3.42]$$

the secondary field is small but depends on the conductivity of the cylinder, whether the ratio γ_2/γ_1 is greater or less than 1. In essence, the inequalities [3.42] are conditions under which an elongated cylinder behaves as an infinite layer with thickness $2b$. Notice that the ratio of the secondary field magnitude of an ideally conducting cylinder to that of an insulating one is the same as the ratio of its semiaxes, a/b. Finally, consider the field inside the cylinder, which can be written as

$$E_2 = \frac{1 + (b/a)}{1 + (\gamma_2 b/\gamma_1 a)} E_0 \quad \text{or} \quad E_2 = \frac{E_0}{1 + [(\gamma_2/\gamma_1) - 1]L}, \qquad [3.43]$$

where the term

$$L = \frac{b/a}{1 + (b/a)}$$

is sometimes called the "depolarization" factor since it characterizes the field decrease inside the conductor when $\gamma_2 > \gamma_1$ (in this case, the word "depolarization" has no particular physical meaning related to the charge distribution). Note the following features of the field E_2:

1. Surface charges are distributed in such a way that the secondary field is uniform inside the cylinder and has either the same direction as the primary field when $\gamma_2 < \gamma_1$, or the opposite direction when $\gamma_2 > \gamma_1$.
2. In the case of the circular cylinder, we have

$$L = \frac{1}{2} \quad \text{and} \quad E_2 = \frac{2E_0}{1 + (\gamma_2/\gamma_1)}.$$

3. As the ratio a/b increases, the influence of charges decreases and we have

$$E_2 \to E_0, \quad \text{if } \frac{\gamma_2}{\gamma_1}\frac{b}{a} \to 0.$$

   This shows that, as the conductivity of the cylinder increases, the influence of surface charges becomes negligible only if the cylinder becomes more elongated along the primary field.
4. As the conductivity of the surrounding medium γ_1 decreases, the field inside the cylinder decreases as well. As γ_1 approaches zero, the surrounding medium becomes an insulator, and the secondary field of induced surface charges cancels the primary field (principle of electrostatic induction).
5. If the elliptical cylinder is an insulator, we have

$$E_2 = \frac{E_0}{1 - L},$$

   and correspondingly the secondary field E_2 has a maximum value for the circular cylinder when

$$L = \frac{1}{2} \quad \text{and} \quad E_2 = 2E_0.$$

6. According to Ohm's law, the current density inside the elliptical cylinder is

$$j_2 = \frac{\gamma_2 E_0}{1 + [(\gamma_2/\gamma_1) - 1]L}.$$

Here, it is appropriate to distinguish two cases:

(a) *A cylinder highly elongated along the primary field*. In this case, the secondary field is negligible, and the current density becomes directly proportional to the cylinder conductivity $j_2 = \gamma_2 E_0$.

(b) *A highly conductive cylinder for which the product* $(\gamma_2/\gamma_1)L$ *is much greater than unity, even when* $a \gg b$. The current density then becomes independent of the cylinder conductivity and proportional to the conductivity of the surrounding medium,

$$j_2 = \frac{\gamma_1 E_0}{L}.$$

Nevertheless, when $a \gg b$ the current density $\mathbf{j}_2$ can exceed the value of the density $\mathbf{j}_0$ when the cylinder is absent. This shows that measurements of the magnetic field generated by currents flowing in a cylinder can be useful for detecting conductors strongly elongated along the primary field (method of charged body).

3.3.5. The Primary Field Directed Along the *Z*-Axis

We now solve for the field in the same model when the primary field is directly along the z-axis (Fig. 3.4B). Then,

$$E_0 = -\frac{\partial U_0}{\partial z} \quad \text{and} \quad U_0 = -E_0 z = -E_0 c \sinh \xi \sin \eta. \qquad [3.44]$$

By analogy with the previous case, we can write for the potential outside and inside the cylinder

$$U_1 = -E_0 c[\sinh \xi + A_* \exp(-\xi)] \sin \eta$$

and [3.45]

$$U_2 = -E_0 c B_* \sinh \xi \sin \eta.$$

Continuity of the potential and normal component of the current density at the surface of the cylinder gives two equations for the unknown constants:

$$\sinh \xi_0 + A_* \exp(-\xi_0) = B_* \sinh \xi_0$$

and [3.46]

$$\gamma_1[\cosh \xi_0 - A_* \exp(-\xi_0)] = \gamma_2 B_* \cosh \xi_0.$$

The solution is

$$B_* = \frac{1 + \tanh \xi_0}{\tanh \xi_0 + (\gamma_2/\gamma_1)}$$

and [3.47]

$$A_* = \frac{[1 - (\gamma_2/\gamma_1)] \sinh \xi_0 \exp(\xi_0)}{\tanh \xi_0 + (\gamma_2/\gamma_1)}.$$

Comparison of Eqs. [3.34] and [3.47] shows that they differ only slightly from each other, and similar to Eqs. [3.35] and [3.36], we have for the field outside the cylinder

$$E_{1\xi} = \frac{E_0}{(\cosh^2\xi - \cos^2\eta)^{1/2}} [\cosh \xi - A_* \exp(-\xi)] \sin \eta$$

and [3.48]

$$E_{1\eta} = \frac{E_0}{(\cosh^2\xi - \cos^2\eta)^{1/2}} [\sinh \xi + A_* \exp(-\xi)] \cos \eta.$$

Inside the cylinder, the field is again constant:

$$E_{2z} = \frac{1 + \tanh \xi_0}{(\gamma_2/\gamma_1) + \tanh \xi_0} E_0.$$

For illustration, consider one case when $\gamma_2/\gamma_1 \gg b/a$. It follows from Eqs. [3.39] that

$$A_* = 0 \quad \text{and} \quad B_* = \frac{\gamma_1}{\gamma_2} E_0. \qquad [3.49]$$

Thus, in the limit when the cylinder is almost transformed into the layer with thickness $2b$, the secondary field outside vanishes, but the total field inside remains uniform and directed along the z-axis. This result can be easily explained, since induced charges are distributed uniformly at two parallel planes and their density differs in sign only. The charges on each plane create uniform fields that differ only in direction and cancel each other outside the cylinder. Inside the cylinder, the fields of the charges on the two planes have the same direction, which is along the primary field if $\gamma_2/\gamma_1 < 1$ or opposite when $\gamma_2/\gamma_1 > 1$. Finally, notice that the solution of the problem for a uniform primary field oriented in an arbitrary direction with respect to the cylinder (Fig. 3.4C) can be represented as a sum of fields along the two axes, using the formulas derived in this section. In the next example, we focus on the behavior of the field in the transition from a 3D model of an inhomogeneity to a 2D cylinder.

3.4. THE SPHEROID IN A UNIFORM ELECTRIC FIELD

3.4.1. Influence of Surface Charges

In this section, we construct the solution for a general ellipsoid of revolution (spheroid) with semiaxes a and b and conductivity γ_2 located in a uniform medium with conductivity γ_1. We first consider some general features of the solution. Assume again that the primary field $\mathbf{E}_0$ is uniform and directed along the major axis a (Fig. 3.5A). As we know, charge will appear on the ellipsoid surface with density

$$\sigma = 2\varepsilon_0 \frac{\gamma_1 - \gamma_2}{\gamma_1 + \gamma_2} E_n^{\text{av}},$$

where $E_n^{\text{av}}(p)$ is the average of the normal component of the electric field on the internal and external sides of the surface at the point p. As the major semiaxis a increases, the charge becomes concentrated at is ends, $\pm a$. The field of these charges becomes smaller with increasing length $2a$ and, when observations are made close to the origin, the total field is practically the same as the primary field $\mathbf{E}_0$. In other words, a marked elongation of the inhomogeneity in the direction of the primary field does not significantly

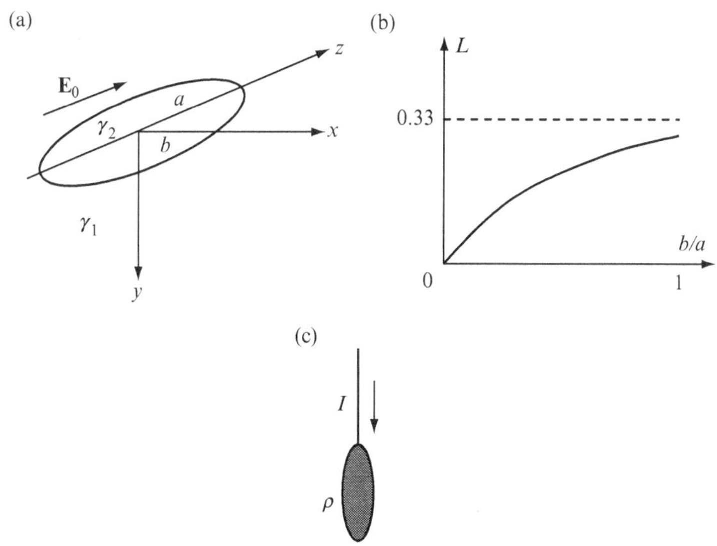

Figure 3.5 (A) Conducting spheroid of revolution in a uniform field directed along the major axis. (B) Function L characterizing the effects of charge on the spheroid surface. (C) Model of a spheroidal electrode for calculating grounding resistance.

distort the electric field if observations are made near the origin at a distance that is much smaller than the length of the semiaxis, a. As the ellipsoid elongates indefinitely, it approaches the model of an infinitely long cylinder. However, the applicability of such a model not only depends on the linear dimensions of the inhomogeneity but also on the ratio of conductivities. Let us consider this question in detail.

3.4.2. The Potential of the Secondary Field

To construct the solution for the spheroid model, we introduce a prolate spheroidal system of coordinates (ξ, ϕ, η) as in Fig. 3.5A, related to the cylindrical system by

$$z = c\xi\eta \quad \text{and} \quad r = c\left[(1-\eta^2)(\xi^2-1)\right]^{1/2}, \qquad [3.50]$$

where $c = \left(a^2 - b^2\right)^{1/2}$. The surface of the spheroid with semiaxes a and b is a coordinate surface ξ_0 = constant, with

$$a = c\xi_0, \quad b = c(\xi_0^2 - 1)^{1/2}. \qquad [3.51]$$

The metric coefficients have the form

$$h_\xi = c\left(\frac{\xi^2-\eta^2}{\xi^2-1}\right)^{1/2}, \quad h_\eta = c\left(\frac{\xi^2-\eta^2}{1-\eta^2}\right)^{1/2}$$

and

$$h_\phi = r.$$

The potential inside and outside a spheroid is again written as

$$U_1 = U_0 + U_s \text{ if } \xi \geq \xi_0, \qquad U_1 = U_2, \text{ if } \xi \leq \xi_0,$$

where the potentials have to satisfy Laplace's equation and boundary conditions at the interface,

$$U_1 = U_2 \quad \text{and} \quad \gamma_1 \frac{\partial U_1}{\partial \xi} = \gamma_2 \frac{\partial U_2}{\partial \xi}, \quad \text{if } \xi = \xi_0, \qquad [3.52]$$

and at infinity

$$U_1 \to U_0. \qquad [3.53]$$

In spheroidal coordinates, a solution of Laplace's equation can be represented as a product of Legendre's functions of the first and second kinds. For the potential of the primary field, we have

$$E_0 = -\frac{\partial U_0}{\partial z} \quad \text{or} \quad U_0 = -E_0 z,$$

since the constant term can be ignored. From Eqs. [3.50], we have

$$U_0 = -E_0 c \xi \eta = -E_0 c P_1(\xi) P_1(\eta), \qquad [3.54]$$

where $P_1(x) = x$ is Legendre's function of the first kind and order one. We make two assumptions: first, that the electric field remains uniform inside the spheroid and directed along the z-axis, and second, that outside the secondary field is given by the product of Legendre's functions of both kinds, but only first order. Then,

$$\begin{aligned} U_1 &= -E_0 c[P_1(\xi) + A Q_1(\xi)] P_1(\eta) \quad \text{and} \\ U_2 &= -E_0 c B P_1(\xi) P_1(\eta). \end{aligned} \qquad [3.55]$$

Here, $Q_1(\xi)$ is Legendre's function of the first order and second kind:

$$Q_1(\xi) = \frac{\xi}{2} \ln \frac{\xi + 1}{\xi - 1} - 1. \qquad [3.56]$$

Now, we attempt to satisfy all boundary conditions with this trial solution. Substitution of Eqs. [3.55] into Eqs. [3.52] gives the system of equations to determine A and B:

$$\xi_0 + A Q_1(\xi_0) = B_1 \xi_0 \quad \text{and} \quad \gamma_1 [1 + A O_1'(\xi_0)] = \gamma_2 B, \qquad [3.57]$$

since

$$P_1(\xi_0) = \xi_0, \quad P_1'(\xi_0) = 1, \quad Q_1'(\xi_0) = \frac{dQ_1}{d\xi}.$$

Equations [3.57] give

$$A = \frac{(\gamma_1 - \gamma_2)\xi_0}{\gamma_2 Q_1(\xi_0) - \xi_0 \gamma_1 Q_1'(\xi_0)}, \quad B = \frac{\gamma_1 [Q_1(\xi_0) - \xi_0 Q_1'(\xi_0)]}{\gamma_2 Q_1(\xi_0) - \xi_0 \gamma_1 Q_1'(\xi_0)}. \qquad [3.58]$$

More explicitly, using the derivative of $Q_1(\xi)$,

$$Q_1'(\xi_0) = \frac{1}{2} \ln \frac{\xi_0 + 1}{\xi_0 - 1} - \frac{\xi_0}{\xi_0^2 - 1}, \qquad [3.59]$$

we have

$$A = \frac{\left(\frac{\gamma_2}{\gamma_1} - 1\right)\xi_0(\xi_0^2 - 1)}{1 + \left(\frac{\gamma_2}{\gamma_1} - 1\right)(\xi_0^2 - 1)\left[\frac{\xi_0}{2}\ln\frac{\xi_0 + 1}{\xi_0 - 1} - 1\right]} \qquad [3.60]$$

and

$$B = \frac{1}{1 + \left(\frac{\gamma_2}{\gamma_1} - 1\right)(\xi_0^2 - 1)\left[\frac{\xi_0}{2}\ln\frac{\xi_0 + 1}{\xi_0 - 1} - 1\right]}.$$

Since we are interested in the transition from the spheroid to the circular cylinder, when $a/b \to \infty$, consider only the field inside the spheroid, which can be written as

$$E_2 = BE_0 = \frac{E_0}{1 + [(\gamma_2/\gamma_1) - 1]L}, \qquad [3.61]$$

where

$$L = (\xi_0^2 - 1)\left[\frac{\xi_0}{2}\ln\frac{\xi_0 + 1}{\xi_0 - 1} - 1\right]. \qquad [3.62]$$

From Eqs. [3.51], we have

$$\xi_0 = \frac{1}{e}, \; e = \left(1 - \frac{b^2}{a^2}\right)^{1/2}. \qquad [3.63]$$

The function L can be expressed in terms of e as

$$L = \frac{1 - e^2}{2e^3}\left(\ln\frac{1 + e}{1 - e} - 2e\right), \qquad [3.64]$$

and it characterizes an influence of surface charges on the field inside a spheroid. Again, consider two limiting cases. If $a \to b$, then the parameter e tends to zero and

$$\ln\frac{1 + e}{1 - e} \to 2e - \frac{2}{3}e^3,$$

and we arrive at the previous expression for the spherical body, $L \to 1/3$. In the opposite case when the spheroid is strongly elongated along the z-axis ($e \to 1$), Eq. [3.64] gives

$$L \approx -\frac{1-e^2}{2} \ln(1-e).$$

Since

$$1-e^2 = \frac{b^2}{a^2} \quad \text{and} \quad \ln\left[1-\left(1-\frac{b^2}{a^2}\right)^{1/2}\right] \approx \ln\frac{1}{2}\frac{b^2}{a^2},$$

we have

$$L \approx \frac{b^2}{a^2} \ln\frac{a}{b} \ll 1 \quad \text{if } e \to 1. \qquad [3.65]$$

It follows from Eq. [3.61] that, as the major axis a increases ($L \to 0$), the electric field approaches the primary field $\mathbf{E}_0$. However, as the conductivity contrast γ_2/γ_1 increases, it requires increasingly more elongate spheroids to approach the limit of an infinitely long cylinder. Of course, when the surrounding medium is insulating, the secondary field is equal to and opposite the primary field inside the spheroid independently of its linear dimensions; that is, we again see a phenomenon of electrostatic induction. The function L is shown in Fig. 3.5B. Taking into account Eq. [3.61], we have for the current density in the spheroid

$$j = \frac{\gamma_2 E_0}{1 + [(\gamma_2/\gamma_1) - 1]L}. \qquad [3.66]$$

For relatively small ratios of conductivities and $L \ll 1$, the current density increases in direct proportion to γ_2 as the spheroid conductivity increases, just as in the case of an infinitely long cylindrical conductor. However, with further increase of the conductivity γ_2, the secondary electric field increases, but the total field becomes smaller. When $(\gamma_i/\gamma_e)L \gg 1$, the current density is practically independent of the conductivity of the spheroid:

$$j = \frac{\gamma_1 E_0}{L} \gg \gamma_1 E_0. \qquad [3.67]$$

Note that if conductivity of the surrounding medium is relatively large such that $\gamma_2/\gamma_1 \ll 1$, for sufficiently small values of a/b, the current density is proportional conductivity of the spheroid, too.

3.5. GROUNDING RESISTANCE OF A STRONGLY ELONGATED SPHEROID

The spheroid model can be used to derive an expression for the grounding resistance of a thin rod, which is often used as an electrode in surface and borehole electric methods. Suppose that an elongated spheroid is surrounded by a uniform medium with resistivity ρ and serves as an electrode through which the current I goes into a medium (Fig. 3.5C). Since the electrode has high conductivity and relatively small dimensions, we can assume that the potential U_s is constant on its surface. The grounding resistance is equal to

$$R = \frac{U_s}{I}, \tag{3.68}$$

since at infinity, the potential vanishes. To determine the potential, we look for a function $U(\xi, \eta)$ that satisfies Laplace's equation; it is constant on the electrode surface and tends to zero at infinity. In addition, in accordance with Ohm's law, we have at the equipotential surface of the electrode $(\xi = \xi_0)$,

$$I = \oint_S j_\xi \, dS = -\oint_S \gamma \frac{\partial U}{\partial l_\xi} \, dS, \tag{3.69}$$

where

$$\partial l_\xi = h_1 \partial \xi \quad \text{and} \quad dS = h_2 h_3 d\eta \, d\phi. \tag{3.70}$$

Also, the metric coefficients are

$$h_1 = c\frac{(\xi^2 - \eta^2)^{1/2}}{(\xi^2 - 1)^{1/2}}, \quad h_2 = c\frac{(\xi^2 - \eta^2)^{1/2}}{(1 - \eta^2)^{1/2}}, \quad \text{and} \quad h_3 = c(\xi^2 - 1)^{1/2}(1 - \eta^2)^{1/2}. \tag{3.71}$$

As before, we assume that the solution of Laplace's equation can be represented as a product of Legendre's functions of the first and second kind. But, the constant potential is independent on the coordinate η, and we assume that this is also true of the full solution. This suggests that the potential is described by just the zero-order harmonic $n = 0$, since $P_0(\eta) = 1$, while Legendre functions of higher order all depend on η. We thus write the potential in the form

$$U(\xi) = AQ_0(\xi), \tag{3.72}$$

where

$$Q_0(\xi) = \frac{1}{2}\ln\frac{\xi+1}{\xi-1}. \tag{3.73}$$

It is easy to show that the function $U(\xi)$ satisfies the condition at infinity. In fact, from Eq. [3.73], it follows that

$$Q_0(\xi) \to \frac{1}{\xi}.$$

Moreover, Eq. [3.50] gives

$$R = (r^2 + z^2)^{1/2} \approx c\xi, \quad \text{if } \xi \gg 1.$$

Thus, far away from the electrode, the potential decays inversely with distance. This behavior is expected since the electrode carries charge of one sign only. To determine the unknown A, we first substitute Eqs. [3.70]–[3.72] into Eq. [3.69]. This gives

$$-I\rho = A\oint_S \frac{h_2 h_3}{h_1}\frac{\partial Q_0}{\partial \xi}\, d\eta\, d\phi = 2\pi c A \int_{-1}^{+1} (\xi_0^2 - 1)\frac{\partial Q_0}{\partial \xi}\, d\eta.$$

Since

$$\frac{\partial Q_0}{\partial \xi} = -\frac{1}{\xi^2 - 1},$$

we have

$$A = \frac{\rho I}{4c\pi}. \tag{3.74}$$

This shows that the potential caused by the surface charges of the electrode is

$$U(\xi) = \frac{\rho I}{8c\pi}\ln\frac{\xi+1}{\xi-1}. \tag{3.75}$$

We have proved that the equipotential surfaces in this case are spheroidal; of course, far away from the electrode, they asymptotically approach spheres. By definition, the grounding resistance of the spheroidal electrode is (from Eq. [3.75])

$$R = \frac{U}{I} = \frac{\rho}{8c\pi} \ln\frac{\xi_0 + 1}{\xi_0 - 1}, \qquad [3.76]$$

or, using Eq. [3.63],

$$R = \frac{\rho}{8\pi ae} \ln\frac{1+e}{1-e}. \qquad [3.77]$$

As usual, it is instructive to consider two limiting cases. First, suppose that the spheroid becomes a sphere, $a \to b$. Then, e tends to zero and we arrive at the known expression of the grounding resistance of a spherical electrode,

$$R = \frac{\rho}{4\pi a}.$$

Next, consider the opposite case, when $b/a \ll 1$ As shown above, if $e \to 1$, we have

$$\ln\frac{1+e}{1-e} \approx 2\ln\frac{2a}{b} \quad \text{and} \quad R \approx \frac{\rho}{4\pi a}\ln\frac{2a}{b}. \qquad [3.78]$$

3.6. PLANE INTERFACE

3.6.1. Distribution of Surface Charge

Suppose that there is one planar interface between two uniform media with resistivities ρ_1 and ρ_2, and that the current electrode A is located in a medium with resistivity ρ_1 (Fig. 3.6A). As current I flows into the medium, the electrode acquires the surface charge

$$e_A = \varepsilon_0 \rho_1 I. \qquad [3.79]$$

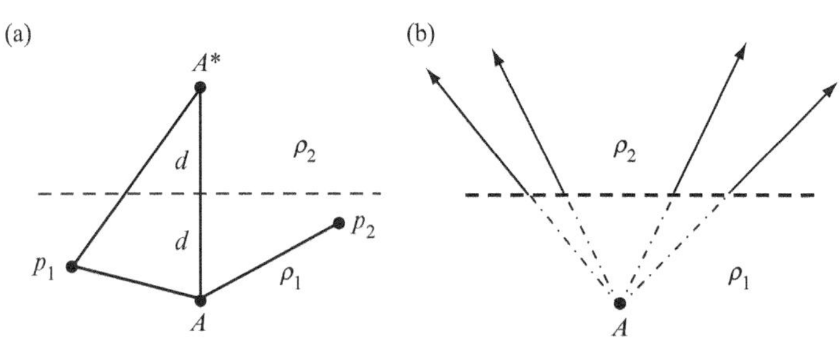

Figure 3.6 (A) Current electrode (A) in the presence of plane interface between uniform media, with mirror reflection source (A^*). (B) Electric field in the second medium.

The electric field of this charge,

$$\mathbf{E}_A(p) = \frac{e_A \mathbf{L}_{Ap}}{4\pi\varepsilon_0 \mathbf{L}_{Ap}^3}, \qquad [3.80]$$

induces charge on the interface and, as was shown in Chapter 2, its density at any point q is

$$\sigma(q) = \frac{K_{12} e_A d}{2\pi (d^2 + r^2)^{3/2}}. \qquad [3.81]$$

Here, d is the distance from the current electrode to the interface, $r = (x^2 + y^2)^{1/2}$, and

$$K_{12} = \frac{\rho_2 - \rho_1}{\rho_2 + \rho_1} = \frac{\gamma_1 - \gamma_2}{\gamma_1 + \gamma_2}. \qquad [3.82]$$

The density $\sigma(q)$ depends critically on the distance d from the electrode to the interface.

For instance, when the electrode is close to the interface, charge is mainly concentrated near the point $\sigma = 0$, and as the electrode moves away, the charge density decays only at relatively large lateral distances, $r \gg d$. The total charge e_S is, however, independent of the distance d and equal to

$$e_S = K_{12} e_A. \qquad [3.83]$$

3.6.2. Potential of Electric Field in the Presence of a Plane Interface

We now formulate the boundary-value problem for two media separated by a plane interface. Given the axial symmetry, we use cylindrical coordinates with the z-axis perpendicular to the boundary $z = d$ (Fig. 3.6A). It is assumed that the electrode A is located at the origin of the system and is so small that its field coincides with that of an elementary charge (Eq. [3.80]). As usual, we write the potential in the half-space containing the electrode as the sum of the primary and secondary fields, and in the other half-space as just one term describing the total field:

$$\begin{aligned} &U_1(r, z) = U_A + U_S, \quad \text{if } z \leq d \\ &U_2(r, z), \quad \text{if } z \geq d. \end{aligned} \qquad [3.84]$$

Here,

$$U_A = \frac{e_A}{4\pi\varepsilon_0 L_{Ap}}, \qquad [3.85]$$

and U_S is the potential of the electric field caused by the electrode charge e_S. The following conditions for the potential define the boundary-value problem:

1. At regular points,

$$\Delta U_1 = 0, \ \Delta U_2 = 0.$$

2. At the interface $z = d$,

$$U_1 = U_2, \quad \gamma_1 \frac{\partial U_1}{\partial z} = \gamma_2 \frac{\partial U_2}{\partial z}.$$

3. At infinity ($L_{Ap} \to \infty$), the potential tends to zero.

To simplify the task of finding a solution satisfying these conditions, it is appropriate to notice that the secondary potential caused by charge at the boundary is an even function with respect to the interface:

$$U_{2S}(z^*) = U_{1S}(z^*). \tag{3.86}$$

Here, z^* is the distance from the interface, and U_{1S}, U_{2S} are potentials in the lower and upper half-spaces caused by charge at the interface. This result follows from the fact that the potential of the given distribution of charge does not depend on the resistivity of a medium. To proceed, we make a key assumption that the potential U_{1S} caused by charge at the interface coincides with the potential of the elementary charge e_S (Eq. [3.83]) located at the point A^* which is the mirror reflection of the point A across the interface (Fig. 3.6A). The expression for the potential of the secondary field in the lower half-space becomes

$$U_{1S}(r, z) = \frac{e_S}{4\pi\varepsilon_0 L_{A^*p}}, \quad \text{if } z \le d,$$

and, using equality [3.86], we have in the upper half-space

$$U_{2S} = \frac{e_S}{4\pi\varepsilon_0 L_{Ap}}, \quad z \ge d.$$

These assumptions allow us to represent the total potential in the form

$$\begin{aligned} U_1(r, z) &= \frac{e_A}{4\pi\varepsilon_0 L_{Ap}} + \frac{K_{12} e_A}{4\pi\varepsilon_0 L_{A^*p}}, \quad \text{if } z \le d \\ U_2(r, z) &= \frac{(1 + K_{12}) e_A}{4\pi\varepsilon_0 L_{Ap}}, \quad \text{if } z \ge d, \end{aligned} \tag{3.87}$$

where

$$L_{Ap} = (r^2 + z^2)^{1/2}, \quad L_{A^*p} = [r^2 + (z - 2d)^2]^{1/2},$$

and K_{12} is given by Eq. [3.82]. It is clear that these functions satisfy Laplace's equation at regular points and the condition at infinity. Let us check the boundary conditions at $z = d$, where $L_{Ap} = L_{A^*p}$. First, it is clear that the potential is continuous; in fact, Eqs. [3.87] give directly

$$\frac{e_A}{L_{Ap}} + \frac{K_{12}e_A}{L_{Ap}} = \frac{(1 + K_{12})e_A}{L_{Ap}},$$

since at points on the interface $L_{Ap} = L_{A^*p}$. Taking derivatives with respect to z, we have for the condition of continuity of the normal component of the current:

$$\gamma_1\left(\frac{d}{L_{Ap}^3} - K_{12}\frac{d}{L_{Ap}^3}\right)e_A = \gamma_2(1 + K_{12})\frac{d}{L_{Ap}^3}, \quad \text{or}$$

$$\gamma_1(1 - K_{12}) = \gamma_2(1 + K_{12}).$$

It follows from Eq. [3.82] that this condition is also satisfied. This shows that our assumptions were correct, and Eqs. [3.87] gives the potential of the electric field of an elementary charge in the presence of the plane interface. The result is the foundation of the "method of images": the charge induced at the plane boundary is assumed to generate the same field (in the first half-space) as that of a point source located at the mirror refection A^* of the true source location A. Certainly, this equivalence was not obvious, and it was necessary to show that the solution constructed this way actually satisfies the boundary-value problem. The point A^* is called a "fictitious source" because, of course, there is no real charge inside the second medium. The field in the second medium is equivalent to that of the elementary charge $(1 + K_{12})e_A$ located at the true source point A, Fig. 3.6B. This is simply the charge on the electrode, augmented by the induced charge on the interface $e_S = K_{12}e_A$. The surface charge always satisfies the condition

$$e_S \leq e_A, \tag{3.88}$$

and is positive when $\rho_2 > \rho_1$ and negative when $\rho_2 < \rho_1$.

3.6.3. Behavior of the Electric Field in the Presence of Plane Interface

According to Eqs. [3.87], the electric field has two nonzero components in cylindrical coordinates (E_r, E_z), while $E_\phi = 0$. By definition,

$$E_r = -\frac{\partial U}{\partial r}, \quad E_z = -\frac{\partial U}{\partial z},$$

or

$$E_{1r} = \frac{e_A r}{4\pi\varepsilon_0}\left\{\frac{1}{(r^2+z^2)^{3/2}} + \frac{K_{12}}{[r^2+(z-2d)^2]^{3/2}}\right\}$$
$$E_{1z} = \frac{e_A}{4\pi\varepsilon_0}\left\{\frac{z}{(r^2+z^2)^{3/2}} - \frac{(2d-z)K_{12}}{[r^2+(z-2d)^2]^{3/2}}\right\}. \qquad [3.89]$$

Also,

$$E_{2r} = \frac{e_A(1+K_{12})r}{4\pi\varepsilon_0(r^2+z^2)^{3/2}}, \quad E_{2z} = \frac{e_A(1+K_{12})z}{4\pi\varepsilon_0(r^2+z^2)^{3/2}}. \qquad [3.90]$$

First, consider the field at the interface $z=d$. By construction, the potential is continuous, which means that the tangential component of the electric field is continuous, $E_{1r}(r,d) = E_{2r}(r,d)$. It follows from the symmetry that this component is equal to zero at the point $z=d$, $r=0$ and increases with r. It reaches a maximum when $r=2^{-1/2}d$, and then decreases to zero at large lateral distances. The normal component E_z is discontinuous, because the normal component of the current density must be continuous,

$$\frac{E_{2z}(r,d)}{E_{1z}(r,d)} = \frac{\gamma_1}{\gamma_2} = \frac{\rho_2}{\rho_1}. \qquad [3.91]$$

This shows that the presence of media with different resistivities across an interface can be detected by measuring the electric field perpendicular to the interface. Since the tangential component of the electric field is continuous, the tangential current density changes abruptly:

$$\frac{j_{2r}(r,d)}{j_{ir}(r,d)} = \frac{\gamma_2}{\gamma_1}.$$

Next, we consider two limiting cases.

3.6.4. Second Medium is an Ideal Conductor: $\rho_2=0$

When the second medium is an ideal conductor, the surface charge is negative and its magnitude is equal to e_A, the charge on the electrode. We have

$$U_1(r,z) = \frac{e_A}{4\pi\varepsilon_0}\left(\frac{1}{L_{Ap}} - \frac{1}{L_{A^*p}}\right), \quad U_2(r,z) = 0.$$

The equipotential surfaces are almost spherical near the current electrode but change as

L_{Ap} increases until their shape blends with the plane interface, which is an equipotential surface at zero potential. From Eqs. [3.89] and [3.90], the electric field is

$$E_{1r} = \frac{e_A r}{4\pi\varepsilon_0}\left(\frac{1}{L_{Ap}^3} - \frac{1}{L_{A^*p}^3}\right), \quad E_{1z} = \frac{e_A}{4\pi\varepsilon_0}\left(\frac{z}{L_{Ap}^3} + \frac{2d - z}{L_{A^*p}^3}\right), \quad \mathbf{E}_2 = 0. \tag{3.92}$$

The electric field lines are directed radially near the electrode, but eventually turn to be everywhere perpendicular to the ideal conductor at $z = d$. As is seen from Eqs. [3.92], when $r \gg (2d - z)$, the vertical component E_{1z} dominates, and the field resembles that of an electric dipole with moment $p = e_A 2d$ directed along the z-axis. Note that if ρ_2 has a nonzero value, but the resistivity in the lower medium ρ_1 tends to infinity (e.g., in air), we again observe the phenomenon of electrostatic induction: a charge in free space cannot create an electric field in a conducting medium.

3.6.5. Second Medium is Insulator: $\gamma_2 = 0$

When the second medium becomes an insulator (infinite resistivity or zero conductivity), the induced surface charge equals that on the electrode and expressions for the total field becomes

$$E_{1r} = \frac{e_A r}{4\pi\varepsilon_0}\left(\frac{1}{L_{Ap}^3} + \frac{1}{L_{A^*p}^3}\right), \quad E_{1z} = \frac{e_A}{4\pi\varepsilon_0}\left(\frac{z}{L_{Ap}^3} + \frac{z - 2d}{L_{A^*p}^3}\right), \tag{3.93}$$

and

$$E_{2r} = \frac{e_A r}{2\pi\varepsilon_0 L_{Ap}^3}, \quad E_{2z} = \frac{e_A z}{2\pi\varepsilon_0 L_{Ap}^3}.$$

We see that the field in an insulator is essentially twice that of a uniform medium; this happens because of the induced surface charges. Notice that the electric field in an insulator contains information about the charge e_A and its location. It is obvious that at both sides of the interface, we have

$$E_{1z}(r,d) = 0, \quad E_{2z} = \frac{e_A d}{2\pi\varepsilon_0 (r^2 + d^2)^{3/2}}. \tag{3.94}$$

Correspondingly, charge is distributed on the surface in such a way that the vector of current density $\mathbf{j}_1(r, d)$ has only a component tangent to this boundary. It is clear that with an increase of the distance from the current electrode, the field $\mathbf{E}_1$ tends to that of an elementary charge equal to $2e_A$.

3.6.6. Current Electrode at the Interface

Returning to the general case, suppose that the current electrode is situated at the boundary. From the physical point of view, it is natural to expect that the radial component of the field will be a symmetric (even) function across the interface, while the z-component will be asymmetric (odd). In a fact, Eqs. [3.89] and [3.90] have the same functional form in both media, but the radial component involves only the square of the z-coordinate, whereas the normal component involves z directly, which changes sign across the interface,

$$E_r = \frac{e_A r}{4\pi\varepsilon_0 (r^2 + z^2)^{3/2}}(1 + K_{12}), \quad E_z = \frac{e_A(1 + K_{12})z}{4\pi\varepsilon_0 (r^2 + z^2)^{3/2}}. \qquad [3.95]$$

In particular, if an observation point is located at the interface, there is only a radial component E_r because all of the charge

$$e = e_A(1 + K_{12}) \qquad [3.96]$$

is situated at the interface (on the electrode). We will use these formulas later to study the behavior of "apparent resistivity" curves near a vertical contact between media with different resistivities. We next consider several more complex models where it is relatively simple to take into account the effect of the earth–air interface.

3.7. VERTICAL CONTACT BENEATH THE EARTH'S SURFACE

3.7.1. Formulation of Boundary-Value Problem

We now consider a model often used to study the effect of geologic faults on electric surveys of the earth's subsurface. Suppose that a vertical plane (the contact plane or "fault") divides the medium beneath the earth's surface into two parts with uniform resistivities ρ_1 and ρ_2. Assume that a current electrode with charge

$$e_A = \varepsilon_0 \rho_1 I$$

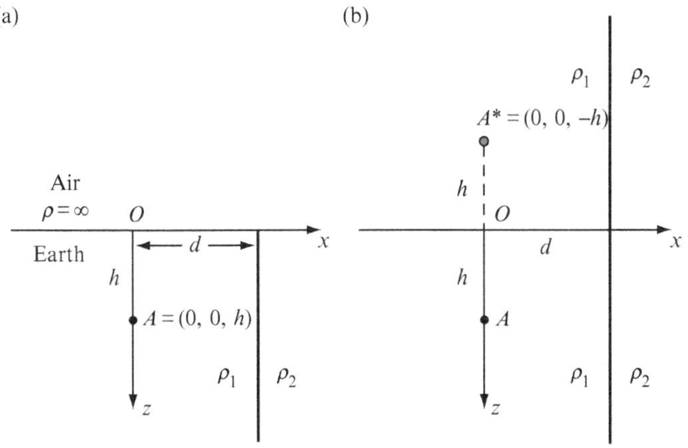

Figure 3.7 (A) Vertical contact or "fault" plane between two uniform media beneath the earth. (B) "Equivalent" model obtained by mirror reflection of medium and charge across the earth–air interface.

is located at the point $A(0, 0, h)$ of the first medium, where h is depth of location of the electrode and d is the distance of the electrode from the fault (Fig. 3.7A). To determine the field, we use Cartesian coordinates (x, y, z) with the origin O at the earth's surface. We assume that the region above the earth's surface is air, which effectively has zero electrical conductivity (infinite resistivity). Since measurements of the electric field are performed either at or beneath the earth's surface, we can focus on the field in the region $z \geq 0$. It is clear that the potential should satisfy the following conditions:

1. At regular points,

$$\Delta U = 0.$$

2. Near the current electrode A, the potential is defined by its charge

$$U_1 = U_A + U_s \to U_A, \quad \text{if } L_{Ap} \to 0.$$

3. At infinity,

$$U_1 \to 0, \; U_2 \to 0, \quad \text{if } L_{Ap} \to \infty.$$

4. At points of the vertical contact,

$$U_1 = U_2, \quad \gamma_1 \frac{\partial U_1}{\partial x} = \gamma_2 \frac{\partial U_2}{\partial x}, \quad \text{if } x = d.$$

5. At points of the earth's surface,

$$\frac{\partial U_1}{\partial z} = \frac{\partial U_2}{\partial z} = 0, \quad \text{if } z = 0, \tag{3.97}$$

since the normal component of the current density is equal to zero at the earth–air interface. Certainly, this boundary-value problem is more complex than the previous ones, as it involves two separate interfaces. Before describing an elegant method for its solution, note that we were able to formulate appropriate boundary conditions at each surface surrounding the conducting media, and so it is possible to find the field in the earth without solving for the field in air.

3.7.2. Potential in a Conducting Medium

To determine the field, we imagine "mirroring" the earth model across the earth–air interface (Fig. 3.7B) and assume that the charge e_A on the electrode is mirrored by charge at the point $A_0(0, 0, -h)$. Thus, we obtain a new model of a space with one vertical boundary ($x = d$) dividing the media with resistivities ρ_1 and ρ_2, and two equal charges located at points A and A_0. We have already solved for the electric field in such model with a single charge. Applying the principle of superposition, we have for the potential caused by two equal charges:

$$\begin{aligned} U_1 &= \frac{e_A}{4\pi\varepsilon_0}\left[\frac{1}{L_{Ap}} + \frac{K_{12}}{L_{A^*p}} + \frac{1}{L_{A_0p}} + \frac{K_{12}}{L_{A_0^*p}}\right], \quad \text{if } x \le d, \\ U_2 &= \frac{e_A(1 + K_{12})}{4\pi\varepsilon_0}\left[\frac{1}{L_{Ap}} + \frac{1}{L_{A_0p}}\right], \quad \text{if } x \ge d. \end{aligned} \tag{3.98}$$

Here,

$$L_{Ap} = [x^2 + y^2 + (z - h)^2]^{1/2}, \quad L_{A_0p} = [x^2 + y^2 + (z + h)^2]^{1/2}$$

and

$$L_{A^*p} = [(x - 2d)^2 + y^2 + (z - h)^2]^{1/2}, \quad L_{A_0^*p} = [(x - 2d)^2 + y^2 + (z + h)^2]^{1/2}. \tag{3.99}$$

Equations [3.98] are assumptions only. We now show that the potential given by these equations solves the original boundary-value problem. The functions U_1 and U_2 obviously satisfy Laplace's equation at regular points; the boundary conditions also hold at the vertical contact; and the conditions at infinity and near the current electrode A are satisfied. We now demonstrate that the condition of no current flow across the earth's surface ($z=0$) is also met. In fact, the derivatives with respect to z give

$$\frac{\partial U_1}{\partial z}=-\frac{e_A}{4\pi\varepsilon_0}\left[\frac{z-h}{L_{Ap}^3}+\frac{K_{12}(z-h)}{L_{A^*p}^3}+\frac{z+h}{L_{A_0p}^3}+\frac{K_{12}(z+h)}{L_{A_0^*p}^3}\right]$$

and [3.100]

$$\frac{\partial U_2}{\partial z}=-\frac{e_A(1+K_{12})}{4\pi\varepsilon_0}\left[\frac{z-h}{L_{Ap}^3}+\frac{z+h}{L_{A_0p}^3}\right].$$

Clearly, in each case, $\partial U/\partial z=0$ at $z=0$. Therefore, all conditions of the boundary-value problem are satisfied and, by the uniqueness theorem, the functions U_1 and U_2 describe the potential in the earth in the presence of the vertical contact. Of course, Eqs. [3.98] cannot be applied above the earth surface since the boundary-value problem was formulated only for points below this surface. Summarizing, we see that, because the lower part of the medium together with the given charges is reflected with respect to the earth's surface into the upper space, a symmetrical distribution of resistivity and charges is created such that the earth's surface, $\partial U/\partial z=0$, is always met. It is important that the potential is known in the new model of the medium, and it satisfies all conditions of the boundary value problem in the conducting part of the original model of the medium which includes the earth's surface.

3.7.3. Potential and Horizontal Component of Electric Field at the Earth's Surface

Now, suppose that an observation point p is located at the earth's surface, $z=0$. From Eqs. [3.98], we have

$$U_1(x,y,0)=\frac{e_A}{2\pi\varepsilon_0}\left(\frac{1}{L_{Ap}}+\frac{K_{12}}{L_{A^*p}}\right),\quad U_2(x,y,0)=\frac{e_A(1+K_{12})}{2\pi\varepsilon_0 L_{Ap}}. \quad [3.101]$$

This shows that the potential is twice as large as in the case when the earth's surface is absent. The same happens for the horizontal components of the electric field. For example, the component E_x at the earth's surface when $y=0$ is

$$E_{1x}(x,0,0) = \frac{e_A}{2\pi\varepsilon_0}\left[\frac{x}{(x^2+h^2)^{3/2}} + \frac{x-2d}{[(x-2d)^2+h^2]^{3/2}}\right]$$
$$E_{2x}(x,0,0) = \frac{e_A(1+K_{12})x}{2\pi\varepsilon_0(x^2+h^2)^{3/2}}. \qquad [3.102]$$

In particular, when the current electrode and observation point are located at the earth's surface, the potential and the component E_x are

$$U_1(x,0,0) = \frac{e_A}{2\pi\varepsilon_0}\left(\frac{1}{x} + \frac{K_{12}}{2d-x}\right), \quad U_2(x,0,0) = \frac{e_A(1+K_{12})}{2\pi\varepsilon_0 x}, \qquad [3.103]$$

$$E_{1x}(x,0,0) = \frac{e_A}{2\pi\varepsilon_0}\left[\frac{1}{x^2} - \frac{K_{12}}{(2d-x)^2}\right],$$
$$E_{2x}(x,0,0) = \frac{e_A(1+K_{12})}{2\pi\varepsilon_0 x^2}. \qquad [3.104]$$

Note that since the distance is positive, we have to represent the denominator in the first equation of the set [3.103] as $2d-x$. Equations [3.103] and [3.104] will allow us to describe the behavior of apparent resistivity curves near a vertical contact for different types of electrode arrays (Chapter 4).

3.7.4. The Potential and Horizontal Electric Field Near a Vertical Contact Plane

It is interesting to examine the fields when the separation between the current electrode and observation point remains constant but the distance to the vertical plane changes (Fig. 3.8A and B). Three distinct cases are shown in Fig. 3.8C. First, recall that the secondary field is caused by charges that appear at the vertical contact and at the earth–air interface. If the second medium is more resistive, $K_{12}>0$ and surface charges e_S have the same sign as the electrode charge (e.g., positive). When the distance from the contact is very large compared to the separation between the electrode and the observation point, $d \gg x$, the influence of charge induced at the contact is negligible,

$$U_1(x,0,0) \to \frac{e_A}{2\pi\varepsilon_0 x}, \quad E_{1x}(x,0,0) \to \frac{e_A}{2\pi\varepsilon_0 x^2},$$

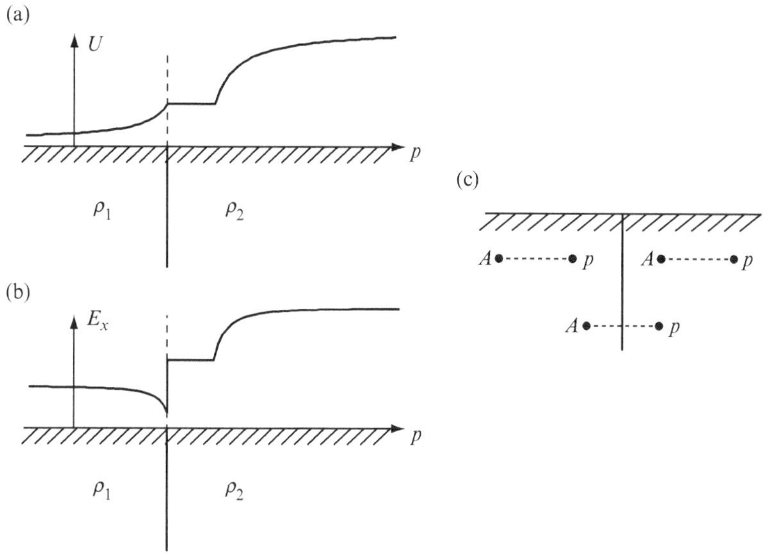

Figure 3.8 (A) Potential and (B) electric field as a two-electrode array crosses a vertical plane intersecting the earth–air interface. (C) Three arrangements of the current electrode and observation point with respect to the contact.

and the field tends to that in a uniform half-space with resistivity ρ_1. Near the contact, the influence of its charge increases and, therefore, the potential U_1 is increased while the electric field E_{1x} is decreased compared to the field in a homogeneous medium. This happens because the induced charge on the contact plane is positive when $\rho_2 > \rho_1$, and its electric field opposes the field of the electrode. If the point of observation p is located close to the contact, $x = d$, we have

$$U_1(d,0,0) = \frac{e_A(1+K_{12})}{2\pi\varepsilon_0 d}, \quad E_{1x}(d,0,0) = \frac{e_A(1-K_{12})}{2\pi\varepsilon_0 d^2}. \qquad [3.105]$$

Next, suppose that the point p is located on the opposite side of the contact, while $x = d$. From Eqs. [3.103] and [3.104], we have

$$U_2(d,0,0) = \frac{e_A(1+K_{12})}{2\pi\varepsilon_0 d}, \quad E_{2x}(d,0,0) = \frac{e_A(1+K_{12})}{2\pi\varepsilon_0 d^2}. \qquad [3.106]$$

Since the charges on the electrode and on the contact create an electric field in the same direction, the horizontal electric field E_{2x} exceeds that in a uniform half-space with resistivity ρ_1. From Eqs. [3.105] and

[3.106], we see again the component of the field normal to the vertical contact is discontinuous,

$$\frac{E_{2x}}{E_{1x}} = \frac{1+K_{12}}{1-K_{12}} = \frac{\rho_2}{\rho_1} = \frac{\gamma_1}{\gamma_2}. \qquad [3.107]$$

Thus, passage of the observation point over the contact, while the electrode remains fixed on one side, shows a sudden change of the component E_x (an increase if the second medium is more resistive). Note also that the potential and electric field stay constant as long as the electrode and observation point remain on opposite sides of the contact and their separation is constant. This behavior can be qualitatively understood as follows. The total charge induced on this surface is constant, but as the current electrode approaches the boundary, the charge density increases near the electrode, and this leads to an increase of the field in this region. At the same time, the observation electrode moves farther from the boundary where the field decreases. As the equations show, these two effects cancel and the field remains constant. Finally, consider the field after the current electrode crosses the contact. Since it is now surrounded by the second medium, its charge is equal to

$$2\varepsilon_0\rho_2 I = \frac{\rho_2}{\rho_1} e_A \quad \text{and} \quad e_A = 2\varepsilon_0\rho_1 I,$$

and as follows from Eqs. [3.103],

$$U_2(x,0,0) = \frac{\rho_2 e_A}{2\pi\varepsilon_0\rho_1}\left(\frac{1}{x} + \frac{K_{21}}{2d+x}\right) = \frac{\rho_2 e_A}{2\pi\varepsilon_0\rho_1}\left(\frac{1}{x} - \frac{K_{12}}{2d+x}\right)$$

$$E_{2x}(x,0,0) = \frac{\rho_2 e_A}{2\pi\varepsilon_0\rho_1}\left[\frac{1}{x^2} - \frac{K_{12}}{(2d+x)^2}\right]. \qquad [3.108]$$

At the moment when the current electrode intersects the contact ($d=0$), we have

$$U_2(x,0,0) = \frac{\rho_2 e_A}{\rho_1 2\pi\varepsilon_0 x}(1-K_{12}), \quad E_{2x}(x,0,0) = \frac{\rho_2 e_A(1-K_{12})}{\rho_1 2\pi\varepsilon_0 x^2}. \qquad [3.109]$$

Comparing with Eqs. [3.105] and [3.109] and recalling that

$$1 + K_{12} = \frac{\rho_2}{\rho_1}(1-K_{12}),$$

we see that the movement of the current electrode over the contact causes a continuous change of the potential and electric field. With further increase of the distance from the contact, its influence becomes smaller and the field gradually approaches that in a uniform half-space with the resistivity ρ_2. It is useful to note that when the current electrode crosses the contact, the surface charges becomes negative ($\rho_2 > \rho_1$). Thus, its electric field is directed opposite to that of the current electrode and tends to decrease the total field. At the same time, as soon as the current electrode crosses the contact and is situated in the second medium, its charge becomes larger by the factor ρ_2/ρ_1. These two effects, the increase of the electrode charge and the change of sign of the surface charge, cancel and the field stays continuous. Similar arguments apply to the field behavior: if the second medium is more conductive than the first ($\rho_2 < \rho_1$).

3.8. METHOD OF MIRROR REFLECTION

In deriving formulas for the potential and the electric field, we used the "method of images" or "mirror reflection." It relies on replacing the original configuration having an infinite plane boundary with a new one in which the model and known charges are mirrored symmetrically with respect to the boundary. It is, in fact, the symmetrical distribution of charge that causes the derivative of the potential normal to the boundary to vanish, thus enforcing the condition that no current flows across the boundary. If the field in this new model can be computed analytically, then this procedure amounts to a method for calculating the field (as above, in the case of the vertical contact). But even if the field cannot be written explicitly in the new model, replacement of the original model by a new one can still can be worthwhile, for example, if it allows the use of different methods to calculate the field or provides physical insight into the behavior of the solution. Let us consider several more examples.

Example 1

The simplest model is a current electrode at the point A with charge e_A located beneath the earth's surface. The electrode charge induces charge on the earth–air interface, whose total amount is also equal to e_A. Then, considering a uniform medium and placing the fictitious charge e_A at the mirror reflection point A_0, we proved that the field caused by these two charges in the conducting part of the original model obeys all conditions of the boundary-value problem. This is the simplest example, in which the method of images allows one to determine the field more quickly than any other method. In spite of its simplicity, this example illustrates one feature that is seen in all cases: namely, every charge inside the conducting medium creates exactly the same charge on the earth surface.

Example 2

If the contact between two media beneath the earth's surface is inclined, the method of mirror reflection shows that such a model is equivalent to that of a wedge (Fig. 3.9A and B). The solution for the boundary-value problem of a conducting wedge in the presence of a current electrode can be represented in the form of integrals. In this case, application of the method of the images avoids the use of purely numerical methods, such integral equation or finite-difference methods, to solve for the field.

Example 3

The value of the method becomes evident in trying to determine the field of a point charge in the presence of a vertical layer intersecting the earth's

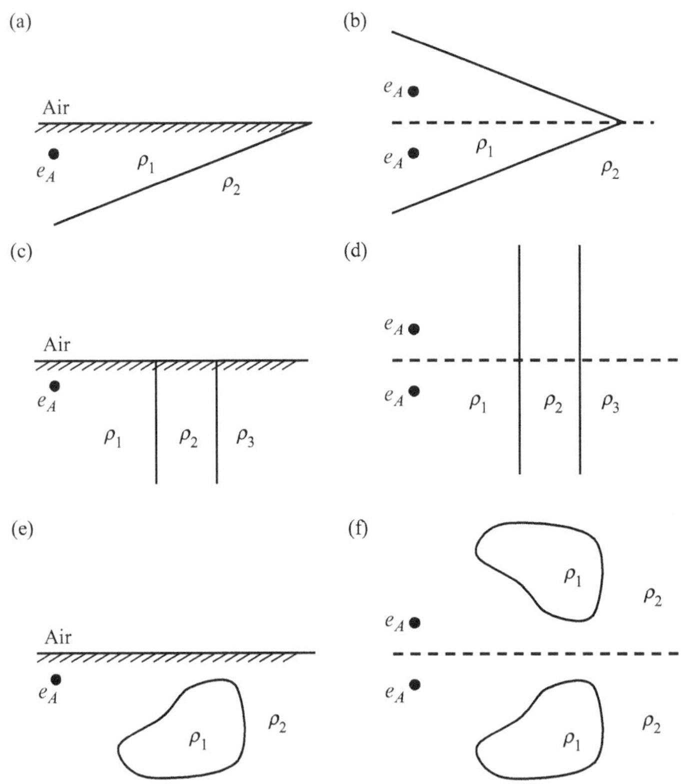

Figure 3.9 Models and mirror-reflection equivalents: (A, B) dipping contact and wedge; (C, D) vertical layer; and (E, F) inhomogeneity beneath the earth's surface.

surface (Fig. 3.9C and D). Mirror reflection of the model with respect to the surface produces a much simpler model with just two parallel plane interfaces. We will show in the next chapter that a solution of this boundary problem for the electric field and its potential can be constructed from relatively simple integrals, whose evaluation has become a standard numerical procedure.

Example 4

Suppose that an arbitrary inhomogeneity is located beneath the earth's surface. The method of mirror reflection shows that the field in a half-space (earth) bounded by an insulator (air) is the same as that caused by two identical inhomogeneities in a whole space (Fig. 3.9E and F). Numerical methods are generally needed to compute the field in this example. At the same time, there are several exceptions which are useful for understanding the influence of inhomogeneities located near the earth's surface, and two of them are considered here.

Case 1: Hemispherical body

If a hemispherical body intersects the earth–air interface along the body's equator, then its mirror refection gives a new model with a full sphere located in a uniform medium (Fig. 3.10A and B). The boundary-value problem for the sphere in a uniform medium was solved earlier for the case when the primary field is uniform over the sphere. In the case where the source is a point electrode e_A, the solution can be represented in the form of an infinite series. However, if the source is located far from the sphere,

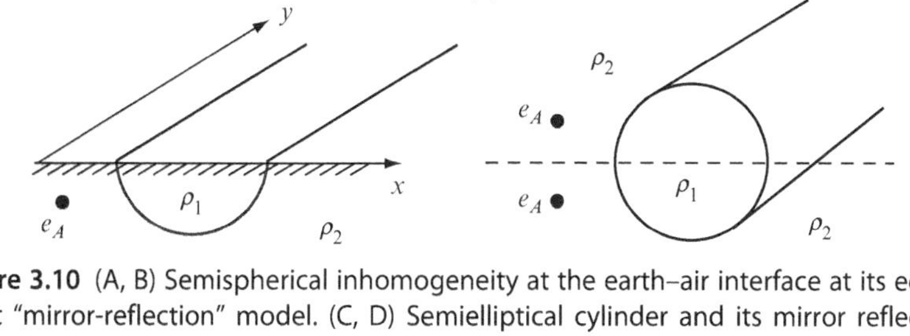

Figure 3.10 (A, B) Semispherical inhomogeneity at the earth–air interface at its equivalent "mirror-reflection" model. (C, D) Semielliptical cylinder and its mirror reflection model.

at a distance two to three times its radius, the primary field is approximately uniform inside the sphere and the solution derived earlier is a good approximation: The total field inside the sphere is also uniform, and the secondary field outside behaves as that of the electric dipole: that is, inside we have

$$E_2 \approx \frac{3\gamma_1}{2\gamma_1 + \gamma_2} \frac{e_A}{2\pi\varepsilon_0 d^2}, \qquad [3.110]$$

where d is the distance between the current electrode and the center of the sphere, while outside we have

$$E_x = \frac{e_A}{2\pi\varepsilon_0 d^2}\left(1 + 2\frac{\gamma_2 - \gamma_1}{\gamma_2 + 2\gamma_1}\frac{a^3}{x^3}\right), \qquad [3.111]$$

where x is the distance between the sphere center and an observation point. To analyze the field in this model, suppose that the observation point moves along the x-axis and the sphere is more resistive than the background, $\rho_2 > \rho_1$. In this case, positive charge appears on the back side of the hemisphere and an equal amount of negative charge arises on the front side. Since the x-component of the current density must be continuous along the earth's surface (where it is normal to the hemisphere), the horizontal component of the electric field experiences a discontinuity at the boundary of the hemisphere, and then, remains constant inside the body. When the point of observation crosses the front side of the hemispherical surface, there is again a sudden change of the field. As the observation point moves away from the sphere, the field gradually approaches the primary field caused by the charge on the current electrode. Inside a resistive inhomogeneity, the ratio E_x/E_{0x} of the total field to the primary field can be as high as 1.5, whereas in a conductive body the total field decreases (because the secondary field opposes the primary field). In this case, the influence of the inhomogeneity can be very large and, as the body conductivity increases, the ratio E_x/E_{0x} tends to zero. Since the total induced charge is equal to zero (the body stays neutral), its field at large distances decays as that of an electric dipole. Correspondingly, the ratio between the secondary and primary fields becomes relatively small as the observation point moves far away from the body.

Case 2: Semielliptical cylinder

The method of mirror reflection can also handle the case of a semielliptical cylinder that intersects the earth–air interface perpendicular to its axis. Mirror reflection across the interface gives a full cylinder in a uniform medium and allows use of the results derived earlier. In particular, if the source electrode is far away, so that its field is approximately uniform over the inhomogeneity, the total field inside the cylinder is constant:

$$E_{2x} = \frac{1 + (b/a)}{1 + (\gamma_2 b/\gamma_1 a)} E_{0x}, \qquad [3.112]$$

where a and b are semiaxes of the cylinder. Note again that as the ratio a/b increases (b/a decreases), the influence of the inhomogeneity becomes smaller and $E_{2x} \rightarrow E_{0x}$. This means that the electric field in the surrounding medium near the interface is

$$E_{1x} \approx \frac{\gamma_2}{\gamma_1} E_{0x}. \qquad [3.113]$$

If measurements are performed along a line parallel to the cylinder axis (e.g., along the y-axis in Fig. 3.10C), then as distance from the electrode increases, the relative contribution of the secondary field becomes smaller. The explanation of this behavior is similar to that for the hemisphere. At large distances from the current electrode, the positive and negative surface charges induced on the cylinder look like an electric dipole, whose field decays more rapidly than the primary field. It is important to point out that in all of these examples the method of mirror reflection does not give the field in the upper half-space.

3.9. DISTRIBUTION OF CHARGES IN A LAYERED MEDIUM AND ASYMPTOTIC FIELD BEHAVIOR

In the next chapters, we will describe different methods of electrical surveying in surface and borehole geophysics, beginning from the "geometrical sounding" method proposed at the beginning of the last century by Conrad Schlumberger (who was later joined by his brother Marcel). Conrad Schlumberger correctly assumed that measuring the voltage on the earth's surface at different distances from current electrodes (also located on the earth's surface) allows determination of the distribution of resistivity with depth in the earth. The description "geometrical sounding" meant that the distance between the current and receiver electrodes was the geometrical factor controlling the depth of investigation of the survey, while the word "sounding" was borrowed from the nautical phrase for determining the depth of water. The Schlumberger method became and remains one of the standard techniques of electrical surveying. Here, we focus on its physical principles by examining the charge distribution in a layered medium with electrodes at the surface.

3.9.1. Distribution of Charge in Layered Media

First, suppose that the subsurface consists of uniform regions with different (finite) resistivities. A current electrode A is located in the innermost region enclosed by the surface S_1 (Fig. 3.11A). Since surface charges create the electric field, it is natural to expect that their distribution defines the most important features of the soundings. First, let us calculate the amount of charge appearing at each interface. The electrode charge e_A is simply

$$e_A = \varepsilon_0 \rho_1 I, \qquad [3.114]$$

where ρ_1 is the resistivity of the medium that surrounds the electrode (whose resistivity is assumed negligible). In Chapter 2, we found that the density of charges at any point q of an interface S_i between media with resistivity ρ_{i-1} and ρ_i is

$$\sigma(q) = \varepsilon_0 (E_n^i - E_n^{i-1}),$$

Ohm's law gives

$$\sigma(q) = \varepsilon_0 (\rho_i - \rho_{i-1}) j_n(q), \qquad [3.115]$$

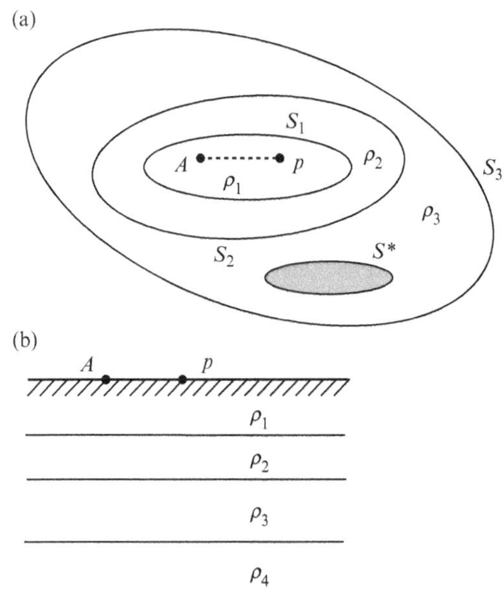

Figure 3.11 (A) Current and measurement electrode enclosed by quasi-concentric layers. The surface S^* encloses a confined inhomogeneity interior to one of the layers. (B) Current and measurement electrodes on the surface of horizontally plane-layered medium.

where q is an arbitrary point of the surface S_i. Since all interfaces are closed around the electrode, the same amount of current I passes through each of them. Thus, the total charge on any surface S_i can be calculated as

$$e_i = \oint_{S_i} \sigma(q)\, \mathrm{d}S = \varepsilon_0(\rho_i - \rho_{i-1}) \oint_{S_i} j_n\, \mathrm{d}S,$$

or

$$e_i = \varepsilon_0(\rho_i - \rho_{i-1})I. \qquad [3.116]$$

This simple expression shows that the total charge on the interface S_i is proportional to the product of the current I and the change in resistivity across the interface. Before we continue, it is appropriate to make several comments: (1) Equations [3.115] and [3.116] do not apply at the interface with an insulator where $\rho_i \to \infty$. This limiting case will be examined in detail later. (2) The sign of the charge e_i is determined by the resistivity difference crossing the layer in the direction of positive current flow. It is positive when $\rho_i > \rho_{i-1}$ and $I > 0$, but negative if $\rho_i < \rho_{i-1}$. Of course, the density of charge $\sigma(q)$ is a function of all parameters of the model, and its magnitude and sign may vary from point to point of an interface. (3) Equation [3.116] determines the total amount of charge at each boundary, but its distribution still remains unknown. Now, we write down expressions for the total charge arising at each interface S_i including the current electrode:

$$\begin{aligned} e_1 &= e_A = \varepsilon_0\rho_1 I, \\ e_2 &= \varepsilon_0(\rho_2 - \rho_1)I, \\ e_3 &= \varepsilon_0(\rho_3 - \rho_2)I, \\ &\vdots \\ e_i &= \varepsilon_0(\rho_i - \rho_{i-1})I, \\ &\vdots \\ e_n &= \varepsilon_0(\rho_n - \rho_{n-1})I. \end{aligned}$$

Here, ρ_n is the resistivity of the final region, which extends to infinity. Adding up all the charge that appears in the medium, we obtain

$$e = \sum_{i=1}^{n} e_i = \varepsilon_0\rho_n I. \qquad [3.117]$$

This remarkable equation shows that the total charge arising at all interfaces is the same as the charge that would appear on the electrode surface if it were surrounded by a uniform medium with the resistivity of the external region ρ_n. We carried out summation over all boundaries, but it is obvious from the nature of the "telescoping" summation that the same result holds for the intermediate interfaces. For instance, summing over only the first two interfaces, we have

$$\varepsilon_0 I[\rho_1 + (\rho_2 - \rho_1) + (\rho_3 - \rho_2)] = \varepsilon_0 \rho_3 I. \qquad [3.118]$$

3.9.2. Asymptotic Field Behavior and Charge Distribution

As we have pointed out before, the field of any distribution of surface charge on a finite body tends to that of an elementary or point charge (of the same total charge) as the observation point moves far away from the body. In the case of a plane-layered medium (Fig. 3.11B), charge may be distributed over an infinite surface, but, as shown earlier, its density decays rapidly with distance from the current electrode. Correspondingly, we expect from Eq. [3.117] that the asymptotic behavior of the potential in the medium shown in Fig. 3.11A will be

$$U(p) \to \frac{\rho_n I}{4\pi L_{Ap}}, \quad \text{if } L_{Ap} \to \infty. \qquad [3.119]$$

In contrast, the potential approaching the current electrode is dominated by the charge on the electrode,

$$U(p) \to \frac{\rho_1 I}{4\pi L_{Ap}}, \quad \text{if } L_{Ap} \to 0. \qquad [3.120]$$

This simple observation allows us to arrive at a result of great importance. Indeed, it shows that if the current and receiver electrodes are situated in a layered medium and the distance between electrodes increases, then the depth of investigation with such an array also increases. For example, in Fig. 3.11B, even if the electrodes remain solely in the medium with resistivity ρ_1, we expect that the measured voltage will approach that of the bottom (uniform) medium with resistivity ρ_4 as the current and receiver electrodes move farther and farther apart. It is not at all obvious that this behavior should hold, but it can be proven rigorously using the arguments outlined above. Equations [3.119] and [3.120], therefore, embody one of the most important features of geometric sounding in surface and borehole geophysics. We note the following additional features: (1) With a plane-layered medium (Fig. 3.11B), as well as with a coaxial system of cylindrical surfaces, each region is not completely enclosed by its two surfaces. The analysis remains valid, however, since we can imagine two additional surfaces "at infinity" that enclose each region, while the amount of current passing through them equals zero. (2) The charges induced at each surface do not have to be confined within a finite area for the conclusions to hold, provided that the density $\sigma(q)$ decreases quickly enough with distance from the current electrode. (3) The total amount of charge in a "layer" does not change if there is also some confined inhomogeneity surrounded by a surface S^* and the current electrode is located outside the inhomogeneity (Fig. 3.11A). In fact, the total charge arising on this surface S^* is

$$e^* = \varepsilon_0(\rho_e - \rho_i)\oint_{S^*} \mathbf{j}\, d\mathbf{S},$$

where ρ_i, ρ_e are the resistivities of the inhomogeneity and surrounding medium, respectively. By the principle of charge conservation, the integral at the right-hand side vanishes, which means that an equal amount of positive and negative charge appears on the surface S^*. Far away from the body, its field tends to that of the electric dipole, which falls off faster than L_{Ap}^{-1}, so the asymptotic behavior of the field is still defined by Eq. [3.119]. (4) To understand in more detail the principle of Schlumberger sounding at the earth's surface, consider the model of a horizontally plane-layered medium (Fig. 3.11B), where all layers have finite resistivity. Assume that the current electrode A is situated slightly beneath the earth's interface; therefore, its charge is given by Eq. [3.114]. Since air is an insulator, all of the current goes into the earth and the amount passing through each interface below the surface is I. The total charge induced on any interface between successive conducting layers is then equal to

$$e_i = \varepsilon_0(\rho_i - \rho_{i-1})I.$$

Earlier, we demonstrated that an elementary charge de_i located anywhere in the earth induces the same amount of charge at the earth–air interface, $de_{i0} = de_i$. This holds for each elementary charge of the interfaces S_i, and we conclude that the charge e_i on each interface is accompanied by the charge e_{i0} at the earth–air interface. Note that charges e_i and e_{i0} are distributed differently. In particular, the charge density σ_{i0} at the earth–air interface decays laterally more slowly than the charge density on the surface S_i, and only at very large distances from the current electrode they decay in the same manner. Thus, the total charge at all boundaries is equal to

$$2\varepsilon_0\rho_1 I + 2\varepsilon_0 I\sum_{i=1}^{n}(\rho_i - \rho_{i-1}) = 2\varepsilon_0\rho_n I. \qquad [3.121]$$

The total induced charge is therefore the same as the charge that would be induced on the current electrode if it were situated on the surface of a uniform half-space with resistivity of the deepest formation. We see that with increasing separation between the current electrode and observation point, all layers become "transparent," and the field is defined by a resistivity ρ_n of the bottom medium. In contrast, close to the current electrode, the field depends only on the resistivity of the upper layer ρ_1. (5) The model of Fig. 3.11A can be used to explain the principle of what is called "lateral sounding" in borehole electrical measurements, which we will study in detail in Chapter 7.

3.10. BOUNDARY CONDITIONS FOR ELECTRIC FIELD IN THE VICINITY OF THIN LAYERS

The earth's subsurface often contains a layer that is thin compared to the size of other layers and to the dimensions of an electrical survey (e.g., the distance between current and receiver electrodes). Such a situation can arise in land or marine electrical prospecting, as well as in borehole surveys. To simplify solution of the forward problem, it is convenient to replace this layer by an idealized surface of vanishing thickness with approximate boundary conditions for the potential and electric field across the surface. As we will see, study of the field behavior near a thin layer is also useful for solving inverse problems because it allows one to identify some stable parameters of a geoelectric section. We will first consider the case of a thin resistive layer (Fig. 3.12A) and then consider the opposite case of a thin conductive layer.

3.10.1. The First Boundary Condition at the Surface of a Thin Layer

We first make several assumptions that greatly facilitate solution of a boundary-value problem with a thin resistive layer. These assumptions are as follows:

1. The conductivity of the thin layer γ_i is much less than the conductivities of the flanking media:

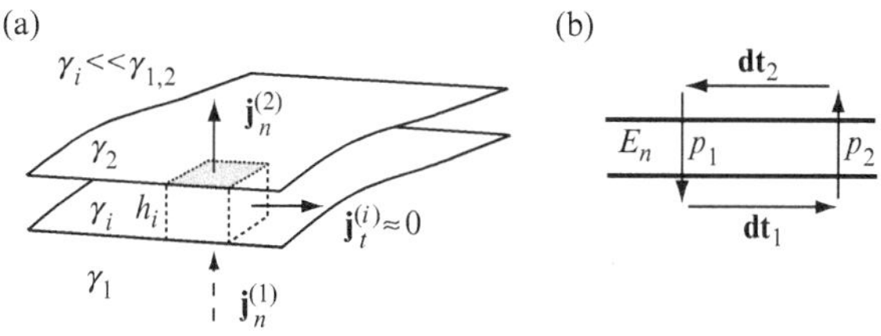

Figure 3.12 (A) Thin resistive layer and geometry for the first boundary condition. Tangential current is negligible in the layer, so the amount of current entering and leaving the layer through elements of its top and bottom surfaces is equal. (B) Geometry of second boundary condition for a thin resistive layer.

$$\gamma_i \ll \gamma_1, \quad \gamma_i \ll \gamma_2. \tag{3.122}$$

2. The thickness of the thin layer h_i is so small that the normal component of the electric field E_n^i inside it changes linearly between the top and bottom of the layer.
3. Because of its high resistivity and small thickness, the tangential component of the current density and electric field inside the layer can be neglected:

$$E_t^i = 0, \quad j_t^i = 0. \tag{3.123}$$

This means that the amount of charges that arrives and leaves through two elements of the surface located opposite each other (Fig. 3.12A) are equal:

$$j_n^{(1)} = j_n^i = j_n^{(2)}. \tag{3.124}$$

Applying Ohm's law gives

$$\gamma_1 E_n^{(1)} = \gamma_i E_n^i = \gamma_2 E_n^{(2)}$$

or

$$\gamma_1 E_n^{(1)} = \gamma_2 E_n^{(2)}. \tag{3.125}$$

Equation [3.125] is the first boundary condition across a thin resistive layer, and does not require knowledge of the field inside the layer. In particular, if the media above and beneath the layer have the same conductivity, then

$$E_n^{(1)} = E_n^{(2)}, \tag{3.126}$$

and the normal component of the electric field is continuous through the layer. As we know, the sources of a constant electric field in a piecewise uniform medium are surface charges; their densities on the top and bottom of the resistive layer are

$$\sigma(p) = 2\varepsilon_0 \frac{\rho_i - \rho_1}{\rho_i + \rho_1} E_n^{S-p}(p), \quad \sigma(q) = 2\varepsilon_0 \frac{\rho_2 - \rho_i}{\rho_2 + \rho_1} E_n^{S-q}(q). \qquad [3.127]$$

Here, p and q are points of the layer boundary located opposite each other, and E_n^{S-p}, E_n^{S-q} are normal components of the field caused by all charges except the charge in the vicinity of points p or q, respectively. Since this part of the normal component of the electric field is a continuous function, we see that the density of charges at points p and q are equal in magnitude but different in sign if $\rho_1 = \rho_2$:

$$\sigma(q) = -\sigma(p). \qquad [3.128]$$

If the layer is thin compared to distance from the layer to an observation point, the field of these charges can be treated as a double layer. In general, when $\rho_2 \neq \rho_1$, the field generated by charge on a thin layer is equivalent to the field of a double layer plus a distribution of surface charge (simple layer).

3.10.2. The Second Boundary Condition Across a Thin Resistive Layer

We next derive the second boundary condition proceeding from the first equation of the electric field

$$\oint \mathbf{E} \cdot \mathbf{dl} = 0. \qquad [3.129]$$

Applying Eq. [3.129] to the closed elementary path shown in Fig. 3.12B gives

$$E_t^{(1)} dt + E_n^i(p_2) h_i - E_t^{(2)} dt - E_n^i(p_1) h_i = 0, \qquad [3.130]$$

where points p_1 and p_2 are inside the layer and close to each other. By definition

$$E_n^i(p_2) - E_n^i(p_1) = \frac{\partial E_n^i}{\partial t} dt,$$

and in place of Eq. [3.130], we have

$$E_t^{(2)} - E_t^{(1)} = h_i \frac{\partial E_n^i}{\partial t}. \quad [3.131]$$

This equation still contains a term involving the field inside the layer. To eliminate it, we will make use of Eq. [3.124] and obtain the secondary boundary condition:

$$E_t^{(2)} - E_t^{(1)} = \gamma_2 T \frac{\partial E_n^{(2)}}{\partial t}, \quad [3.132]$$

where

$$T = \rho_i h_i \quad [3.133]$$

is the transverse resistance of the thin resistive layer. Like the first condition, Eq. [3.133] relates field components on opposite sides of the layer. In the limit when $\rho_i \to \infty$ and $h_i \to 0$, but the product $T = \rho_i h_i$ has a finite and nonzero value $(T \neq 0)$, we arrive at the concept of an idealized surface T that represents a thin resistive layer.

3.10.3. Boundary Conditions at the Surface *T*

Thus, we have derived approximate boundary conditions at the surface T from the principle of charge conservation and the first equation of the electric field:

$$\gamma_1 E_n^{(1)} = \gamma_2 E_n^{(2)}, \quad E_t^{(2)} - E_t^{(1)} = T\gamma_2 \frac{\partial E_n^{(2)}}{\partial t}. \quad [3.134]$$

It is convenient to represent these conditions in terms of the potential U. For instance, in the cylindrical system of coordinates where $t = r, n = z$, we have

$$\gamma_1 \frac{\partial U_1}{\partial z} = \gamma_2 \frac{\partial U_2}{\partial z}, \quad U_2 - U_1 = T\gamma_2 \frac{\partial U_2}{\partial z} \quad [3.135]$$

where the last equality is obtained after integration with respect to r. It clearly shows that the density of the double layer is equal to

$$\eta = \varepsilon_0 T\gamma_2 \frac{\partial U_2}{\partial z}. \quad [3.136]$$

The discontinuity of the tangential component of the electric field across the layer is caused by a change in the density of the dipole moments η.

Thus, the electric field in the vicinity of the surface T is characterized by continuity of the normal component of the current density and discontinuity of the tangential component of the electric field.

3.10.4. The First Boundary Condition on the Surface *S*

Now, we assume that the thin layer has much higher conductivity than that of the surrounding media:

$$\gamma_i \gg \gamma_1, \quad \gamma_i \gg \gamma_2. \qquad [3.137]$$

This allows us to assume that the tangential component of the electric field prevails inside the layer and dominates the normal component, which can be neglected:

$$E_t^i \gg E_n^i. \qquad [3.138]$$

As before, surface charge appears on both sides of the layer, but must be distributed in such a way that the normal component of the primary field nearly compensates the normal component caused by induced charge. Since the resistivity of the thin layer is less than that of the media on either side, it is obvious that a sign of surface charges on each side are opposite to each other. Applying the first equation of the electric field to an elementary contour crossing the layer (Fig. 3.12B) and using Eqs. [3.130] and [3.138] gives

$$E_t^{(2)} = E_t^{(1)} = E_t^i. \qquad [3.139]$$

Thus, the tangential component of the electric field is continuous across a thin conducting layer. This is the first approximate boundary condition.

3.10.5. The Second Boundary Condition

To derive the second boundary condition, consider an elementary volume of the thin conducting layer (Fig. 3.12C). Applying the principle of charge conservation to the surface surrounding this volume, gives

$$\begin{aligned}&[j_z^{(2)}(p_2) - j_z^{(1)}(p_1)]\mathrm{d}x\,\mathrm{d}y + [j_x^i(p_4) - j_x^i(p_3)]h_i\,\mathrm{d}y+\\&[j_y^i(p_6) - j_y^i(p_5)]h_i\,\mathrm{d}x = 0,\end{aligned} \qquad [3.140]$$

where the points p_i are midpoints of surface faces. Since

$$j_x^i(p_4) - j_x^i(p_3) = \frac{\partial j_x^i}{\partial x}\,\mathrm{d}x, \quad j_y^i(p_6) - j_y^i(p_5) = \frac{\partial j_y^i}{\partial y}\,\mathrm{d}y,$$

Eq. [3.140] becomes

$$j_z^{(2)} - j_z^{(1)} = -S\left(\frac{\partial E_x^i}{\partial x} + \frac{\partial E_y^i}{\partial y}\right), \qquad [3.141]$$

where

$$S = \gamma_i h_i \qquad [3.142]$$

is called the longitudinal conductance of the layer. This parameter often plays an important role in the geoelectric section. The right-hand side of Eq. [3.141] contains the tangential component of the electric field inside the layer and, for this reason, is not yet a true boundary condition. Since the horizontal components E_x and E_y of the field are continuous and the derivatives are taken tangent to the layer, we have

$$j_z^{(2)} - j_z^{(1)} = -S\left(\frac{\partial E_x^{(2)}}{\partial x} + \frac{\partial E_y^{(2)}}{\partial y}\right).$$

The right-hand side of this equation can be simplified using div $\mathbf{E}=0$. This gives the second boundary condition,

$$j_z^{(2)} - j_z^{(1)} = S\frac{\partial E_z^{(2)}}{\partial z},$$

or

$$j_z^{(2)} - j_z^{(1)} = S\frac{\partial E_z^{(1)}}{\partial z}. \qquad [3.143]$$

3.10.6. Boundary Conditions on the Surface *S*

Thus, we have introduced the concept of an idealized surface S that replaces a thin conducting layer with the following boundary conditions:

$$E_t^{(1)} = E_t^{(2)}, \quad j_z^{(2)} - j_z^{(1)} = S\frac{\partial E_z^{(1)}}{\partial z}. \qquad [3.144]$$

The first condition holds that the tangential component of the electric field is continuous across the surface, while the second condition shows that the discontinuity of the normal component of current density depends on the conductance of the layer and on the derivative of the normal component of the field in the direction perpendicular to the layer. In terms of the potential, these conditions are

$$U_1 = U_2, \quad \gamma_2\frac{\partial U_2}{\partial z} - \gamma_1\frac{\partial U_1}{\partial z} = S\frac{\partial^2 U_2}{\partial z^2}. \qquad [3.145]$$

3.11. THEOREM OF RECIPROCITY

3.11.1. Principle of Interchange of Current and Receiver Electrodes

We now study a general feature of field behavior that is important for comparison of measurements with different arrays. Suppose that a four-electrode array, with two current (source) and two voltage (receiver) electrodes, is placed in an inhomogeneous conducting medium (Fig. 3.13A). Here, I_{AB} is the current in the line AB and $V^{AB}_{MN} = U_M - U_N$ is the voltage measured in the receiver line MN. Now, interchange the roles of the source and receiver electrodes and assume that the line MN serves as the current line with current equal to I_{MN}, while the voltage V^{MN}_{AB} is measured in the line AB (Fig. 3.13B). Our goal is to prove that the voltages and currents in these two configurations are related as follows:

$$\frac{V^{AB}_{MN}}{I_{AB}} = \frac{V^{MN}_{AB}}{I_{MN}}. \qquad [3.146]$$

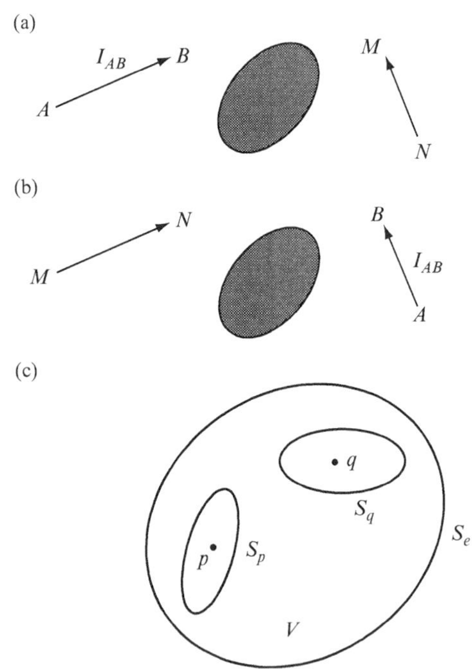

Figure 3.13 (A) Source–receiver array *AMNB* and (B) its reciprocal array *MABN*. (C) Surfaces and volumes involved in proof of the reciprocity theorem.

This equality is hardly obvious and certainly requires proof. For example, the primary field in the first measurement is caused by charges at electrodes A and B, and the secondary field will of course depend on a position of these electrodes. Also, the difference of potentials, $U_M - U_N$, will depend on a location of the current and receiver electrodes, and so will the resistance of the medium between equipotential surfaces passing through receiver electrodes:

$$R_{MN} = \frac{V_{MN}^{AB}}{I_{AB}}.$$

Nevertheless, we will demonstrate that Eq. [3.146] is correct. This equation can also be written in terms of resistances as

$$R_{MN}^{AB} = R_{AB}^{MN}. \qquad [3.147]$$

The proof of this equality is called the theorem of reciprocity.

3.11.2. The Theorem of Reciprocity

Suppose that there are two electrodes P and Q, where each can be either a current or voltage electrode (Fig. 3.13A and B). Let us introduce the following notation: I_p is the current through the electrode P and $U_P(p)$, $U_P(q)$ are potentials at electrodes P and Q due to the current I_p. Similarly, I_Q is the current through the electrode Q and $U_Q(q)$, $U_Q(p)$ are potentials at electrodes Q and P caused by this current. Our goal is to prove the equality,

$$\frac{U_P(q)}{I_P} = \frac{U_Q(p)}{I_Q}. \qquad [3.148]$$

Assuming that the potential is zero at infinity, the latter can be rewritten as

$$R_q^P = R_p^Q. \qquad [3.149]$$

That is, we have to show that the resistance of the medium between the equipotential surface passing point q and infinity when P is the current electrode, is equal to the resistance between the equipotential surface passing electrode p and infinity when Q is the current electrode. In essence, this part of the theorem of reciprocity shows that the principle of interchange of source and receiver electrodes is valid for a two-electrode array. In the case of a uniform medium, Eq. [3.148] is obvious. Indeed, as we know

$$U_P(q) = \frac{\rho I_P}{4\pi L_{pq}}, \quad U_Q(p) = \frac{\rho I_Q}{4\pi L_{qp}},$$

and these two equations give Eq. [3.148]. To prove the principle of reciprocity for an inhomogeneous medium and electrodes of finite dimensions, we will take three surfaces: S_p, S_q, and S_e. The first two surround electrodes P and Q, respectively, while S_e is an external surface; V is the volume enclosed by the three surfaces (Fig. 3.13C). Introduce the vector

$$\mathbf{X} = \gamma(U_Q \nabla U_P - U_P \nabla U_Q), \qquad [3.150]$$

where U_P and U_Q are potentials at any point of a medium when the current electrode is P and Q, respectively. It is useful to notice that each term of the vector $\mathbf{X}$ is a product of the potential and electric field caused by different current electrodes. Gauss's theorem gives

$$\oint_S \mathbf{X}\cdot d\mathbf{S} = \int_V \operatorname{div} \mathbf{X}\, dV,$$

or

$$\oint_S \gamma \left(U_Q \frac{\partial U_P}{\partial n} - U_p \frac{\partial U_Q}{\partial n} \right) dS = \int_V [U_Q \operatorname{div}(\gamma \nabla U_P) - U_P \operatorname{div}(\gamma \nabla U_Q)]\, dV.$$

Since

$$\operatorname{div}(\gamma \nabla U) = -\operatorname{div}(\gamma \mathbf{E}) = -\operatorname{div} \mathbf{j} = 0,$$

we have

$$\oint_S \gamma \left(U_Q \frac{\partial U_P}{\partial n} - U_P \frac{\partial U_Q}{\partial n} \right) dS = 0. \qquad [3.151]$$

The integral actually covers all three surfaces S_q, S_p, and S_e. Since the surface S_e expands to infinity to include all of space, the integral over S_e vanishes (because the integrand falls over faster than the surface area grows). Thus, we obtain in place of Eq. [3.151]

$$\oint_{S_p} \gamma U_Q \frac{\partial U_P}{\partial n}\, dS - \oint_{S_p} \gamma U_P \frac{\partial U_Q}{\partial n}\, dS + \oint_{S_q} \gamma U_Q \frac{\partial U_P}{\partial n}\, dS - \oint_{S_q} \gamma U_P \frac{\partial U_Q}{\partial n}\, dS = 0.$$

Since these surfaces are equipotentials, we have

$$U_Q(p) \oint_{S_p} \gamma \frac{\partial U_P}{\partial n}\, dS - U_P(p) \oint_{S_p} \gamma \frac{\partial U_Q}{\partial n}\, dS + U_Q(q) \oint_{S_q} \gamma \frac{\partial U_P}{\partial n}\, dS - U_P(q) \oint_{S_q} \gamma \frac{\partial U_Q}{\partial n}\, dS = 0.$$

From the principle of charge conservation, we also have

$$\oint_{S_p} \gamma \frac{\partial U_P}{\partial n}\, dS = -\oint \gamma E_{Pn}\, dS = -I_P, \quad \oint_{S_q} \gamma \frac{\partial U_Q}{\partial n}\, dS = -I_Q,$$

and

$$\oint_{S_p} \gamma \frac{\partial U_Q}{\partial n}\, dS = \oint_{S_q} \gamma \frac{\partial U_P}{\partial n}\, dS = 0.$$

Therefore,

$$U_Q(p) I_P = U_P(q) I_Q,$$

and the theorem is proved. We now show that the principle of interchange of current and receiver lines of any array follows from Eq. [3.148]. First, consider the array shown in Fig. 3.13A. By definition, we have

$$V_{AB}^{MN} = U_{AB}^{M} - U_{AB}^{N} = U_A^M + U_B^M - U_A^N - U_B^N.$$

Thus,

$$\frac{V_{AB}^{MN}}{I_{AB}} = \frac{U_A^M}{I_A} - \frac{U_B^M}{I_B} - \frac{U_A^N}{I_A} + \frac{U_B^N}{I_B}, \qquad [3.152]$$

since $I_{AB} = I_A$ and $I_{AB} = -I_B$. By analogy, for an array $MABN$, Fig. 3.13C, we have

$$V_{MN}^{AB} = U_{MN}^{A} - U_{MN}^{B} = U_M^A + U_N^A - U_M^B - U_N^B,$$

and

$$\frac{V_{MN}^{AB}}{I_{MN}} = \frac{U_M^A}{I_M} - \frac{U_N^A}{I_N} - \frac{U_M^B}{I_M} + \frac{U_N^B}{I_N}, \qquad [3.153]$$

because $I_M = I_{MN}$ and $I_N = -I_{MN}$. Applying the theorem of reciprocity, Eq. [3.148], we conclude that the principle interchange of the current and receiver parts of any array is valid. The following features of the principle and its proof are worth noting:

1. Although the proof assumed that the conductivity is a scalar function (isotropic conductivity), the principle of reciprocity turns out to be valid in the general case of anisotropic medium.
2. In deriving Eq. [3.148], we assumed that

$$\operatorname{div}(\gamma \nabla U) = 0,$$

that is, an extraneous (non-Coulomb) field is absent. As we know, in general, Ohm's law is

$$\mathbf{j} = \gamma(\mathbf{E}^c + \mathbf{E}^{\text{ext}}),$$

where γ and $\mathbf{E}^{\text{ext}}$ are independent on the field $\mathbf{E}^c$. The latter gives

$$\operatorname{div}\mathbf{j} = -\operatorname{div}(\gamma\nabla U) + \operatorname{div}\gamma\mathbf{E}^{\text{ext}} = 0.$$

This means that

$$\operatorname{div}(\gamma\nabla U) \neq 0, \qquad [3.154]$$

and the proof does not go through. It is possible to imagine one exception. Suppose that the extraneous field is directly proportional to the Coulomb field:

$$\mathbf{E}^{\text{ext}}(p) = \alpha(p)\mathbf{E}^c(p), \qquad [3.155]$$

where $\alpha(p)$ is a parameter characterizing a medium in the vicinity of the point p, but is independent of the field $\mathbf{E}^c$. In such case, Ohm's law can be written as

$$\mathbf{j}(p) = \gamma^* E^c = -\gamma^*\nabla U, \qquad [3.156]$$

where

$$\gamma^* = \gamma[1 + \alpha(p)]. \qquad [3.157]$$

The theory of induced polarization is based on the assumption that the extraneous field is directly proportional to the Coulomb field (Eq. [3.155]), and correspondingly, the principle of interchange of the current and receiver electrodes applies.

3.12. RELATIONSHIP BETWEEN POTENTIAL AND BOUNDARY CONDITIONS

One other interesting general feature of the field follows from the theorem of uniqueness, namely, that the field inside some volume V is defined by the boundary conditions at the surface S surrounding this volume. It is natural to derive an explicit equation establishing this link proceeding naturally from Gauss's theorem. Let the vector $\mathbf{X}(q)$ be constructed from two scalar functions $\phi(q)$ and $\psi(q,p)$:

$$\mathbf{X} = \phi(q)\nabla\psi(q,p) - \psi(q,p)\nabla\phi(q), \qquad [3.158]$$

where ϕ and ψ are continuous functions with continuous first and second derivatives; p is an arbitrary observation point, where the potential is to be determined; and q is an arbitrary point that varies inside the volume. (The derivatives in Eq. [3.158] are taken with respect to the variable q.) Then, Gauss theorem can be written as

$$\int_V (\phi\nabla^2\psi - \psi\nabla^2\phi)\,\mathrm{d}V = \oint_S \left(\phi\frac{\partial\psi}{\partial n} - \psi\frac{\partial\phi}{\partial n}\right)\mathrm{d}S, \qquad [3.159]$$

since

$$\mathbf{X}\cdot\mathrm{d}\mathbf{S} = X_n\mathrm{d}S = \left(\phi\frac{\partial\psi}{\partial n} - \psi\frac{\partial\phi}{\partial n}\right)\mathrm{d}S$$

and

$$\mathrm{div}\,\mathbf{X} = \nabla\phi\cdot\nabla\psi + \phi\nabla^2\psi - \nabla\phi\cdot\nabla\psi - \psi\nabla^2\phi.$$

Equation [3.159] is called the Green's second formula (or theorem), and it expresses Gauss's theorem when the vector $\mathbf{X}$ is given by Eq. [3.158]. Letting $\psi =$ const. gives Green's first formula

$$\int_V \nabla^2\phi\,\mathrm{d}V = \oint_S \frac{\partial\phi}{\partial n}\,\mathrm{d}S. \qquad [3.160]$$

Our goal is to express the potential of the field $U(p)$ within the volume V in terms of the potential and its derivatives on the boundary S $(= S_1 + S_2)$, as in Fig. 3.14A. To solve this problem, we will make use of Eq. [3.159], assuming that the function ϕ is the potential of the electric field U, whereas the function $\psi = G(q, p)$ obeys the following conditions:

1. It is a solution of Laplace's equation:

$$\nabla^2 G = 0, \qquad [3.161]$$

everywhere except the observation point p.

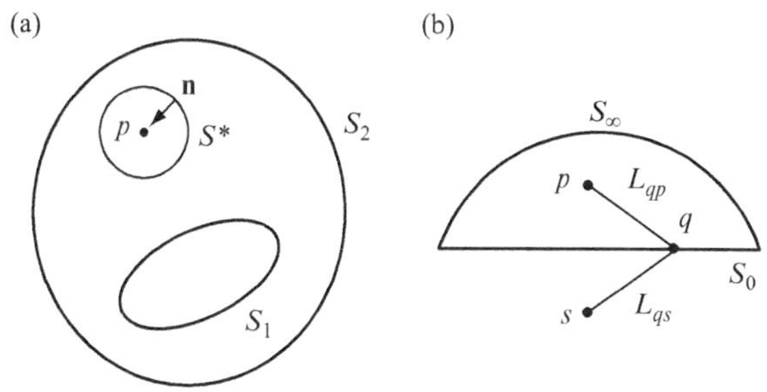

Figure 3.14 (A) Illustration of Eq. [3.160]. (B) Geometry of Eq. [3.169].

2. In approaching the point p, this function tends to infinity as

$$G(q,p) \to \frac{1}{L_{qp}}, \quad \text{as } L_{qp} \to 0. \qquad [3.162]$$

Here, L_{qp} is the distance between the points q and p. The function $G(q,p)$ is called Green's function. There are an infinite number of such functions and all of them have a singularity at the observation point p. The second Green's formula was been derived assuming that singularities of the functions U and G are absent within volume V; therefore, we cannot directly use G in Eq. [3.159]. To avoid the singularity, we surround the point p by a small spherical surface S^* and apply Eq. [3.159] to the volume enclosed by surfaces S and S^* (Fig. 3.14A). We will then take the limit as the surface S^* shrinks down to the point p. Since charges are absent in V, we have

$$\nabla^2 U = 0, \qquad [3.163]$$

and using Eq. [3.161], we see that the volume integral in the second Green's formula disappears. Then, we obtain in place of Eq. [3.159],

$$\oint_S \left(U\frac{\partial G}{\partial n} - G\frac{\partial U}{\partial n} \right) \mathrm{d}S + \oint_{S^*} \left(U\frac{\partial G}{\partial n} - G\frac{\partial U}{\partial n} \right) \mathrm{d}S = 0. \qquad [3.164]$$

To derive an explicit expression for the potential U, we have to reduce the second integral. With this purpose in mind, consider the limiting value of this integral when the radius of the spherical surface r tends to zero. Since both the potential and its derivative are continuous functions inside the volume, we have

$$U(q) \to U(p), \quad \frac{\partial U(q)}{\partial n} \to \frac{\partial U(p)}{\partial n},$$

that is, functions $U(q)$ and $\partial U/\partial n$ on the surface S^* approach their values at the observation point, respectively. As seen from Fig. 3.14A, it follows that the normal $\mathbf{n}$ and radius vector $\mathbf{r}$ on this surface are opposite to each other, and therefore, for points on this surface, we have

$$G \to \frac{1}{L_{qp}} = \frac{1}{r} \quad \text{and} \quad \frac{\partial G}{\partial n} \to -\frac{\partial}{\partial r}\frac{1}{r} = \frac{1}{r^2}, \quad \text{as} \quad r \to 0.$$

The surface integral around the observation point p can be represented as

$$\oint_{S^*}\left[U(q)\frac{\partial}{\partial n}\frac{1}{L_{qp}}-\frac{1}{L_{qp}}\frac{\partial U}{\partial n}\right]\mathrm{d}S=\frac{1}{r^2}\oint_{S^*}U(q)\,\mathrm{d}S-\frac{1}{r}\oint_{S*}\frac{\partial U}{\partial n}\,\mathrm{d}S, \quad [3.165]$$

and, applying the mean value theorem, we have

$$\frac{1}{r^2}\oint_{S^*}U(q)\,\mathrm{d}S=\frac{U(p)}{r^2}4\pi r^2=4\pi U(p)$$

and

$$\frac{1}{r}\oint_{S^*}\frac{\partial U(q)}{\partial n}\,\mathrm{d}S=\frac{1}{r}\frac{\partial U(p)}{\partial n}4\pi r^2\to 0.$$

Therefore, in the limit, the integral over the surface S^* is

$$\oint_{S^*}\left[U(q)\frac{\partial G}{\partial n}-G\frac{\partial U(q)}{\partial n}\right]\mathrm{d}S=4\pi U(p). \quad [3.166]$$

This is the most important result of the derivation, since it gives an expression for the potential at any point of the volume in terms of boundary conditions. Substitution of Eq. [3.166] into Eq. [3.164] gives

$$U(p)=\frac{1}{4\pi}\oint_{S}\left[G(q,p)\frac{\partial U(q)}{\partial n}\right]-U(q)\frac{\partial G(q,p)}{\partial n}\,\mathrm{d}S. \quad [3.167]$$

Here, q is the point of the boundary surface and p is the observation point. To determine the potential from Eq. [3.167], we have to know both the potential and its normal derivative on the surface S. The uniqueness theorem for the first and second boundary-value problems showed, however, that it is sufficient to know either one of these quantities on S.

This apparent contradiction can be resolved by the appropriate choice of Green's function. We will illustrate this approach by calculating the field in the upper half-space; that is, above the earth's surface. In this case, the volume V is a half-space surrounded by a plane (earth's surface) and a semispherical surface of infinitely large radius, where the field decays to zero (Fig. 3.14B). Then, in place of Eq. [3.167] we have

$$U(p)=\frac{1}{4\pi}\int_{S_0}\left[G(q,p)\frac{\partial U(q)}{\partial z_q}-U(q)\frac{\partial G(q,p)}{\partial z_q}\right]\mathrm{d}S, \quad [3.168]$$

where the z-axis is directed downward outside the volume V. The value of the normal derivative (the vertical component of the electric field) is discontinuous at the earth's surface. Its value inside the earth is zero and characterizes the condition of zero current flow across the interface, but Eq. [3.168] contains the term with the vertical component

$$E_z = -\frac{\partial U}{\partial z},$$

at the other side of the surface S_0 (in the air) where the field differs from zero and is unknown (there is also an atmospheric electric field normal to the earth's surface, which is in many orders greater than the field caused by charges inside the earth). But if we can find a function $G(q, p)$ such that the first term of the integrand in Eq. [3.168] vanishes, the need to know the vertical component of the electric field is eliminated. Consider the function

$$G(q, p) = \frac{1}{L_{qp}} - \frac{1}{L_{qs}}, \qquad [3.169]$$

where s is the mirror reflection of the point p with respect of the earth's surface and

$$\begin{aligned} L_{qp} &= \left[(x_p - x_q)^2 + (y_p - y_q)^2 + (z_p - z_q)^2\right]^{1/2}, \\ L_{qs} &= \left[(x_s - x_q)^{1/2} + (y_s - y_q)^{1/2} + (z_s - z_q)^2\right]^{1/2}. \end{aligned} \qquad [3.170]$$

By definition, when point q is located on the earth's surface, the Green's function is equal to zero and Eq. [3.168] gives

$$U(p) = -\frac{1}{4\pi}\int_{S_0} U(q)\frac{\partial G(q, p)}{\partial z_q}\,\mathrm{d}S. \qquad [3.171]$$

Performing the differentiation and taking into account that $z_q = 0$ and $z_p = -z_s$, we obtain

$$U(p) = \frac{z_p}{2\pi}\int_{S_0} \frac{U(q)}{L_{qp}^3}\,\mathrm{d}S. \qquad [3.172]$$

It is a simple matter to express the horizontal components of the electric field in the upper half-space by differentiating this expression with respect to x_p and y_p. Note that the function $U(q)$ is the potential on the earth's surface and can be measured directly. Equation [3.172] can actually be used to extrapolate the field measured at the earth's surface upward into the air layer. This "upward continuation" attenuates most rapidly the electric field caused by small, relatively shallow inhomogeneities that are close to the surface and, therefore, may help to reduce geologic noise in field surveys.

PART TWO

Electric Methods of Surface Geophysics

CHAPTER FOUR

Electric Soundings

In general, we will treat a medium beneath the earth's surface as piecewise uniform, where each uniform region is characterized by a constant resistivity. For instance, in the horizontally layered medium, the resistivity and thickness of these layers

$$\rho_1, h_1 \quad \rho_2, h_2 \quad \rho_3, h_3 \quad \rho_4, h_4 \ldots \rho_n, h_n$$

form the set of parameters which describes the geoelectric section. Examples of different geoelectric sections are shown in Fig. 4.1. Also, there are cases, such as permafrost, when resistivity is a continuous function of a point, but they are exceptions. In order to study the distribution of resistivity beneath the earth's surface, two main groups of methods were introduced. One of them defines the resistivity as a function of the distance downward from the earth's surface and, correspondingly, it is called *soundings*. The second group is designed to trace the lateral changes of the resistivity of formations located approximately at the same depth, and it is naturally called *profiling*. Note that this classification is equally applied to both the surface and borehole methods, as well as to electromagnetic methods. Inasmuch as the goal of soundings is the determination of parameters of a geoelectric section, the quantitative interpretation of results of measurements is the important part of this approach. On the contrary, in most methods of profiling, it is sufficient to perform a qualitative or semiquantitative interpretation.

4.1. ARRAYS OF THE ELECTRIC METHODS

Earlier, we demonstrated that a two-electrode array has very small depth of investigation, and for this reason, in order to perform measurements for either the soundings or profiling, four-electrode arrays are always used in the surface geophysics. Examples of such arrays are shown in Fig. 4.2. Each of them consists of two parts, namely, the current and receiver lines.

The current line AB includes a source of current and an ammeter which measures the current I. The role of this part of the array is to create an electric field in the medium where charges arise at interfaces between

Methods in Geochemistry and Geophysics, Volume 44
ISSN 0076-6895, DOI: 10.1016/S0076-6895(10)44004-4
© 2010 Elsevier B.V.
All rights reserved.

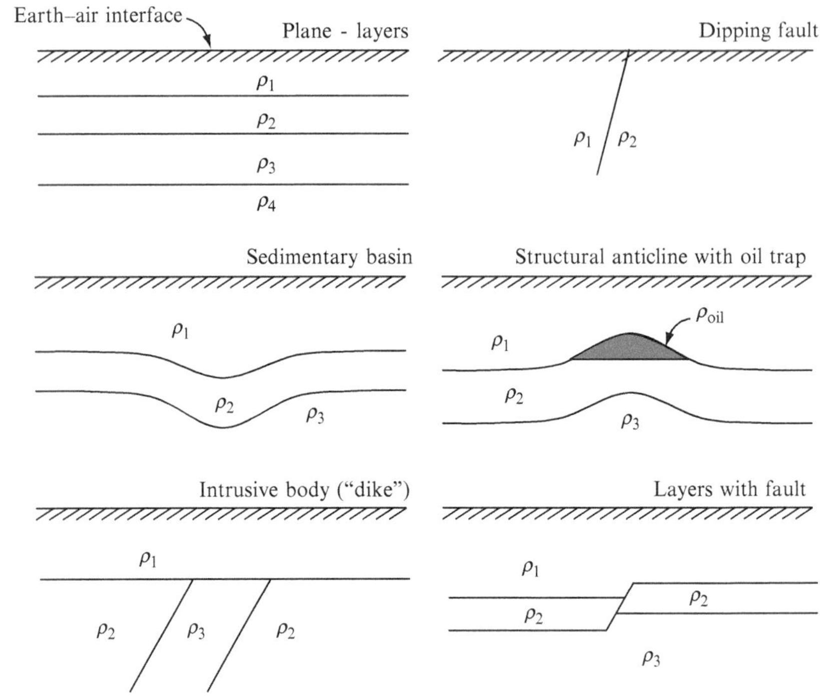

Figure 4.1 Models of geoelectric sections.

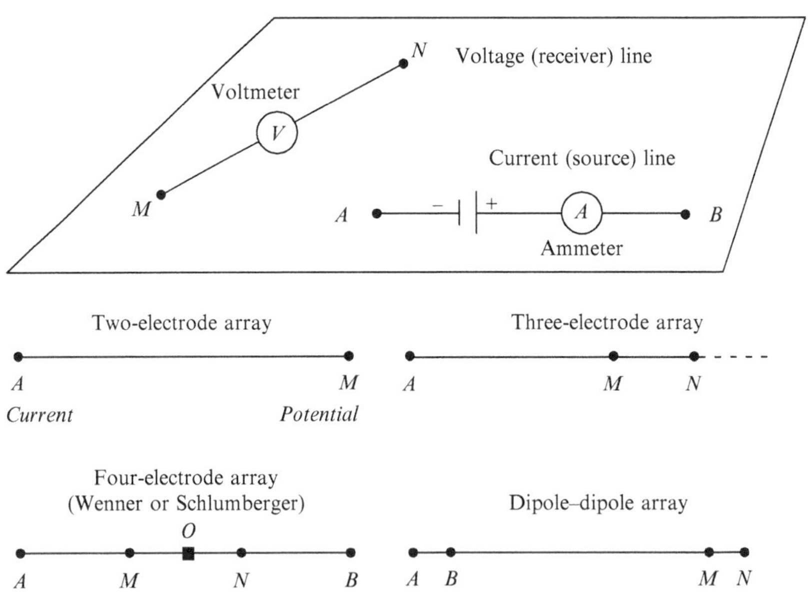

Figure 4.2 Different electric arrays.

regions with different resistivity, including the surface of current electrodes. All these charges depend on geoelectric parameters of a medium and, in accordance with Coulomb's law, they create an electric field at any point, in particular, near the earth's surface. The role of the receiver line MN is to measure the voltage between electrodes M and N caused by the field $\mathbf{E}$, and the latter contains information about a geoelectric section. By definition,

$$V_{MN} = \int_M^N \mathbf{E} \cdot \mathbf{dl} = \int_M^N E_l dl. \qquad [4.1]$$

Here, E_l is the scalar component of the field along the element of the receiver line dl. In many practical cases, the line MN is straight and so small that at its points, the field $\mathbf{E}$ is constant. Then, in place of Eq. [4.1], we have

$$V_{MN} = E_{MN} MN, \qquad [4.2]$$

that is, by measuring the voltage, the electric field can be calculated. Here, E_{MN} is the component of the field $\mathbf{E}$ along the receiver line and MN is its length. Thus, an array performs two measurements: a, the current I in the line AB and b, the voltage V_{MN} in the line MN. There are several factors, including the ground resistance of electrodes A and B, which change the current I and, therefore, the voltage V_{MN} at different positions of the array. In order to eliminate the influence of the current variations, the ratio V_{MN}/I is calculated. As was shown earlier, it describes the resistance of a medium confined by the earth's surface and two equipotential surfaces passing through electrodes M and N, respectively. The function

$$R_{MN} = \frac{V_{MN}}{I} \qquad [4.3]$$

contains information about the geoelectric section, and it can be used for either quantitative (soundings) or qualitative (profiling) interpretation: that is, for a solution of the inverse problem.

4.2. APPARENT RESISTIVITY AND ARRAY COEFFICIENT

4.2.1. Relation Between Resistance R_{MN} and Apparent Resistivity

At the same time, it is very convenient to use the so-called apparent resistivity ρ_a in place of the resistance. Such transition from R_{MN} to ρ_a neither increases nor decreases information about the parameters of a medium, but usually it greatly facilitates the study of the geoelectric section. Let us discuss this subject in some detail. First, it is natural to introduce a quantity which shows how a given medium differs from a uniform half-space with resistivity of the upper layer ρ_1. With this purpose in mind, consider the ratio

$$\frac{V_{MN}}{V_{MN}(\rho_1)}. \qquad [4.4]$$

Here, V_{MN} is the voltage measured with the given array, while $V_{MN}(\rho_1)$ corresponds to that measured in a uniform half-space with the resistivity ρ_1. It is clear that this ratio characterizes the degree with which a medium differs from a homogeneous half-space with resistivity ρ_1. For instance, if regardless of separation between the current and receiver electrodes

$$\frac{V_{MN}}{V_{MN}(\rho_1)} = 1,$$

then an array is located on the earth's surface of a medium which creates the same signal as that in the case of a uniform half-space. Taking into account this important property of the ratio [4.4], the apparent resistivity is introduced as

$$\frac{\rho_a}{\rho_1} = \frac{V_{MN}}{V_{MN}(\rho_1)}, \qquad [4.5]$$

and the quantity ρ_a/ρ_1 can be treated as a measure of a deviation of a medium from a uniform half-space. The apparent resistivity ρ_a has a dimension of ohm m, and it can be either positive, negative, or zero. Certainly, it does not have the meaning of an average value of resistivity of the different regions of a medium. It is proper to notice that, in the case when induced polarization effect is taken into account, the apparent resistivity can be complex. As will be seen many times, the behavior of the apparent resistivity as a function of separation between the current and receiver electrodes resembles that of a medium resistivity, with depth much better than with the voltage V_{MN}. Besides, it is more convenient to compare the results of measurements with different arrays using the apparent resistivity, because in a uniform medium, $\rho_a = \rho_1$ for all different arrays.

4.2.2. The Array Coefficient

Next, we describe the relationship between the apparent resistivity and measured voltage and current. As follows from Eq. [4.5], we have

$$\rho_a = \frac{\rho_1}{V_{MN}(\rho_1)} V_{MN} = \frac{\rho_1 I}{V_{MN}(\rho_1)} \frac{V_{MN}}{I}$$

or

$$\rho_a = kR_{MN} = k\frac{V_{MN}}{I}, \qquad [4.6]$$

where

$$k = \frac{\rho_1 I}{V_{MN}(\rho_1)} \qquad [4.7]$$

is called the array coefficient which depends on geometric parameters of an array only: that is, of the mutual positions of the current and receiver electrodes. Note that since $V_{MN}(\rho_1)$ is also directly proportional to ρ_1, the coefficient k is independent of the resistivity. Bearing in mind that, when current electrodes are situated at the surface of a uniform half-space, charges arise only on the surface of the current electrodes:

$$e_A = 2\varepsilon_0 \rho_1 I, \quad e_B = -2\varepsilon_0 \rho_1 I,$$

it is simple matter to derive an expression for an array coefficient. In fact, the influence of charges that appear on the surface of the receiver electrodes can be taken into account during measurements since the potentials of electrodes M and N are

$$U_M = \frac{\rho_1 I}{2\pi}\left(\frac{1}{AM} - \frac{1}{BM}\right), \quad U_N = \frac{\rho_1 I}{2\pi}\left(\frac{1}{AN} - \frac{1}{BN}\right). \qquad [4.8]$$

Therefore,

$$V_{MN}(\rho_1) = \frac{\rho_1 I}{2\pi}\left(\frac{1}{AM} - \frac{1}{AN} + \frac{1}{BN} - \frac{1}{BM}\right), \qquad [4.9]$$

and the coefficient of any array is

$$k = 2\pi\left(\frac{1}{AM} - \frac{1}{AN} + \frac{1}{BN} - \frac{1}{BM}\right)^{-1}. \qquad [4.10]$$

In performing this simple algebra, we used the principle of superposition and presented the four-electrode array *AMNB* as the sum of four-electrode arrays: *AM*, *AN*, *BM*, and *BN*.

4.2.3. Arrays in Electric Methods

There are several types of four-electrode arrays often used in electric methods, and their application depends on the geological problems that have to be solved as well as the conditions on the earth's surface. Now, we describe some of them.

4.2.3.1 Two-Electrode Array

The receiver electrode M is located at a distance AM from the electrode A, but electrodes B and N are situated far away from each other and from the electrodes A and M. Correspondingly, the potential at the point N is practically equal to zero and the influence of the electrode B can be neglected. This means that the voltage between electrodes M and N coincides with the potential at the point M. At the surface of a uniform half-space, we have

$$U_M(\rho_1) = \frac{\rho_1 I}{2\pi AM}.$$

Thus, the coefficient of the two-electrode array is

$$k = \frac{\rho_1 I}{U_M(\rho_1)} = 2\pi AM, \qquad [4.11]$$

and the apparent resistivity is expressed as

$$\rho_a = 2\pi AM \frac{U_M}{I}, \qquad [4.12]$$

where U_M is numerically equal to the voltage measured between the receiver electrodes M and N. We see that, in this simplest case of an array, the coefficient k is directly proportional to the separation AM, and the ratio U_M/I is equal to the resistance of the medium enclosed by an equipotential surface passing through an electrode M and that located at infinity.

4.2.3.2 Hummel Three-Electrode Array

In this array, the current line is the same as in the previous case, but the receiver electrodes M, N, and the electrode A are located in a straight line and $AM=MN$. Then, Eq. [4.9] gives

$$V_{MN}(\rho_1) = \frac{\rho_1 I}{2\pi}\left(\frac{1}{AM} - \frac{1}{AN}\right) = \frac{\rho_1 I \cdot MN}{2\pi AM \cdot AN} = \frac{\rho_1 I}{2\pi AN}.$$

Thus,

$$k = 2\pi AN = 4\pi AM. \qquad [4.13]$$

4.2.3.3 Schlumberger Three-Electrode Array

As follows from Eq. [4.13], the coefficient of the three-electrode array is

$$k = 2\pi \frac{AM \cdot AN}{MN} \qquad [4.14]$$

and

$$\rho_a = 2\pi \frac{AM \cdot AN}{MN} \frac{V_{MN}}{I} \qquad [4.15]$$

or

$$\rho_a = 2\pi \frac{AM \cdot AN}{MN} R_{MN}, \qquad [4.16]$$

where R_{MN} is the resistance of a medium bounded by the earth's surface and two equipotential surfaces passing through receiver electrodes. In particular, if $MN \ll AM$, we have

$$\rho_a = \frac{2\pi (A0)^2}{MN} \frac{V_{MN}}{I}. \qquad [4.17]$$

Here, $A0$ is the distance between the current electrode A and the middle point of the receiver line. In this case, the component of the electric field E_{MN} does not practically change along the receiver line:

$$V_{MN} = E_{MN} \cdot MN,$$

and in place of Eq. [4.17], we obtain

$$\rho_a = 2\pi (A0)^2 \frac{E_{MN}}{I}. \qquad [4.18]$$

4.2.3.4 Four-Electrode Array

The current and receiver electrodes are placed along the same line, and as follows from Eq. [4.10],

$$k = \frac{2\pi}{MN} \left(\frac{1}{AM \cdot AN} + \frac{1}{BM \cdot BN} \right)^{-1}. \qquad [4.19]$$

In the case of Wenner array, when $AM=MN=NB$, Eq. [4.19] becomes

$$k = 2\pi MN \qquad [4.20]$$

and

$$\rho_a = 2\pi MN \frac{V_{MN}}{I}. \qquad [4.21]$$

4.2.3.5 Symmetrical Schlumberger Four-Electrode Array

This is a special case of the four-electrode array when $AM=BN$ and $AN=BM$. Correspondingly, Eq. [4.10] yields

$$k = \pi \frac{AM \cdot AN}{MN}. \qquad [4.22]$$

For instance, if the separation MN is relatively small, we have

$$k = \pi \frac{(A0)^2}{MN} \qquad [4.23]$$

and

$$\rho_a = \pi (A0)^2 \frac{E_{MN}}{I} \qquad [4.24]$$

and the last relation is often used in practice.

4.2.3.6 Dipole Arrays

Suppose that the distance r between the current electrodes and an observation point located on the surface of a uniform half-space is much greater than the separation between the current electrodes: $r \gg AB$. Then, the system of charges on these electrodes can be treated as an electric dipole. Its moment is directed from the electrode B to A and it is equal to

$$p = 2\varepsilon_0 \rho_1 IAB. \qquad [4.25]$$

As shown in Chapter 1, the potential of this dipole is

$$U = \frac{\rho_1 I}{2\pi} \frac{AB \cos\theta}{r^2}, \qquad [4.26]$$

where r and θ are coordinates of the observation point. Taking into account that

$$E_r = -\frac{\partial U}{\partial r}, \quad E_\theta = -\frac{\partial U}{r\partial \theta},$$

for the radial and azimuthal components of the electric field, we have

$$E_r = \frac{\rho_1 I}{\pi}\frac{AB\cos\theta}{r^3}, \quad E_\theta = \frac{\rho_1 I}{2\pi}\frac{AB\sin\theta}{r^3}. \qquad [4.27]$$

Assuming that $MN \ll r$, the apparent resistivity for the axial ($\theta = 0$) and equatorial ($\theta = \pi/2$) dipole–dipole arrays is defined as

$$\rho_a = \frac{\pi r^3}{AB}\frac{E_r}{I} \quad \text{if } \theta = 0$$

and [4.28]

$$\rho_a = \frac{2\pi r^3}{AB}\frac{E_\theta}{I} \quad \text{if } \theta = \pi/2.$$

4.3. THE ELECTRIC FIELD ON THE SURFACE OF A TWO-LAYERED MEDIUM

4.3.1. Introduction

Now, we begin to describe the theory of Schlumberger soundings and start from the simplest model of a two-layered medium when the current electrode A is located on the earth's surface (Fig. 4.3A). The resistivity and thickness of the upper layer is ρ_1 and h_1, respectively, and the resistivity of the underlying medium (basement) is ρ_2. Before we begin to describe the boundary-value problem, it may be appropriate to notice the following. Imagine that at some instant $t = 0$, the source of current is turned on. It does not mean that at each point of the medium, including the line AB, the current instantly becomes constant. On the contrary, at the beginning, the electromagnetic field propagates through air and the conducting medium with a finite velocity, and until it reaches an observation point, the electric and magnetic fields are equal to zero at this point. It is also proper to point out that this process of propagation is accompanied by diffusion. During this first time interval, $t < t_0$ (Fig. 4.3B), the field behavior is described by Maxwell's equations. The movement of an electromagnetic field from the current line to the point of observation can take place along different paths. For instance, in the case of a two-layered medium, it may occur through the air and the first layer. Also, an electromagnetic wave may travel from the current line through the upper layer into a formation and back to the observation point.

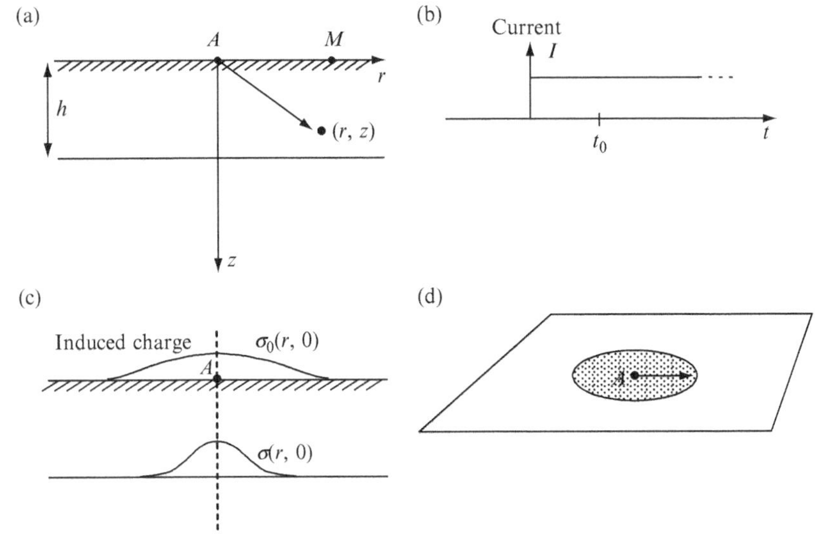

Figure 4.3 (A) A model of a two-layered medium, (B) transient and stationary ranges, (C) density of charges as a function of the distance *r*, and (D) electric field as a difference of two fields.

In general, the contribution of signals arriving along these paths can significantly differ from each other. Finally, when the propagation and diffusion effects are established, we observe the stage at which the current is constant. This is the range in which the direct current methods are used, while measurements in electromagnetic methods are performed in the previous time interval, which rarely exceeds a few seconds, and this happens only when the separation of arrays is several kilometers. Thus, we may say that to some extent, the main features of the constant electric field reflect the behavior of an electromagnetic field propagating through a conducting medium and an upper half-space.

4.3.2. Charges on the Surfaces of a Two-Layered Medium

First, consider again a distribution of charges at the boundaries of a two-layered medium. As was shown earlier, the charge on the electrode surface is

$$e_A = 2\varepsilon_0 \rho_1 I$$

and, correspondingly, its potential and electric field are

$$U_0(p) = \frac{\rho_1 I}{2\pi L_{Ap}}, \quad \mathbf{E}_0(p) = \frac{\rho_1 I}{2\pi L_{Ap}^2} \mathbf{L}_{Ap}^0. \qquad [4.29]$$

Here, L_{Ap} is the distance between the current electrode A and the observation point p, $\mathbf{L}_{Ap}^0$ is the unit vector along $\mathbf{L}_{Ap}$, and I is the current passing through the electrode A. Due to this primary electric field $\mathbf{E}_0$, surface charges arise at the boundary between the upper layer and the base (underlying medium) and their density is given by

$$\sigma(q) = \varepsilon_0(\rho_2 - \rho_1)j_z,$$

where j_z is the vertical component of the current density. Taking into account the fact that the current through this interface equals I, the total charge at the plane boundary $z=h_1$ is

$$e_S = \int_S \sigma(q)\mathrm{d}S = \varepsilon_0(\rho_2 - \rho_1)\int_S j_z(q)\mathrm{d}S = \varepsilon_0(\rho_2 - \rho_1)I. \qquad [4.30]$$

Note that the value of j_z is defined by the total electric field but not by the primary field only; that is, the interaction of charges is taken into account. This charge (Eq. [4.30]) is positive when $\rho_2 > \rho_1$ or negative if $\rho_2 < \rho_1$, provided that $e_A > 0$. Elementary charges are distributed over the interface and create everywhere the electric field and, therefore, they induce charges on the earth's surface. As was pointed out earlier, each elementary charge located in a conducting medium produces at the boundary with an insulator a surface charge with the same magnitude and sign. For this reason, the induced charge on the earth's surface e_0 coincides with that of the lower boundary, e_S:

$$e_0 = \varepsilon_0(\rho_2 - \rho_1)I, \qquad [4.31]$$

but it is distributed in a different manner. Qualitatively, the distribution of charges e_S and e_0 is illustrated in Fig. 4.3C. It is essential that on the earth's surface, the density of charges decays more slowly than at the interface $z=h_1$. Thus, the total charge situated at the surfaces is

$$e = 2\varepsilon_0\rho_1 I + 2\varepsilon_0(\rho_2 - \rho_1) = 2\varepsilon_0\rho_2 I, \qquad [4.32]$$

and it coincides with the charge of an electrode located at the surface of a uniform half-space with the resistivity of the underlying medium. As was shown in Chapter 3, this analysis of the charge distribution allows us in a very simple manner to establish the asymptotic behavior of the field as a function of the distance between the current electrode and an observation point. In fact, let us represent the electric field at the point of observation 0 on the earth's surface as the sum:

$$E_r(0) = \frac{\rho_1 I}{2\pi r^2} + E_{rs}(0). \quad [4.33]$$

Here, $r = A0$, and $E_{rs}(0)$ is the tangential component of the secondary electric field caused by all charges except the electrode charge. Inasmuch as this component has everywhere a finite value, in approaching the current electrode, the field is mainly defined by the charge at the electrode surface; that is,

$$E_r(r) \to \frac{\rho_1 I}{2\pi r^2}. \quad [4.34]$$

This means that, with a decrease of the separation between the current and receiver electrodes r, the depth of investigation also decreases; that is, the field is practically defined by the resistivity of the upper layer ρ_1 only. In addition, let us point out that near the current electrode, the density σ_0 of induced charges on the earth's surface decreases very slowly and, correspondingly, its tangential component of the electric field is small. Besides, charges at the low boundary mainly create the vertical component of the field if the separation r is sufficiently small. For this reason, there is some interval of distances near the current source where the field is defined by Eq. [4.34]. In the opposite case, when the point of observation is far away from the current electrode, the influence of both surface charges e_S and e_0 is the same as if the total charge were placed at the current electrode. Then, we have

$$E_r(r) \to \frac{\rho_2 I}{2\pi r^2}. \quad [4.35]$$

Thus, with an increase of the separation r, the depth of investigation increases in spite of the fact that the current and receiver electrodes are located on the surface of the upper layer with resistivity ρ_1. This occurs because the tangential component of electric field as well as potential become functions of the basement resistivity ρ_2 only, and this result does not depend on the thickness of the upper layer. Moreover, as was shown earlier, this asymptotic behavior remains valid regardless of the number of layers and the presence of inhomogeneities of finite dimensions. In other words, these layers become "transparent." In essence, we outlined again the main concept of geometrical soundings based on measuring the voltage at different separations from the current electrodes. This approach has at least two merits. First of all, it is based on the behavior of the field sources (charges), which represents the physical foundation of the geometrical soundings. Secondary, we were able to obtain the asymptotic formulas without complicated derivations. Besides, it is possible to show that the behavior of the field on the earth's surface can be explained by an equivalent distribution of charges at this surface, so that the point of observation is located somewhere between the external and internal charges, as shown in Fig. 4.3D. With an increase of the distance from the current electrode, the relative contribution of external charges becomes smaller, and this means that the field gradually approaches its asymptotic value. We emphasized the role of charges for understanding the geometrical sounding, but at the same time, it is proper to notice that in two limiting cases: $\rho_2=0$ and $\rho_2=\infty$, this approach is hardly useful.

4.3.3. Formulation of the Boundary-Value Problem

Now, we will derive formulas for the potential and electric field at any separation from the current electrode, and with this purpose in mind, formulate the boundary-value problem. In accordance with the theorem of uniqueness, it requires that:

(1) Within the layer and underlying medium, the potential satisfies Laplace's equation:

$$\nabla^2 U_1 = 0 \quad \text{if } 0 \leq z \leq h_1$$

and

$$\nabla^2 U_2 = 0 \quad \text{if } z \geq h_1.$$

(2) In approaching the current electrode, the potential U_1 tends to that caused by the charge of this electrode (Fig. 4.3A):

$$U_1 \to \frac{\rho_1 I}{2\pi R} \quad \text{if } R \to 0,$$

where

$$R = (r^2 + z^2)^{1/2}.$$

(3) On the earth's surface, the normal component of the current density is zero and, therefore, for the normal component of electric field, we have

$$\frac{\partial U_1}{\partial z} = 0.$$

(4) With an increase of the distance R from the current electrode, the electric field and its potential tend to zero:

$$U_1 \to 0, \quad U_2 \to 0, \quad \text{as} \quad R \to \infty.$$

(5) At the interface $z = h_1$, the potential and the vertical component of the current density are continuous functions:

$$U_1 = U_2, \quad \gamma_1 \frac{\partial U_1}{\partial z} = \gamma_2 \frac{\partial U_2}{\partial z}.$$

Here, γ_1 and γ_2 are the conductivities of the upper and underlying layers, respectively. Note that, since the conducting medium is surrounded by surfaces where at each point the boundary conditions are defined, we do not need to determine the field above the earth's surface. In other words, we formulated a boundary-value problem which uniquely defines the potential and electric field. In particular, this means that the position of the current generator, as well as that of a wire connecting current electrodes, does not have any influence on the field behavior beneath the earth's surface. Besides, due to electrostatic induction, all constant charges situated above this surface also do not create the field inside a conducting medium.

4.3.4. Solution of Laplace's Equation

As the first step of a solution of the boundary-value problem, we find the potential that satisfies Laplace's equation. Let us choose the cylindrical system of coordinates r, ϕ, z with the origin at the point A where the current electrode is located. The z-axis is directed downwards. Taking into account the axial symmetry of the field,

$$\frac{\partial U}{\partial \phi} = 0,$$

Laplace's equation takes the form

$$\frac{\partial^2 U}{\partial r^2} + \frac{1}{r}\frac{\partial U}{\partial r} + \frac{\partial^2 U}{\partial z^2} = 0. \qquad [4.36]$$

This is a differential equation of the second order with partial derivatives, since the potential depends on two coordinates: r and ϕ. To solve this equation, we will suppose that its solution can be represented as the product of two functions and each of them depends on one argument only. Consequently, we have

$$U(r,z) = T(r)\Phi(z). \quad [4.37]$$

Substitution of Eq. [4.37] into Eq. [4.36] gives

$$\Phi\frac{d^2T}{dr^2} + \frac{\Phi}{r}\frac{d\Phi}{dr} + T\frac{d^2\Phi}{dz^2} = 0.$$

Dividing both sides of the equation by $T\Phi$, we obtain

$$\frac{1}{T}\frac{d^2T}{dr^2} + \frac{1}{rT}\frac{dT}{dr} + \frac{1}{\Phi}\frac{d^2\Phi}{dz^2} = 0. \quad [4.38]$$

It is natural to distinguish two terms on the left-hand side of Eq. [4.38]:

$$\text{Term}\,1 = \frac{1}{T}\frac{d^2T}{dr^2} + \frac{1}{rT}\frac{dT}{dr}, \quad \text{Term}\,2 = \frac{1}{\Phi}\frac{d^2\Phi}{dz^2}.$$

At first glance, it seems that they depend on the arguments r and z, respectively, and Eq. [4.38] can be written as

$$\text{Term}\,1(r) + \text{Term}\,2(z) = 0.$$

However, such equality is impossible since a change of one argument, for example, r, leads to a change of the first term while the second term remains the same and vice versa. Correspondingly, the sum of these terms cannot be equal to zero for arbitrary values of r and z. Therefore, we have to conclude that every term does not depend on the coordinates and it is a constant. This fact constitutes the key point of the method of separation of variables, allowing us to describe the potential $U(r, z)$ as a product of two functions. For convenience, we represent this constant in the form $\pm m^2$, where m is called a constant of separation. Thus, instead of Laplace's equation, we obtain two ordinary differential equation of the second order:

$$\frac{1}{T}\frac{d^2T}{dr^2} + \frac{1}{rT}\frac{dT}{dr} = \pm m^2$$

and [4.39]

$$\frac{1}{\Phi}\frac{d^2\Phi}{dz^2} = \mp m^2.$$

It may be proper to emphasize that the replacement of a differential equation with partial derivatives by two ordinary differential equations is the main purpose of the method of separation of variables, since their solutions are known. In order to find the function $T(r)$, we have to choose the sign of the right-hand side of the first equation of the set [4.39]. Later, we will demonstrate that if the positive sign is chosen, then, the solutions would be the modified Bessel functions $I_0(mr)$ and $K_0(mr)$, and they have singularities either at infinity or at points of the z-axis, respectively:

$$I_0(mr) \to \infty \quad \text{if } r \to \infty$$

and

$$K_0(mr) \to \infty \quad \text{if } r \to 0.$$

Inasmuch as all these points are located in the upper layer and in the base, functions $I_0(mr)$ and $K_0(mr)$ cannot describe the potential which has finite values everywhere except at the origin of coordinates. For this reason, we will choose the negative sign in equation for $T(r)$ and then obtain

$$\frac{d^2T}{dr^2} + \frac{1}{r}\frac{dT}{dr} + m^2T = 0. \qquad [4.40]$$

At the same time, the second equation of the set [4.39] becomes

$$\frac{d^2\Phi}{dz^2} - m^2\Phi = 0. \qquad [4.41]$$

Its partial solution is a combination of exponential functions:

$$\Phi(m, z) = C_m \exp(mz) + D_m \exp(-mz), \qquad [4.42]$$

where C_m and D_m are unknown coefficients which are independent of the coordinate z.

Introducing a new variable $x = mr$ and performing differentiation in Eq. [4.40],

$$\frac{dT}{dr} = \frac{dT}{dx}\frac{dx}{dr} = m\frac{dT}{dx}, \quad \frac{d^2T}{dr^2} = m^2\frac{d^2T}{dx^2},$$

we obtain

$$\frac{d^2T}{dx^2} + \frac{1}{x}\frac{dT}{dx} + T = 0. \qquad [4.43]$$

The solutions of this equation are Bessel functions of the first and second type but zero order, $J_0(x)$ and $Y_0(x)$, respectively. They were introduced more than hundred years ago and studied thoroughly and widely used in numerous theoretical and engineering problems. Since the functions $Y_0(mr)$ has a

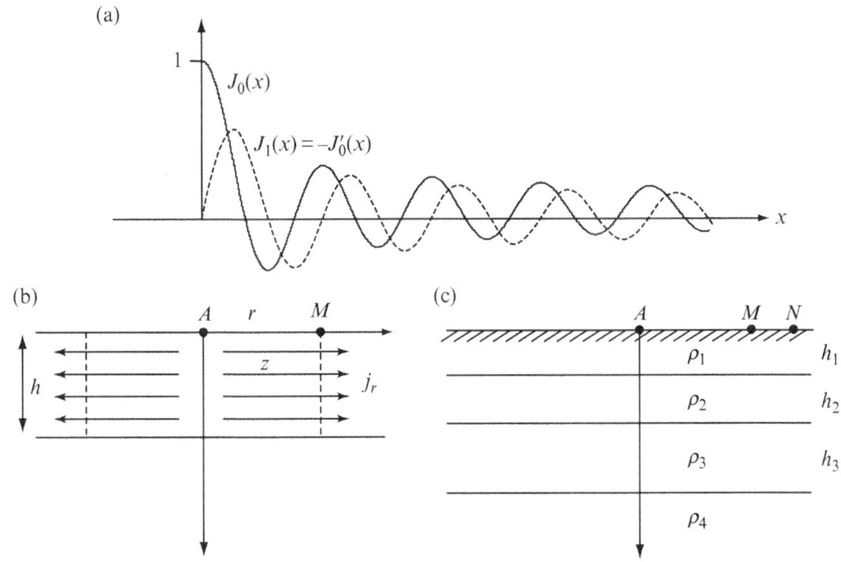

Figure 4.4 (A) Behavior of functions $J_0(x)$ and $J_0'(x)$, (B) distribution of currents when the base is insulator, and (C) illustration of Eq. [4.75].

logarithmic singularity at points of the z-axis ($r=0$), it cannot be used to describe the potential and electric field. The behavior of the function $J_0(x)$, and its first derivative $J_0'(x)$, is shown in Fig. 4.4A. In particular, we have

$$J_0(x) \to 1 \quad \text{if } x \to 0$$

and

$$J_0(x) \to \left(\frac{2}{\pi x}\right)^{1/2} \cos\left(x - \frac{\pi}{4}\right) \quad \text{if } x \to \infty. \qquad [4.44]$$

Thus, for every value of the separation constant, we obtain

$$U_m(r,z,m) = T(r,m)\Phi(z,m) = \left(C_m e^{mz} + D_m e^{-mz}\right)J_0(mr). \qquad [4.45]$$

By definition, the functions $U_m(r,z,m)$ satisfy Laplace's equation, and we may think that the first step of solving the boundary-value problem is accomplished. However, this assumption is incorrect, since the function $U_m(r,z,m)$ depends on m which appears as a result of the transformation of Laplace's equation into two ordinary differential equations. At the same time, the potential U describing the electric field $\mathbf{E}$ in the medium is independent of m. Inasmuch as the function $U_m(r,z,m)$, given by Eq. [4.45], obeys Laplace's equation for any m, we will represent the potential U as a superposition of partial solutions U_m; that is, in the form of a definite integral

$$U(r,z)=\int_0^\infty (C_m e^{mz}+D_m e^{-mz})J_0(mr)\,dm, \qquad [4.46]$$

which is independent of m. In fact, an integrand is a function of m, but after an integration and substitution of the upper and low limits, this dependence disappears. Thus, we have arrived at the general solution of Laplace's equation which includes an infinite number of partial solutions corresponding to different values of m: $0 \leq m < \infty$.

4.3.5. Expressions for the Potential and Electric Field

Representing the potential $U_1(r,z)$ in the upper layer by

$$U_1(r,z)=\frac{\rho_1 I}{2\pi R}+\int_0^\infty (C_m e^{mz}+D_m e^{-mz})J_0(mr)\,dm, \qquad [4.47]$$

we see that the function $U_1(r,z)$ satisfies two conditions of the boundary-value problem: the condition near the source and Laplace's equation. To satisfy the condition on the earth's surface, $\partial U_1/\partial z = 0$, we take the first derivative of the potential with respect to z. This gives

$$\frac{\partial U_1}{\partial z}=-\frac{\rho_1 Iz}{2\pi(r^2+z^2)^{3/2}}+\int_0^\infty m(C_m e^{mz}-D_m e^{-mz})J_0(mr)\,dm. \qquad [4.48]$$

Letting $z=0$, we have

$$\int_0^\infty m(C_m-D_m)J_0(mr)\,dm=0, \qquad [4.49]$$

that is, the normal component of the current density of both the primary and secondary fields vanishes on the earth's surface. The last equality is a very complicated integral equation with respect to unknowns C_m and D_m. Fortunately for us, integrals of the type

$$\int_0^\infty \phi(m)J_0(mr)\,dm=0$$

have one remarkable feature similar to that of Fourier integrals, namely, from the equality

$$\int_0^\infty \phi_1(m)J_0(mr)\mathrm{d}m = \int_0^\infty \phi_2(m)J_0(mr)\mathrm{d}m$$

or

$$\int_0^\infty [\phi_1(m) - \phi_2(m)]J_0(mr)\mathrm{d}m = 0, \qquad [4.50]$$

it follows that

$$\phi_1(m) = \phi_2(m). \qquad [4.51]$$

Thus, Eq. [4.49] is drastically simplified and we have

$$C_m = D_m,$$

and the expression for the function U_1 becomes

$$U_1(r,z) = \frac{\rho_1 I}{2\pi R} + \int_0^\infty C_m(\mathrm{e}^{mz} + \mathrm{e}^{-mz})J_0(mr)\mathrm{d}m. \qquad [4.52]$$

In the underlying medium (base), where the coordinate z unlimitedly increases, the term with $\exp(mz)$ has to be absent. Therefore, a solution of Laplace's equation is represented in the form

$$U_2(r,z) = \int_0^\infty B_m\mathrm{e}^{-mz}J_0(mr)\mathrm{d}m, \qquad [4.53]$$

and the condition at infinity is satisfied. Note that, due to the oscillating character of Bessel function $J_0(mr)$, both functions U_1 and U_2 tend to zero as $r \to \infty$. To satisfy the condition at the bottom of the upper layer, it is necessary to represent the potential of the primary and secondary fields in the same form. With this purpose in mind, we will make use of the Lipschitz's integral

$$\frac{1}{R} = \frac{1}{(r^2+z^2)^{1/2}} = \int_0^\infty \mathrm{e}^{-m|z|}J_0(mr)\mathrm{d}m \qquad [4.54]$$

or

$$\frac{1}{(r^2+z^2)^{1/2}} = \int_0^\infty \mathrm{e}^{-mz}J_0(mr)\mathrm{d}m \quad \text{if } z > 0. \qquad [4.55]$$

Substitution of Eq. [4.55] into Eq. [4.52] gives for potential in the upper medium the following expression:

$$U_1(r, z) = \int_0^\infty [Ce^{-mz} + C_m(e^{mz} + e^{-mz})] J_0(mr)dm. \qquad [4.56]$$

Here,

$$C = \frac{\rho_1 I}{2\pi}.$$

Correspondingly, the continuity of the potential and normal component of the current density take place at the interface $z = h_1$ if

$$\int_0^\infty \{C\exp(-mh_1) + C_m[\exp(mh_1) + \exp(-mh_1)]\}J_0(mr)dm$$
$$= \int_0^\infty B_m \exp(-mh_1)J_0(mr)dm$$

and

$$\gamma_1 \int_0^\infty \{-C\exp(-mh_1) + C_m[\exp(mh_1) - \exp(-mh_1)]\}mJ_0(mr)dm$$
$$= -\gamma_2 \int_0^\infty B_m \exp(-mh_1)mJ_0(mr)dm.$$

[4.57]

Now, making use again Eqs. [4.50] and [4.51], the set [4.57] is greatly simplified, and we obtain for every m two equations with two unknowns:

$$C\exp(-mh_1) + C_m[\exp(mh_1) + \exp(-mh_1)] = B_m \exp(-mh_1)$$

and

$$\gamma_1\{-C\exp(-mh_1) + C_m[\exp(mh_1) - \exp(-mh_1)]\} = -\gamma_2 B_m \exp(-mh_1).$$

[4.58]

Let us note that here we observe one of the most important features of the so-called special functions, such as Bessel functions, namely, their orthogonality. This property makes them extremely useful in solving many boundary-value problems. The system [4.58] gives

$$C_m = \frac{\rho_1 I}{2\pi} \frac{K_{12}\exp(-2mh_1)}{1 - K_{12}\exp(-2mh_1)}$$

and [4.59]

$$B_m = \frac{\rho_1 I}{2\pi} \frac{1 + K_{12}}{1 - K_{12}\exp(-2mh_1)},$$

where

$$K_{12} = \frac{\rho_2 - \rho_1}{\rho_2 + \rho_1} \qquad [4.60]$$

is the contrast coefficient and, as we know, it varies within the range

$$-1 \le K_{12} \le 1.$$

Therefore, the functions $U_1(r,z)$ and $U_2(r,z)$, given by Eqs. [4.52] and [4.53], obey all conditions of the boundary-value problem provided that the coefficients C_m and B_m are defined by Eqs. [4.59]. In other words, these functions describe the potential of the electric field when the current electrode is located at the origin of coordinates on the earth's surface. In particular, if an observation point is also situated at this surface $z=0$, we have

$$U_1(r,0) = \frac{\rho_1 I}{2\pi}\left[\frac{1}{r} + 2K_{12}\int_0^\infty \frac{\exp(-2mh_1)}{1 - K_{12}\exp(-2mh_1)} J_0(mr)\mathrm{d}m\right]$$

and [4.61]

$$E_r(r,0) = \frac{\rho_1 I}{2\pi}\left[\frac{1}{r^2} + 2K_{12}\int_0^\infty \frac{m\exp(-2mh_1)}{1 - K_{12}\exp(-2mh_1)} J_1(mr)\mathrm{d}m\right].$$

Here, $E_r = -\partial U/\partial r$ is the radial component of the electric field, which is tangential to the earth's surface and

$$J_1(x) = -\frac{\partial J_0(x)}{\partial x} = -J_0'(x)$$

is the Bessel function of the first order. Note that the normal component E_z vanishes in the upper layer in the vicinity of the boundary $z=0$, but it differs from zero at the opposite side. We have derived formulas for two- and three-electrode arrays, but applying the principle of superposition, it is a simple matter to obtain an expression of the voltage for an arbitrary system of electrodes. For instance, in the case of the symmetrical array *AMNB*, we have

$$V_{MN} = \frac{\rho_1 IMN}{\pi}\left[\frac{1}{r^2} + 2K_{12}\int_0^\infty \frac{m\exp(-2mh_1)J_1(mr)}{1 - K_{12}\exp(-2mh_1)}\mathrm{d}m\right], \qquad [4.62]$$

since charges at both current electrodes A and B cause the same voltage between the receiver electrodes. Here, $r = AB/2 = A0$ and $MN \ll A0$. As follows from the second equation of the set [4.61], for the voltage of the axial dipole array, we have

$$V_{MN} = \frac{\rho_1 IMN \cdot AB}{\pi}\left[\frac{1}{r^3} - K_{12}\int_0^\infty \frac{m^2\exp(-2mh_1)}{1-K_{12}\exp(-2mh_1)}J_1'(mr)\mathrm{d}m\right], \quad [4.63]$$

since the voltage V_{MN} is the difference of the voltages due to current electrodes A and B and

$$\frac{1}{r_{A0}^2} - \frac{1}{r_{B0}^2} = -AB\frac{\partial}{\partial r}\frac{1}{r^2} = \frac{2AB}{r^3},$$

$$J_1(mr_{A0}) - J_1(mr_{B0}) = -AB\frac{\partial}{\partial r}J_1(mr) = -AB\cdot mJ_1'(mr).$$

Algorithms for calculating integrals in Eqs. [4.61] are well developed and described later in some detail.

4.3.6. Asymptotic Behavior of the Field

Before we describe the results of field calculations, let us use integrals in Eqs. [4.61] and study again the asymptotic behavior of the potential and its electric field. As follows from the first equation of the set [4.61], in approaching the current electrode, $J_0(mr) \to 1$ and the integral has a finite value. Therefore, as was pointed out earlier, the condition in the vicinity of the electrode A is met:

$$U_1(r,z) \to \frac{\rho_1 I}{2\pi r} \quad \text{if } r \to 0.$$

At the same time, the integrand for the electric field E_r tends to zero since $J_1(mr) \to 0$ if $r \to 0$. This result can be expected because of the fact that near the electrode A, the charges are almost uniformly distributed on the earth's surface and the tangential component of the field is mainly defined by the charge e_A:

$$E_r \to E_0 = \frac{\rho_1 I}{2\pi r^2}. \quad [4.64]$$

In the opposite case when the distance r unlimitedly increases and $r/h_1 \gg 1$, we will take into account the fact that, due to the oscillating character of Bessel function $J_0(mr)$, the asymptotic value of the integral

$$\int_0^\infty \phi(m)J_0(mr)\mathrm{d}m$$

is defined by values of the integrand at the initial part of integration, $m \to 0$. Then, assuming that exponents in Eq. [4.61] are equal to unity, we have

$$\frac{m\exp(-2mh_1)}{1-K_{12}\exp(-2mh_1)}\to\frac{m}{1-K_{12}}\quad\text{as}\quad m\to 0$$

and

$$E_r(r,0)=\left[\frac{1}{r^2}+\frac{2K_{12}}{1-K_{12}}\int_0^\infty mJ_1(mr)dm\right]\quad\text{if } r\to 0.$$

Since

$$mJ_1(mr)=-\frac{\partial}{\partial r}J_0(mr),$$

and as it follows from Eq. [4.54]

$$\frac{1}{r}=\int_0^\infty J_0(mr)dm,$$

we have

$$E_r(r,0)=\frac{\rho_1 I}{2\pi}\left(\frac{1}{r^2}-\frac{2K_{12}}{1-K_{12}}\frac{\partial}{\partial r}\frac{1}{r}\right)\quad\text{if } r\to\infty$$

or

$$E_r(r,0)=\frac{\rho_1 I}{2\pi r^2}\frac{1+K_{12}}{1-K_{12}}=\frac{\rho_2 I}{2\pi r^2}. \qquad [4.65]$$

Thus, we see again that, regardless of the ratio of resistivity with an increase of the separation r, the electric field approaches the value of the field in a uniform half-space with the resistivity of the base provided that $\rho_2\neq 0$ or $\rho_2\neq\infty$. Of course, both asymptotes when either $r\to 0$ or $r\to\infty$ have been already derived earlier proceeding from the charge distribution. Next, we will study an asymptotic behavior of the field when the base is an insulator or an ideal conductor.

4.4. ASYMPTOTIC BEHAVIOR OF THE FIELD WHEN A BASE IS AN INSULATOR

In the case when the underlying medium is an insulator, Eq. [4.61] becomes

$$E_r(r,0)=\frac{\rho_1 I}{2\pi}\left[\frac{1}{r^2}+2\int_0^\infty\frac{m\exp(-2mh_1)}{1-\exp(-2mh_1)}J_1(mr)dm\right], \qquad [4.66]$$

since

$$K_{12} = 1.$$

As before, with a decrease of a distance r, the influence of the first term becomes dominant and

$$E_r(r,0) \rightarrow \frac{\rho_1 I}{2\pi r^2}.$$

In deriving the asymptotic formula for the field far away from the current electrode, we will follow two approaches.

4.4.1. The First Approach

When $r \rightarrow \infty$, we assume that the integral is still defined by its initial part ($m \rightarrow 0$) and this gives

$$\frac{m\exp(-2mh_1)}{1-\exp(-2mh_1)} \rightarrow \frac{m}{2mh_1} = \frac{1}{2h_1} \quad \text{as} \quad m \rightarrow 0.$$

Thus, we obtain

$$E_r(r,0) = \frac{\rho_1 I}{2\pi}\left[\frac{1}{r^2} + \frac{1}{h_1}\int_0^\infty J_1(mr)\mathrm{d}m\right] = \frac{\rho_1 I}{2\pi}\left(\frac{1}{r^2} + \frac{1}{rh_1}\right) \quad \text{if } r \rightarrow \infty \tag{4.67}$$

since

$$\int_0^\infty J_v(mr)\mathrm{d}m = \frac{1}{r},$$

and $J_v(mr)$ is the Bessel function of the v order. As follows from Eq. [4.67], far away from the current electrode, the secondary field prevails and, thus, neglecting the first term, we have

$$E_r(r,0) = \frac{I}{2\pi S_1 r} \quad \text{if } r/h_1 \gg 1. \tag{4.68}$$

Here, $S_1 = \gamma_1 h_1$ is the conductance of the upper layer, and, by definition, it is a product of conductivity and thickness of the layer. We see that with an increase of the distance the electric field E_r becomes inversely proportional to the conductance S_1 and it does not depend separately on the thickness and resistivity of the layer. From a practical point of view, this means that by performing measurements far away from the current electrode, $r/h_1 \gg 1$, we can determine the conductance S_1 only. It is useful to point out that far away from the electrode A the measured field decreases inversely proportional to r, that is relatively slowly, and it is caused by the fact that the rate of a decay of the surface charges with the distance r is rather low. Note also that Eq. [4.68], as well as its generalization of an n-layered medium, plays an important role in the interpretation of the geometric soundings.

4.4.2. The Second Approach

It is also instructive to derive this equation in a different way. From the physical point of view, it is clear that with an increase of the separation r, the vector of the current density becomes practically horizontal and independent of both coordinates r and z (Fig. 4.4B). Then, due to the axial symmetry, the total current through any lateral surface of the cylinder with the radius r and height h_1 is

$$I = j_r 2\pi r h_1.$$

Taking into account Ohm's law $j_r = \gamma_1 E_r$, we obtain

$$E_r = \rho_1 j_r = \frac{I}{2\pi S_1 r}$$

which coincides with Eq. [4.68]. It is always useful to derive the same equation from both the mathematical and physical points of view. For instance, in our case the use of both approaches allowed us to understand that the asymptotic behavior of the field, described by Eq. [4.68], corresponds to a uniform distribution of the current density in the upper layer. It turns out that in this limiting case, the density of surface charges varies so slowly that regardless of a distance r, their field cannot be replaced by that of a point charge.

4.4.3. Asymptotic Behavior in an *N*-Layered Medium

Moreover, this physical approach allows us to generalize Eq. [4.68] for an n-layered medium provided that the base is an insulator. In fact, inasmuch as the electric field is horizontal far away from the electrode A, the equipotential surfaces are cylindrical, as it is for the two-layered medium (Fig. 4.4C). We will consider two arbitrary equipotential surfaces located at the distance MN from each other. At any layer, the difference of potentials between them, that is voltage, is the same regardless of the layer, and it can be written as

$$V = I_1 R_1 = I_2 R_2 = I_3 R_3 = \cdots = I_n R_n, \qquad [4.69]$$

where I_i is the current in the ith layer, R_i is the resistance of the cylindrical layer with the thickness MN and height h_i. By definition,

$$V = E_r MN, \quad R_i = \frac{\rho_i MN}{2\pi r h_i} = \frac{MN}{2\pi S_i r}. \qquad [4.70]$$

Here, $S_i = \gamma_i h_i$ is the conductance of the ith layer. Note that the second equality in the set [4.70] implies that $r \gg MN$. Substitution of Eqs. [4.70] into Eq. [4.69] yields the first relation between the measured voltage on the earth's surface and the conductance of layers:

$$2\pi E_r = \frac{I_1}{S_1} = \frac{I_2}{S_2} = \cdots = \frac{I_n}{S_n}. \qquad [4.71]$$

Also, at great distances from the current electrode, it is natural to treat layers as conductors connected in parallel. This gives

$$I = I_1 + I_2 + \cdots + I_n. \qquad [4.72]$$

Note that the current I is measured, while currents I_i are unknowns. From the last two equations, we obtain

$$I = I_1 + \frac{S_2}{S_1} I_1 + \frac{S_3}{S_1} I_1 + \cdots + \frac{S_n}{S_1} I_1.$$

Thus, we found the relation between the total current and that in the first layer:

$$I = I_1 \frac{S}{S_1}, \qquad [4.73]$$

where

$$S = \sum_{i=1}^{n} S_i \qquad [4.74]$$

is called the total conductance of a system of layers, and often it is a very informative parameter of a geoelectric section. Finally, from Eqs. [4.71] and [4.73], we obtain

$$E_r(r, 0) = \frac{I_1}{2\pi S_1 r} = \frac{I}{2\pi S r} \quad \text{if } r/h \gg 1. \qquad [4.75]$$

Here,

$$h = \sum_{1}^{n} h_i \qquad [4.76]$$

is the total thickness of layers above the underlying insulator. Similarity of Eqs. [4.68] and [4.75] is obvious.

4.4.4. The S-Zone

We derived the asymptotic expression for the field, Eqs. [4.75], conditional that the separation between the current and receiver electrodes is much greater than the total thickness of layers and the base is an insulator. At the same time, we may expect such a behavior in real conditions where an underlying layer has a relatively high but finite value of resistivity. For illustration, consider a two-layer medium where $\rho_2 \gg \rho_1$. Intuitively, we can imagine the following behavior of the current: At small distances from the electrode A, the vector of current density has mainly radial direction, and it indicates that the influence of the interface $z = h_1$ is insignificant. With the increase of the distance, the current lines tend to be parallel to this boundary and it happens because the leakage of the current into a base is very small. If measurements of the voltage take place within this range of distances, we may see that this field is defined by the conductance S_1 (Eq. [4.68]). At greater distances, the effect of leakage becomes more noticeable, and the density of charges decreases more rapidly. Correspondingly, the field approaches that of a uniform half-space with resistivity ρ_2. Thus, there is an intermediate range of separations where the field is defined by the longitudinal conductance S_1, and this interval is called the S-zone. A similar behavior can also be observed in the presence of several layers.

4.4.5. Distribution of Charges on the Surface of a Base

We found out that far away from the current electrode, the electric field has only the radial component E_r which is inversely proportional conductance and it is independent of the coordinate z. It is obvious that such peculiar behavior is related to the distribution of charges at the boundary with an insulator as well as on the earth's surface. In this light, let us find the density of charges at distances greatly exceeding the thickness h_1. By definition, at each point of the interface $z = h_1$, we have

$$E_{2z}(q) - E_{1z}(q) = \frac{\sigma(q)}{\varepsilon_0} \quad \text{or} \quad \sigma(q) = \varepsilon_0 E_{2z} \qquad [4.77]$$

since the normal component E_{1z} is equal to zero at the boundary with an insulator. In accordance with Eqs. [4.53] and [4.59], the vertical component of the electric field at this boundary is

$$E_{2z}(h_1, r) = \frac{\rho_1 I}{2\pi}(1 + K_{12}) \int_0^\infty \frac{m \exp(-mh_1)}{1 - K_{12}\exp(-2mh_1)} J_0(mr)\mathrm{d}m. \qquad [4.78]$$

Assuming that the second medium is an insulator and the field is studied far away from the current electrode, we focus only at the initial part of the integration. This gives

$$\sigma(q) = \varepsilon_0 \frac{\rho_1 I}{2\pi h_1} \int_0^\infty J_0(mr)\mathrm{d}m,$$

since $K_{12}=1$, $\exp(-mh_1) \approx 1$, and $\exp(-2mh_1) \approx 1 - 2mh_1$. Bearing in mind that integral is equal to $1/r$, we have

$$\sigma(r) = \varepsilon_0 \frac{I}{2\pi S_1 r}. \qquad [4.79]$$

This shows that a charge density decreases very slowly, which may explain why in this case, surface charges cannot be replaced by the same charge on the electrode A regardless of how far away the field is measured.

4.5. ASYMPTOTIC BEHAVIOR OF THE FIELD WHEN THE BASE IS AN IDEAL CONDUCTOR

4.5.1. Transformation of the Expression of the Potential (Eq. [4.61])

Again, proceeding from Eqs. [4.61], we have on the earth's surface when $\rho_2=0$

$$U_1(r, 0) = \frac{\rho_1 I}{2\pi}\left[\frac{1}{r} - 2\int_0^\infty \frac{\exp(-2mh_1)}{1 + \exp(-2mh_1)} J_0(mr)\mathrm{d}m\right] \qquad [4.80]$$

since $K_{12}=-1$. It is obvious that in approaching the current electrode, the field is mainly defined by its charge

$$U_1(r, 0) \to \frac{\rho_1 I}{2\pi r}.$$

In order to study the asymptotic behavior of the field at large distances r, it is convenient to perform some transformations in Eq. [4.80]. Taking into account the equality

$$\frac{1}{r} = \int_0^\infty J_0(mr)\mathrm{d}m,$$

we can represent Eq. [4.80] as

$$U_1(r,0) = \frac{\rho_1 I}{2\pi}\int_0^\infty \frac{1-\exp(-2mh_1)}{1+\exp(-2mh_1)} J_0(mr)\mathrm{d}m$$

or

$$U_1(mr) = \frac{\rho_1 I}{2\pi}\int_0^\infty \tanh(mh_1)J_0(mr)\mathrm{d}m. \qquad [4.81]$$

First, assume as before that the potential and electric field at large distances from the current electrode are defined by small values of m, that is, the initial part of integration plays the dominant role, and then demonstrate what happens in this case. At the beginning, let us prove the following equality:

$$\int_0^\infty m^p J_0(mr)\mathrm{d}m = 0 \qquad [4.82]$$

if p is an odd number. Performing a differentiation of Eq. [4.55] with respect to z, we have

$$z(r^2+z^2)^{-3/2} = \int_0^\infty m\exp(-mz)J_0(mr)\mathrm{d}m. \qquad [4.83]$$

Therefore,

$$\int_0^\infty mJ_0(mr)\mathrm{d}m = 0.$$

Repeating the differentiation of Eq. [4.83] and letting each time $z=0$, we arrive at Eq. [4.82]. Now, we will try to derive the asymptotic expression for the potential $U_1(r,0)$ for large values of r. As is well known for small values of m, we have

$$\tanh mh_1 \approx mh_1.$$

Then, Eq. [4.81] becomes

$$U_1(r,0) \approx \frac{\rho_1 I h_1}{2\pi}\int_0^\infty mJ_0(mr)\mathrm{d}m$$

or

$$U_1(r,0) \approx 0 \quad \text{if} \ \ r/h_1 \gg 1,$$

that is, in this approximation the potential of the secondary field cancels that of the primary one, and it suggests that the total field is much smaller. Inasmuch as the Mc Laurent series of the function tanh x contains only terms with odd powers of x,

$$\tanh x \approx x - \frac{x^3}{3} + \frac{2}{15}x^5 - \frac{17}{315}x^7 + \cdots,$$

its better approximation still does not allow us to obtain the asymptotic formulas for the potential and its field E_r. In other words, the method of evaluation of the field, based on the assumption that the initial part of the integration plays the dominant role when $r \to \infty$, is invalid. From the physical point of view, this relates to the fact that charges on the interfaces have the same magnitude as the charge on the current electrode but they differ in sign.

4.5.2. Deformation of the Contour of Integration in Eq. [4.81]

Now, let us attempt to change the interval of integration: $0 \le m < \infty$, in such way that it would be possible to derive asymptotic formulas. For this reason, we will apply a method based on Cauchy theorem and also use the equality

$$J_0(mr) = \frac{1}{2}[H_0^{(1)}(mr) + H_0^{(2)}(mr)]. \qquad [4.84]$$

Here, $H_0^{(1)}(mr)$ and $H_0^{(2)}(mr)$ are Hankel's functions of the first and second kind. As is well known from the theory of Bessel functions, their asymptotic expressions for large argument are

$$H_0^{(1)}(mr) \approx \left(\frac{2}{\pi mr}\right)^{1/2} \exp \mathrm{i}\left(mr - \frac{\pi}{4}\right)$$

and [4.85]

$$H_0^{(2)}(mr) \approx \left(\frac{2}{\pi mr}\right)^{1/2} \exp\left[-\mathrm{i}\left(mr - \frac{\pi}{4}\right)\right] \quad \text{if } mr \gg 1.$$

Substitution of Eq. [4.84] into Eq. [4.81] gives

$$U_1(r,0) = \frac{C}{2}\int_0^\infty F(m)H_0^{(1)}(mr)\mathrm{d}m + \frac{C}{2}\int_0^\infty F(m)H_0^{(2)}(mr)\mathrm{d}m, \qquad [4.86]$$

where

$$C = \frac{\rho_1 I}{2\pi}, \quad F(m) = \tanh mh_1. \qquad [4.87]$$

First, consider the behavior of the function $F(m)$ on the complex plane m even though the integration is performed along the real axis of m, $\mathrm{Re}\, m > 0$. Function

$$F(m) = \frac{1 - \exp(-2mh_1)}{1 + \exp(-2mh_1)} \qquad [4.88]$$

has singularities which are called poles, and their position is determined from the condition

$$1 + \exp(-2mh_1) = 0 \quad \text{or} \quad \exp(-2mh_1) = -1 = \exp[\mathrm{i}\pi(2n-1)].$$

This means that

$$-2m_n h_1 = \mathrm{i}\pi(2n-1) \quad \text{or} \quad m_n = -\mathrm{i}\frac{\pi}{2}\frac{2n-1}{h_1}, \qquad [4.89]$$

that is, the poles are located on the imaginary axis, $\mathrm{Im}\, m < 0$. Thus, $F(m)$ is an analytical function on the complex plane m except the infinite set of poles situated from each other at a distance equal to π/h_1. As follows from the theory of function of complex variables, the integral of a function along any closed path is zero if the function is analytical inside the area bounded by this path. This remarkable feature is used in deriving the asymptotic formula. In accordance with Eqs. [4.85], Hankel's functions of the first and second kind decay exponentially with the increase of m if $\mathrm{Im}\, m > 0$ and $\mathrm{Im}\, m < 0$, respectively. Taking into account this fact, consider the integral

$$\oint_{C_1} F(m) H_0^{(1)}(mr) \mathrm{d}m \qquad [4.90]$$

along the closed path C_1 located in the first quadrant of the complex plane m (Fig. 4.5A). Due to the exponential decay of $H_0^{(1)}(mr)$ for large arguments, the integral along the portion of the path with an infinitely large radius tends to zero (Jordan lemma), and we have

$$\int_0^{\infty} F(m) H_0^{(1)}(mr) \mathrm{d}m + \int_{\mathrm{i}\infty}^{0} F(m) H_0^{(1)}(mr) \mathrm{d}m = 0$$

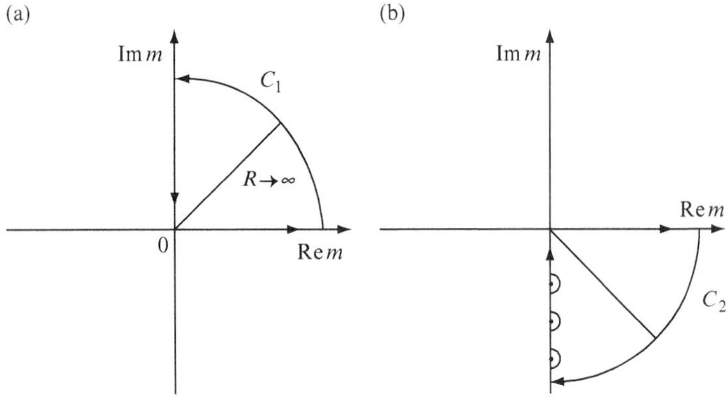

Figure 4.5 Contours of integration.

or

$$\int_0^{\infty} F(m)H_0^{(1)}(mr)\mathrm{d}m = \int_0^{\mathrm{i}\infty} F(m)H_0^{(1)}(mr)\mathrm{d}m, \qquad [4.91]$$

and integration along the real axis of m is replaced by that along the imaginary one. Next, consider the closed path C_2, situated in the fourth quadrant (Fig. 4.5B), and it includes again the real axis of m, the portion of the circle with an infinitely large radius ($R \to \infty$), the axis $\mathrm{Im}\, m < 0$, and finally, the set of semicircles around poles. Inasmuch as inside this closed path C_2, the integrand of the second term in Eq. [4.86] is an analytical function, we have

$$\oint_{C_2} F(m)H_0^{(2)}(mr)\mathrm{d}m = 0, \qquad [4.92]$$

or taking into account Jordan lemma,

$$\int_0^{\infty} F(m)H_0^{(2)}(mr)\mathrm{d}m + \int_{-\mathrm{i}\infty}^{0} F(m)H_0^{(2)}(mr)\mathrm{d}m + \sum = 0$$

or

$$\int_0^{\infty} F(m)H_0^{(2)}(mr)\mathrm{d}m = \int_0^{-\mathrm{i}\infty} F(m)H_0^{(2)}(mr)\mathrm{d}m - \sum, \qquad [4.93]$$

where $\sum$ is a sum of integrals along semicircles in the vicinity of the poles. Thus, making use of Eqs. [4.91] and [4.93] in place of Eq. [4.86], we have

$$U_1(r,0) = \frac{C}{2}\left[\int_0^{\mathrm{i}\infty} F(m)H_0^{(1)}(mr)\mathrm{d}m + \int_0^{-\mathrm{i}\infty} F(m)H_0^{(2)}(mr)\mathrm{d}m - \sum\right]. \qquad [4.94]$$

We derived three expressions for the potential, and it seems the last one is even more complicated than others, but it is an apparent impression. In fact, since

$$F(m) = -F(-m) \quad \text{and} \quad H_0^{(1)}(mr) = -H_0^{(2)}(-mr),$$

we obtain

$$\int_0^{i\infty} F(m)H_0^{(1)}(mr)dm = \int_0^{i\infty} F(-m)H_0^{(2)}(-mr)dm = \int_{-i\infty}^{0} F(m)H_0^{(2)}(mr)dm.$$

This means that the sum of the integrals in Eq. [4.94] is equal to zero and

$$U_1(r,0) = -\frac{\rho_1 I}{4\pi}\sum. \qquad [4.95]$$

Thus, a determination of the potential is reduced to a calculation of residuals of the integrand

$$F(m)H_0^{(2)}(mr),$$

and this result justifies a transition from integration along the real axis to that in the complex plane of m. Bearing in mind that the calculation of the residual is related to an integration only along semicircles, we have

$$U_1(r,0) = -\frac{\rho_1 I}{4\pi} i\pi \sum \text{Res}\, F(m)H_0^{(2)}(mr). \qquad [4.96]$$

Here,

$$\text{Res} = \frac{\phi_1(m_n)}{\phi_2(m_n)}, \qquad [4.97]$$

$$\phi_1(m_n) = [1 - \exp(-2m_n h_1)]H_0^{(2)}(m_n r),$$

and

$$\phi_2(m_n) = [1 + \exp(-2mh_1)]' \quad \text{if } m = m_n,$$

where m_n characterizes a position of the poles along the imaginary axis m. From Eq. [4.89], we have

$$\phi_1(m) = \{1 - \exp[i\pi(2n-1)]\}H_0^{(2)}\left[-\frac{i}{2}(2n-1)\frac{r}{h_1}\right]$$

or

$$\phi_1(m) = 2H_0^{(2)}\left[-\mathrm{i}\frac{\pi}{2}(2n-1)\frac{r}{h_1}\right]$$

and

$$\phi_2(m_n) = -2h_1 \exp(-2m_n h_1) = -2h_1 \exp[\mathrm{i}\pi(2n-1)] = 2h_1.$$

Thus, for each n we have

$$\mathrm{Res} = \frac{1}{h_1} H_0^{(2)}\left[-\mathrm{i}\frac{\pi}{2}(2n-1)\frac{r}{h_1}\right] = \frac{2\mathrm{i}}{\pi h_1} K_0\left[\frac{\pi}{2}(2n-1)\frac{r}{h_1}\right] \qquad [4.98]$$

and, therefore,

$$U_1(r,0) = \frac{\rho_1 I}{2\pi h_1}\sum_{n=1}^{\infty} K_0\left[\frac{\pi}{2}(2n-1)\frac{r}{h_1}\right]. \qquad [4.99]$$

Here, $K_0(x)$ is the modified Bessel function which plays an important role in many applications, in particular, in the theory of borehole geophysics. With the increase of the argument, it exponentially decays:

$$K_0(x) \to \left(\frac{\pi}{2x}\right)^{1/2} \exp(-x) \quad \text{if } x \gg 1. \qquad [4.100]$$

Since we are interested in the asymptotic behavior of the field, it is sufficient to consider only the first term of the sum [4.99], which gives

$$U_1(r,0) = \frac{\rho_1 I}{2\pi h_1} K_0\left(\frac{\pi}{2}\frac{r}{h_1}\right) \qquad [4.101]$$

or

$$U_1(r,0) = \frac{\rho_1 I}{2\pi h_1}\left(\frac{h_1}{r}\right)^{1/2} \exp\left(-\frac{\pi}{2}\frac{r}{h_1}\right) \quad \text{if } r/2h_1 > 1. \qquad [4.102]$$

Correspondingly, for the radial component of the electric field, we have

$$E_\mathrm{r} \approx \frac{\rho_1 I}{4\pi h_1^{1/2} r^{3/2}}\left(1+\pi\frac{r}{h_1}\right)\exp\left(-\frac{\pi}{2}\frac{r}{h_1}\right) \qquad [4.103]$$

or

$$E_\mathrm{r} \approx \frac{\rho_1 I}{4h_1^{3/2} r^{1/2}} \exp\left(-\frac{\pi}{2}\frac{r}{h_1}\right) \quad \text{if } r/2h_1 \gg 1. \qquad [4.104]$$

The latter indicates that the charges are distributed at both interfaces in such a way that the field decays very rapidly with the distance from the current electrode, and it demonstrates the high sensitivity to a change of the thickness as soon as $r/2h_1 > 1$. Note that in principle, the same approach based on the use of the residue theorem can be applied when the resistivity of the base is equal to infinity or, in the general case, an arbitrary value ρ_2.

4.6. APPARENT RESISTIVITY CURVES FOR A TWO-LAYERED MEDIUM

First, we perform a simple transformation in the second equation of the set [4.61]. Introduction of a new variable $x = mh_1$ in the expression for $E_r(r, 0)$ gives

$$E_r(r,0) = E_{0r}\Phi\left(\mu, \frac{r}{h_1}\right), \qquad [4.105]$$

where

$$E_{0r}(r,0) = \frac{\rho_1 I}{2\pi r^2}$$

is the primary electric field caused by the electrode charge only and

$$\Phi\left(\mu, \frac{r}{h_1}\right) = \left[1 + 2K_{12}\left(\frac{r}{h_1}\right)^2 \int_0^\infty \frac{x\exp(-2x)J_1[(r/h_1)x]}{1 - K_{12}\exp(-2x)}\,\mathrm{d}x\right], \qquad [4.106]$$

where

$$K_{12} = \frac{\mu - 1}{\mu + 1} \quad \text{and} \quad \mu = \frac{\rho_2}{\rho_1}.$$

In order to visualize a geoelectric section and observe the influence of the separation on the depth of investigation, it is convenient to use the apparent resistivity, which shows how the field differs from that in a uniform half-space with resistivity ρ_1. By definition,

$$\frac{\rho_a}{\rho_1} = \frac{E_r(\rho_1, \mu, r, h_1, I)}{E_{0r}(\rho_1, r, I)}.$$

Taking into account Eq. [4.105], we have

$$\frac{\rho_a}{\rho_1} = \Phi\left(\mu, \frac{r}{h_1}\right) \quad \text{or} \quad \rho_a = 2\pi r^2 \frac{E_r}{I}. \qquad [4.107]$$

Note that, by applying the log–log scale, Eq. [4.107] can be written as

$$\log \rho_a = \log \rho_1 + \log \Phi\left(\mu, \log\frac{r}{h_1}\right). \qquad [4.108]$$

It is clear that in this scale, a change of the resistivity of the upper layer ρ_1 and its thickness h_1 do not change the shape of the apparent resistivity curve, but it results in a parallel shift which is either vertical or horizontal. This fact is very useful for the interpretation of the geometric soundings. Curves of the apparent resistivity are given in Fig. 4.6, and they form two groups with $\mu > 1$ and $\mu < 1$, correspondingly. In accordance with a behavior of the field E_r, the left asymptote of all curves is equal to unity:

$$\rho_a \to \rho_1.$$

In this case, $r/h_1 < 1$, the field due to an electrode charge gives main contribution, and it happens because surface charges are distributed almost uniformly within this range. Correspondingly, the tangential component of

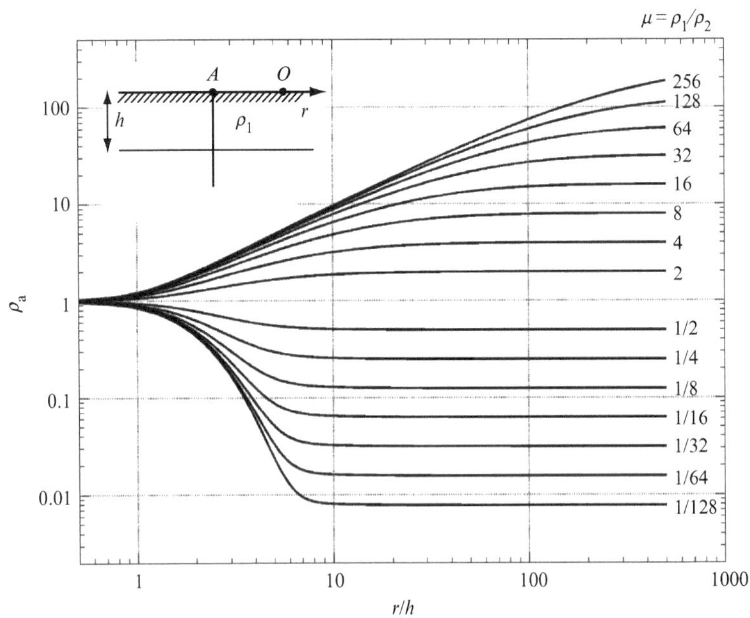

Figure 4.6 Apparent resistivity curves for a two-layer model.

the electric field caused by charges on the earth's surface is negligible, but charges on the base give rise mainly to the normal component. Then, with an increase of r, the influence of the base gradually increases, and, in the limit, the curves approach their right asymptote, equal to ρ_2/ρ_1; that is $\rho_a \rightarrow \rho_2$.

As is seen from these curves, with an increase of the resistivity contrast as well as the thickness h_1, an approach to the right asymptote takes place at greater distances and such behavior is in agreement with the distribution of surface charges. In the case of a much more resistive base, $\mu \gg 1$, there is an intermediate range of distances r where all curves approach that corresponding to a nonconducting base. Therefore, within this range, the apparent resistivity ρ_a depends on one parameter of the medium only, namely, the conductance of the upper layer S_1. As we know, it is related to the fact that the electric field in the upper layer has mainly the tangential component distributed uniformly with depth. It is interesting to note that this field decays with distance slightly more rapidly than $1/r$ because of the leakage of the current into a base. From Eq. [4.68], we have

$$\log \rho_a = \log r - \log S_1, \qquad [4.109]$$

that is, it is a straight line forming with the horizontal axis an angle $\pi/4$. This is a very useful relationship and, as follows from Eq. [4.75], remains valid in an n-layered medium.

Note that, if the slope of the apparent resistivity curve with $\mu \gg 1$ exceeds unity, it may indicate that there are lateral changes of the resistivity ρ_1. In the case of a much more conductive base ($\mu \ll 1$), there is an initial portion of the descending branch where the apparent resistivity curves are very close to their asymptote, $\mu = 0$. In this limiting case, we have (Eq. [4.104]):

$$\frac{\rho_a}{\rho_1} = \frac{\pi}{2}\left(\frac{r}{h_1}\right)^{3/2} \exp\left(-\frac{\pi}{2}\frac{r}{h_1}\right). \qquad [4.110]$$

With a further increase of r, the curves depart from this asymptote and each of them approaches its asymptotic value equal to ρ_2/ρ_1. Such a behavior is explained by the fact that with an increase in the separation, more and more charges are located relatively close to the electrode A, and they can be treated as the point charges. It may be proper to notice that in the case of a horizontally layered medium, the curves of the apparent resistivity curves never intersect their asymptote. As regards a more conductive basement ($\mu \ll 1$), it is possible to observe within some range of r a relatively high sensitivity of the apparent resistivity to the thickness h_1. We outlined some features of the

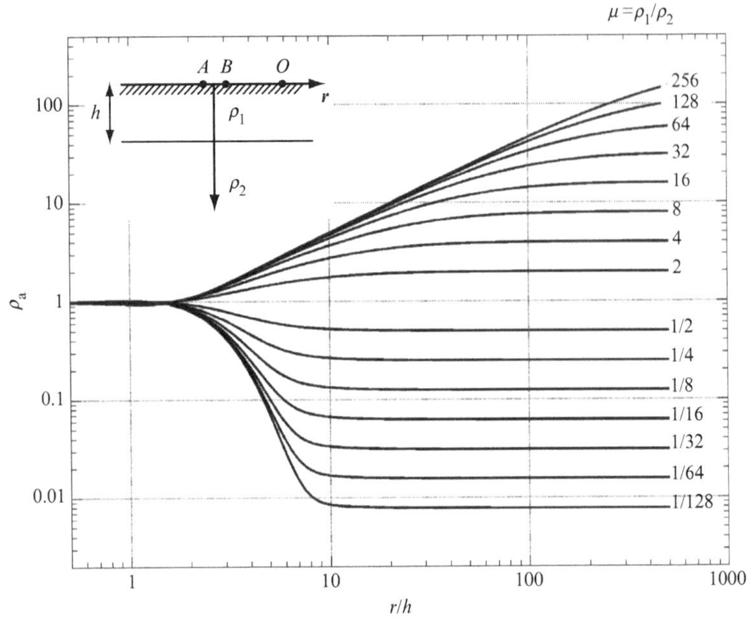

Figure 4.7 Apparent resistivity curves for an axial dipole–dipole array.

apparent resistivity curves which can be very useful for interpretation of the geometric soundings. In addition, let us point out that the apparent resistivity curves, shown in Fig. 4.6, were derived for a two-electrode array $A0$ assuming that the electric field is measured at the point 0. It is obvious that the same curves are observed for the three-electrode array AMN if the distance between the receiver electrodes is much smaller than the distance between the current and receiver electrodes; that is, $V_{MN} \approx E_r MN$. Besides, this result can also be applied for the symmetrical four-electrode array $AMNB$, when $MN \ll AB$. Proceeding from Eqs. [4.61], it is a simple matter to calculate the apparent resistivity curves for the different arrays. For instance, the function ρ_a/ρ_1 for the axial dipole–dipole array is given in Fig. 4.7.

4.7. ELECTRIC FIELD IN THE PRESENCE OF A THIN HORIZONTAL LAYER

Suppose there is a horizontal layer beneath the earth's surface whose thickness is much smaller than its depth of burial (Fig. 4.8). We can sometimes model the response of this layer by a horizontal plane with a transverse resistance T or a longitudinal conductance S. Such models play an

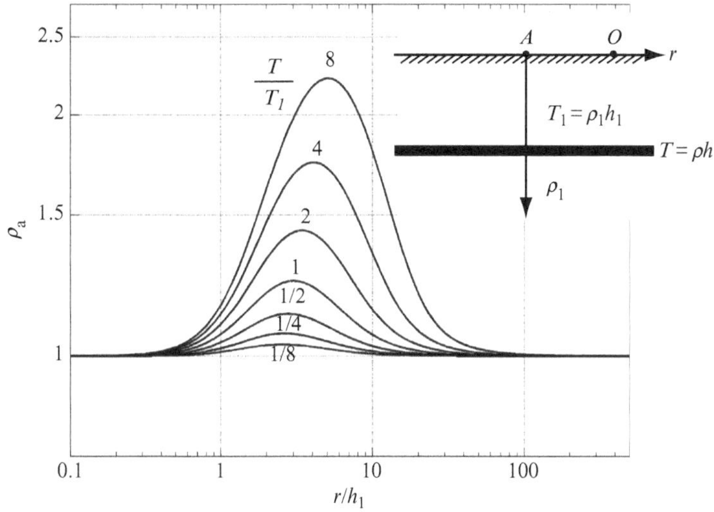

Figure 4.8 Thin layer beneath the earth's surface and apparent resistivity curves.

important role in the solution of inverse problems. We assume that the current electrode A is located at the origin of the cylindrical system of coordinates on the earth's surface and the horizontal plane is at depth h_1. Then, by analogy with the two-layer medium, the potential of the electric field can be represented at regular points as

$$U_1(r,z) = \frac{\rho_1 I}{2\pi}\int_0^\infty \{\exp(-mz) + [A_m \exp(mz) + B_m \exp(-mz)]\}J_0(mr)\mathrm{d}m \quad \text{if } 0 < z < h_1 \qquad [4.111]$$

and

$$U_2(r,z) = \frac{\rho_1 I}{2\pi}\int_0^\infty C_m \exp(-mz)J_0(mr)\mathrm{d}m \quad \text{if } z > h_1. \qquad [4.112]$$

On the earth's surface, we have

$$\frac{\partial U_1}{\partial z} = 0 \quad \text{if } z = 0. \qquad [4.113]$$

To treat the thin layer, we use the boundary conditions derived in Chapter 3. At the surface of a thin layer with the transverse resistance T, we have (Eq. [3.135])

$$\gamma_1 \frac{\partial U_1}{\partial z} = \gamma_2 \frac{\partial U_2}{\partial z}, \quad U_2 - U_1 = T\gamma_2 \frac{\partial U_2}{\partial z}, \qquad [4.114]$$

whereas for a layer characterized by longitudinal conductance S, we have (Eq. [3.145])

$$U_1 = U_2, \quad \gamma_2 \frac{\partial U_2}{\partial z} - \gamma_1 \frac{\partial U_1}{\partial z} = S \frac{\partial^2 U_2}{\partial z^2}. \qquad [4.115]$$

The condition on the earth's surface gives $A_m = B_m$. Therefore,

$$U_1(r,z) = \frac{\rho_1 I}{2\pi} \int_0^\infty \{\exp(-mz) + B_m[\exp(mz) + \exp(-mz)]\} J_0(mr)\mathrm{d}m. \qquad [4.116]$$

We first consider a thin resistive layer.

Case 1 Resistive thin layer

Substituting Eqs. [4.112] and [4.116] into Eqs. [4.114] and for simplicity assuming that $\gamma_2 = \gamma_1$, we obtain a system of equations for B_m and C_m:

$$B_m[\exp(mh_1) - \exp(-mh_1)] + C_m \exp(-mh_1) = \exp(-mh_1)$$

and

$$-B_m[\exp(mh_1) + \exp(-mh_1)] + C_m(1 + T\gamma_1 m)\exp(-mh_1) = \exp(-mh_1). \qquad [4.117]$$

The solution gives

$$B_m = \frac{T\gamma_1 m \exp(-mh_1)}{2\exp(mh_1) + T\gamma_1 m[\exp(mh_1) - \exp(-mh_1)]}, \qquad [4.118]$$

so the potential on the earth's surface is

$$U_1(r,0) = \frac{\rho_1 I}{2\pi}\left\{\frac{1}{r} + 2T\gamma_1 \int_0^\infty \frac{m\exp(-mh_1)J_0(mr)\mathrm{d}m}{2\exp(mh_1) + T\gamma_1 m[\exp(mh_1) - \exp(-mh_1)]}\right\}. \qquad [4.119]$$

The horizontal component of the field is, therefore,

$$E_r(r,0) = \frac{\rho_1 I}{2\pi}\left\{\frac{1}{r^2} + 2T\gamma_1 \int_0^\infty \frac{m^2\exp(-mh_1)J_1(mr)\mathrm{d}m}{2\exp(mh_1) + T\gamma_1 m[\exp(mh_1) - \exp(-mh_1)]}\right\}. \qquad [4.120]$$

Introducing the integration variable $x = mh_1$, the apparent resistivity is given by

$$\frac{\rho_a}{\rho_1} = 1 + 2\frac{T}{T_1}\left(\frac{r}{h_1}\right)^2 \int_0^\infty \frac{x^2 \exp(-x) J_1[(r/h_1)x]\mathrm{d}x}{2\exp(x) + (T/T_1)x[\exp(x) - \exp(-x)]}, \quad [4.121]$$

where

$$T_1 = \rho_1 h_1 \quad [4.122]$$

is the transverse resistance of the upper layer. Figure 4.8 shows typical curves of apparent resistivity. As is seen from these curves, there is an intermediate range of separations where the influence of the thin resistive layer is strong. As the separations increase beyond this range, the effect of charges on the surfaces of a thin layer becomes relatively small, and in the limit, it vanishes.

Case 2 Conductive thin layer

Next, suppose that the thin layer is much more conductive than the surrounding medium (Fig. 4.9). To determine the field, we use again Eqs. [4.112] and [4.116], which define functions that obey the boundary condition near the current electrode and at infinity, as well as on the earth's surface. The unknowns B_m and C_m are now calculated from the boundary conditions for a thin layer with longitudinal conductance S (Eqs. [4.115]). Letting $\gamma_1 = \gamma_2$ and substituting Eqs. [4.112] and [4.116] into the set [4.115], we obtain

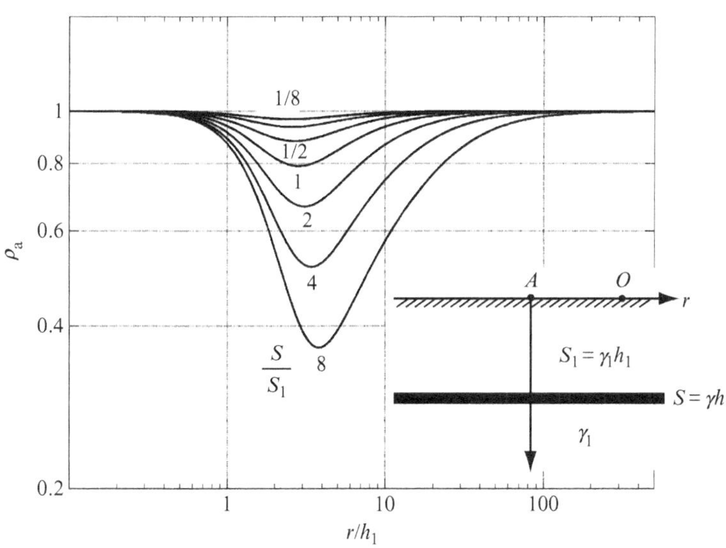

Figure 4.9 Apparent resistivity curves for a thin conductive layer.

$$B_m[\exp(mh_1) + \exp(-mh_1)] - C_m \exp(-mh_1) = -\exp(-mh_1)$$

and

$$B_m[\exp(mh_1) - \exp(-mh_1)] + C_m(1 + Sm\rho_1)\exp(-mh_1) = \exp(-mh_1). \quad [4.123]$$

This gives for the coefficient B_m characterizing the field in the upper layer:

$$B_m = -\frac{Sm\rho_1 \exp(-mh_1)}{2\exp(mh_1) + Sm\rho_1[\exp(mh_1) + \exp(-mh_1)]}. \quad [4.124]$$

Thus, we have for the potential and electric field on the earth's surface

$$U_1(r,0) = \frac{\rho_1 I}{2\pi}\left\{\frac{1}{r} - 2S\rho_1 \int_0^\infty \frac{m\exp(-mh_1)J_0(mr)\mathrm{d}m}{2\exp(mh_1) + Sm\rho_1[\exp(mh_1) + \exp(-mh_1)]}\right\} \quad [4.125]$$

and

$$E_r(r,0) = \frac{\rho_1 I}{2\pi}\left\{\frac{1}{r^2} - 2S\rho_1 \int_0^2 \frac{m^2\exp(-mh_1)J_1(mr)\mathrm{d}m}{2\exp(mh_1) + Sm\rho_1[\exp(mh_1) + \exp(-mh_1)]}\right\}. \quad [4.126]$$

Introducing again a variable $x = mh_1$, the apparent resistivity is

$$\frac{\rho_a}{\rho_1} = 1 - 2\frac{S}{S_1}\frac{r^2}{h_1^2}\int_0^\infty \frac{x^2\exp(-x)J_1[(r/h_1)x]\mathrm{d}x}{2\exp(x) + (S/S_1)x[\exp(x) + \exp(-x)]}, \quad [4.127]$$

where $S_1 = \gamma_1 h_1$ is the longitudinal conductance of the upper layer and the ratio ρ_a/ρ_1 is defined by the two quantities, S/S_1 and r/h_1. As in the case of a thin resistive layer, there is an intermediate range of separations where the influence of the plane S is maximal (Fig. 4.9).

4.8. THE ELECTRIC FIELD ON THE SURFACE OF A THREE-LAYERED MEDIUM

4.8.1. Derivation of Formulas for the Potential in the Upper Layer

Next, we will study the field in the presence of two planar and horizontal boundaries beneath the earth's surface. By analogy with the case of a two-layered medium, let us represent the potential in each layer as

$$U_1(r,z) = \frac{\rho_1 I}{2\pi R} + \int_0^\infty [A_m \exp(mz) + B_m \exp(-mz)] J_0(mr) \mathrm{d}m \quad \text{if} \quad 0 \le z \le h_1,$$

$$U_2(r,z) = \int_0^\infty [C_m \exp(mz) + D_m \exp(-mz)] J_0(mr) \mathrm{d}m \quad \text{if} \quad h_1 \le z \le h_1 + h_2,$$

$$U_3(r,z) = \int_0^\infty F_m \exp(-mz) J_0(mr) \mathrm{d}m \quad \text{if} \quad z \ge h_1 + h_2. \qquad [4.128]$$

In the last integral, the term with $\exp(mz)$ is discarded; otherwise, the potential U_3 would not decrease when $z \to \infty$. Note that, as before, the current electrode A, situated on the earth's surface, is placed at the origin of the cylindrical system of coordinates and its charge is equal to $2\varepsilon_0 \rho_1 I$. From the previous sections, we know that functions U_1, U_2, and U_3 are solutions of Laplace's equation for any values of coefficients A_m, B_m, C_m and D_m, F_m. Besides, the boundary-value problem requires that

$$\frac{\partial U_1}{\partial z} = 0 \quad \text{if } z = 0 \qquad [4.129]$$

and

$$U_1 = U_2, \quad \gamma_1 \frac{\partial U_1}{\partial z} = \gamma_2 \frac{\partial U_2}{\partial z} \quad \text{if } z = h_1, \qquad [4.130]$$

$$U_2 = U_3, \quad \gamma_2 \frac{\partial U_2}{\partial z} = \gamma_3 \frac{\partial U_3}{\partial z} \quad \text{if } z = h_1 + h_2. \qquad [4.131]$$

It is clear that, due to the presence of the exponential terms and oscillating Bessel function, U_1, U_2, and U_3 obey the condition at infinity when either $z \to \infty$ or $r \to \infty$. From the first equation of the set [4.128], we arrive at the known result

$$\frac{\partial U_1}{\partial z} = -\frac{\rho_1 I}{2\pi} \frac{z}{(r^2+z^2)^{3/2}} + \int_0^\infty m[A_m \exp(mz) - B_m \exp(-mz)] J_0(mr) \mathrm{d}m$$

and, as it follows from Eq. [4.129], $A_m = B_m$. Thus, as before, it is convenient for derivations to represent the potential U_1 as

$$U_1(r,z) = \int_0^\infty \{C \exp(mz) + A_m[\exp(mz) + \exp(-mz)]\} J_0(mr) \mathrm{d}m.$$

Here, C is equal to $\rho_1 I / 2\pi$. The boundary conditions [4.130] and [4.131] give the system of four equations with four unknowns and they are

$$
\begin{aligned}
&C\exp(-mh_1) + A_m[\exp(mh_1) + \exp(-mh_1)] \\
&\quad = C_m \exp(mh_1) + D_m \exp(-mh_1), \\
&\gamma_1\{-C\exp(-mh_1) + A_m[\exp(mh_1) - \exp(-mh_1)]\} \\
&\quad = \gamma_2[C_m \exp(mh_1) - D_m \exp(-mh_1)], \\
&C_m \exp[m(h_1+h_2)] + D_m \exp[-m(h_1+h_2)] = F_m \exp[-m(h_1+h_2)], \\
&\gamma_2\{C_m \exp[m(h_1+h_2)] - D_m \exp[-m(h_1+h_2)]\} = -\gamma_3 F_m \exp[-m(h_1+h_2)].
\end{aligned} \qquad [4.132]
$$

Our goal is to find the coefficient A_m, which describes the potential of the secondary field in the upper layer and, in particular, on the earth's surface. Eliminating F_m from the last two equations of the set [4.132], we obtain

$$C_m = K_{23} \exp[-2m(h_1+h_2)] D_m. \qquad [4.133]$$

Here,

$$K_{23} = \frac{\rho_3 - \rho_2}{\rho_3 + \rho_2}.$$

Substitution of Eq. [4.133] into the first two equations of the system [4.132] yields

$$
\begin{aligned}
&C\exp(-mh_1) + A_m[\exp(mh_1) + \exp(-mh_1)] \\
&\quad = \{K_{23}\exp[-m(h_1+2h_2)] + \exp(-mh_1)\} D_m
\end{aligned}
$$

and

$$
\begin{aligned}
&\gamma_1\{-C\exp(-mh_1) A_m[\exp(mh_1) - \exp(-mh_1)]\} \\
&\quad = \gamma_2\{K_{23}\exp[-m(h_1+2h_2)] - \exp(-mh_1)\} D_m,
\end{aligned}
$$

whence

$$\frac{\gamma_1\{-C\exp(-mh_1) + A_m[\exp(mh_1) - \exp(-mh_1)]\}}{\gamma_2\{C\exp(-mh_1) + A_m[\exp(mh_1) + \exp(-mh_1)]\}} = P_m. \qquad [4.134]$$

Here,

$$P_m = \frac{K_{23}\exp(-2mh_2) - 1}{K_{23}\exp(-2mh_2) + 1}. \qquad [4.135]$$

From Eq. [4.134] it follows that

$$A_m = \frac{[(\rho_2/\rho_1) + P_m]\exp(-mh_1)}{(\rho_2/\rho_1)[\exp(mh_1) - \exp(-mh_1)] - [\exp(mh_1) + \exp(-mh_1)]P_m} C \qquad [4.136]$$

and, making use of Eq. [4.128], we found the potential and electric field in the upper layer. In order to examine Eq. [4.136], consider several cases:

(1) $$\rho_1 = \rho_2 = \rho_3.$$

Then, $P_m = -1$, $K_{12} = K_{23} = 0$, and $A_m = 0$ which corresponds to a uniform half-space.

(2) $$\rho_2 = \rho_3.$$

Since $K_{23} = 0$, the function $P_m = -1$ and Eqs. [4.136] give

$$A_m = \frac{K_{12}\exp(-2mh_1)}{1 - K_{12}\exp(-2mh_1)} C$$

which coincides with the expression of this coefficient in a two-layered medium with parameters ρ_1, h_1, and ρ_2.

(3) $$\rho_1 = \rho_2.$$

Then, we have

$$A_m = \frac{(1 + P_m)\exp(-mh_1)C}{(1 - P_m)\exp(mh_1) - (1 + P_m)\exp(-mh_1)}. \qquad [4.137]$$

In as much as

$$1 + P_m = \frac{2K_{23}\exp(-2mh_2)}{K_{23}\exp(-2mh_2) - 1}, \qquad 1 - P_m = \frac{2}{K_{23}\exp(-2mh_2) - 1},$$

we obtain

$$A_m = \frac{K_{13}\exp[-2m(h_1 + h_2)]}{1 - K_{13}\exp[m(h_1 + h_2)]} C \qquad [4.138]$$

and

$$K_{13} = \frac{\rho_3 - \rho_1}{\rho_3 + \rho_1}.$$

It is obvious that A_m given by Eqs. [4.138] corresponds to a two-layered medium with resistivity ρ_1 and ρ_3 and the thickness of the upper layer is equal to $h_1 + h_2$.

(4) $$h_2 \to \infty.$$

Then, as before $P_m \to -1$ and we have the second case when $\rho_2 = \rho_3$.

(5) $$h_2 \to 0.$$

As follows from Eq. [4.135]

$$P_m = \frac{K_{23} - 1}{K_{23} + 1} = -\frac{\rho_2}{\rho_3}$$

and

$$A_m = \frac{K_{13}\exp(-2mh_1)}{1 - K_{13}\exp(-2mh_1)} C,$$

which also corresponds to a two-layered medium.

4.8.2. Asymptotic Behavior of the Field

In accordance with Eq. [4.128], the electric field E_r on the earth's surface is

$$E_r(r,0) = \frac{\rho_1 I}{2\pi r^2} + 2\int_0^\infty m A_m J_1(mr)\mathrm{d}m. \qquad [4.139]$$

As in the case of a two-layered medium, the secondary field has everywhere a finite value and near the current electrode the primary field prevails:

$$E_r(r,0) \to \frac{\rho_1 I}{2\pi r^2}. \qquad [4.140]$$

In order to study a field in the opposite case when $r \to \infty$, we make use again the oscillating character of Bessel function and consider the behavior of the function A_m as $m \to 0$. First of all, we see that

$$P_m \to -\frac{\rho_2}{\rho_3}, \quad \exp(\pm mh_1) \to 1, \quad \exp(\pm mh_2) \to 1$$

and this gives

$$A_m = \frac{C}{2}\left(\frac{\rho_3}{\rho_1} - 1\right). \qquad [4.141]$$

Its substitution into Eq. [4.139] yields

$$E_r(r,0) = \frac{\rho_1 I}{2\pi r^2} + \frac{\rho_1 I}{2\pi}\left(\frac{\rho_3}{\rho_1} - 1\right)\int_0^\infty m J_1(mr)\mathrm{d}m$$

or

$$E_r(r,0) \to \frac{\rho_3 I}{2\pi r^2} \quad \text{if } r \to \infty. \qquad [4.142]$$

Thus, in the far zone when $r \gg h_1 + h_2$, the field tends to that of a uniform half-space with resistivity of the underlying medium, ρ_3. In other words, the two upper layers become transparent, and we again observe this remarkable feature of the geometric soundings. Also, it is useful to consider the case when the base is an insulator, $\rho_3 \to \infty$. Preserving only the two first terms of the power series of exponents in the expression of P_m, we have

$$P_m \approx -mh_2 \quad \text{if } m \to 0$$

and

$$A_m \approx \frac{(\rho_2/\rho_1) - mh_2}{(\rho_2/\rho_1)2mh_1 + 2mh_2} C \quad \text{or} \quad A_m = \frac{I}{4\pi m(S_1 + S_2)}, \qquad [4.143]$$

that is, in the far zone the secondary field prevails as in the case of a two-layered medium, and it is defined by the total conductance of the two upper layers. This means that we confirmed the results derived earlier from the distribution of the current density. Note that if the resistivity of the base is relatively low ($\rho_3 \ll \rho_1$, $\rho_3 \ll \rho_2$), it is possible to show with the help of the residual theorem that there is a range of separations over which the potential and electric field decay almost exponentially.

4.8.3. Apparent Resistivity Curves

As in the case of a two-layered medium, we represent the field (Eq. [4.139]) in the form

$$E_r(r, 0) = \frac{\rho_1 I}{2\pi r^2} F(\mu_1, \mu_2, v_1, A0/h_1) \qquad [4.144]$$

or

$$E_r(r, 0) = \frac{\rho_1 I}{2\pi r^2} \frac{\rho_a}{\rho_1}. \qquad [4.145]$$

Here, $\mu_1 = \rho_2/\rho_1$, $\mu_2 = \rho_3/\rho_1$, $v_1 = h_2/h_1$, and $r = A0$ is the distance between the current electrode and an observation point; ρ_a is the apparent resistivity. Behavior of the ratio ρ_a/ρ_1 is shown in Fig. 4.10A–D. It is convenient to consider separately the apparent resistivity curves for four types of geoelectric section and they are:

(1) A type $\rho_1 < \rho_2 < \rho_3$.
(2) H type $\rho_1 > \rho_2 < \rho_3$.

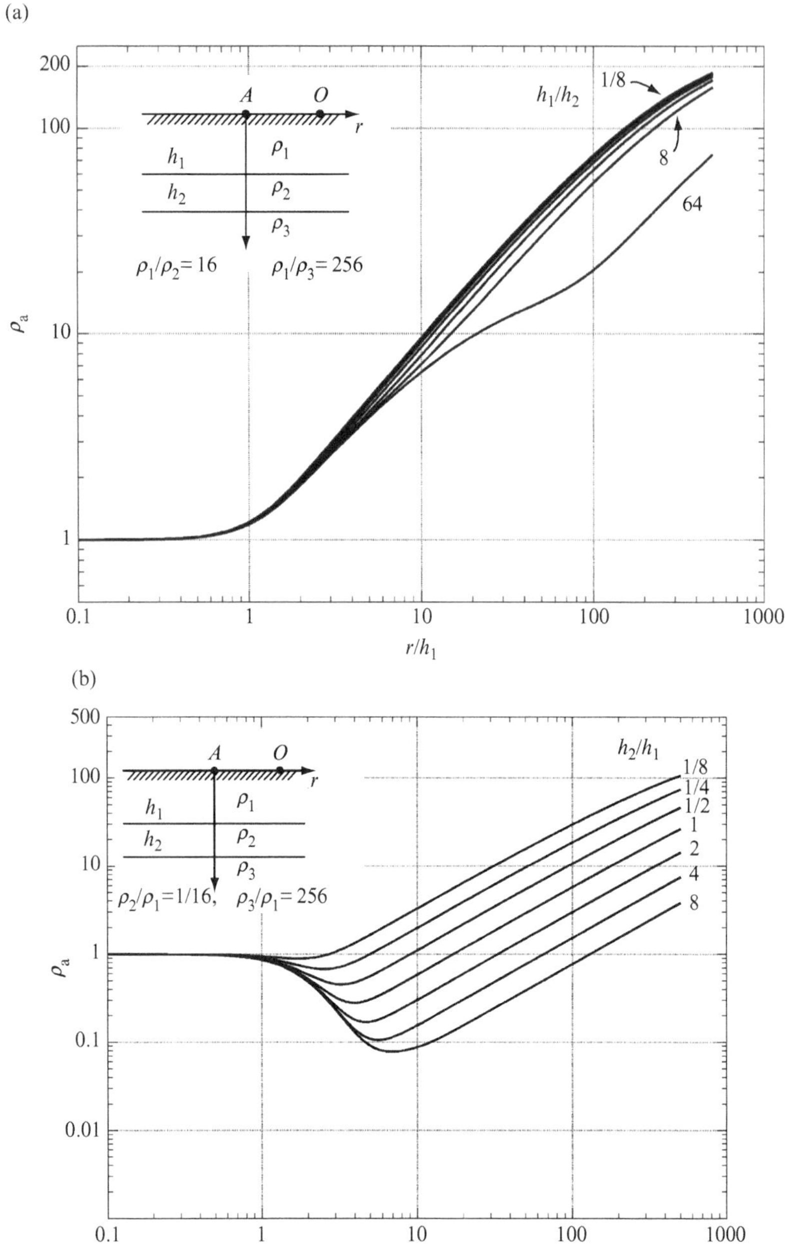

Figure 4.10 Illustration of apparent resistivity curves for a three-layer model: (A) Type A: resistivity increases with depth, (B) Type H: conductive middle layer,

Continued

(c)

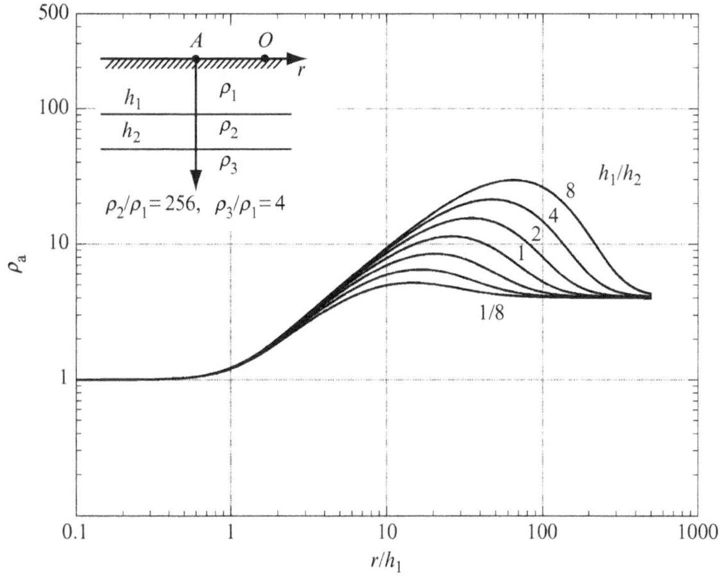

(d)

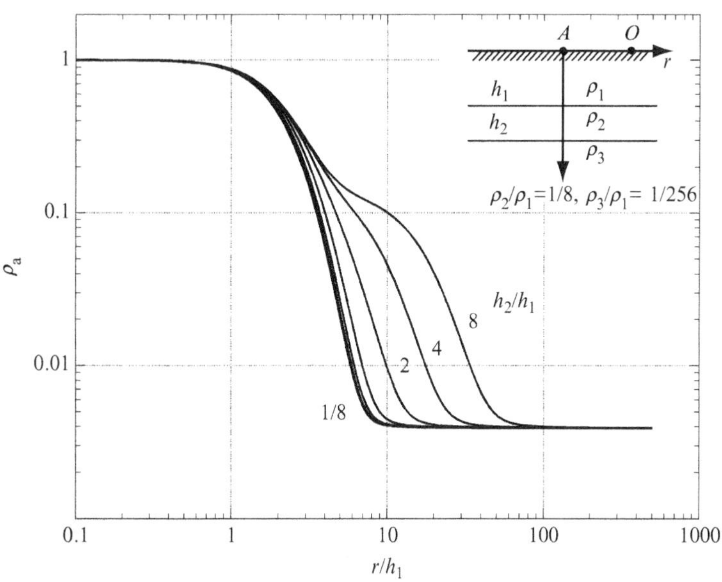

Figure 4.10—Cont'd (C) Type K: resistive middle layer, and (D) Type Q: resistivity decreases with depth.

(3) K type $\rho_1 < \rho_2 > \rho_3$.
(4) Q type $\rho_1 > \rho_2 > \rho_3$.
Parameters μ_1 and μ_2 characterize each set of curves, while the curve's index is the ratio of the layer thickness ν_1. As in the case of the two-layered medium, from Eqs. [4.144] and [4.145], we also have

$$\log \rho_a = \log \rho_1 + \log F(\mu_1, \mu_2, \nu_1, \log r - \log h_1).$$

These curves clearly show that, with an increase of the thickness of an intermediate layer, there is a range of distances over which the apparent resistivity approaches ρ_2. Also, we may see that, in the case of a strong contrast of resistivity, it is possible sometimes to observe the S-zone, as well as an almost exponential decay of the field.

4.9. ELECTRIC FIELD ON THE EARTH'S SURFACE OF AN *N*-LAYERED MEDIUM

Now, we derive formulas for the potential for an arbitrary number of the horizontal layers, and with this purpose in mind, consider again the expressions of the potential in several known cases and represent them in a different form.

4.9.1. Uniform Half-Space

In this simplest model, we have

$$U_1(r,z) = \frac{\rho_1 I}{2\pi}\left[\frac{1}{R} + \int_0^\infty A_m \exp(-mz) J_0(mr)\mathrm{d}m\right].$$

The condition on the earth's surface $\partial U_1/\partial z = 0$ gives $A_m = 0$, and

$$U_1(r,0) = \frac{\rho_1 I}{2\pi}\int_0^\infty R_1 J_0(mr)\mathrm{d}m, \quad \text{where } R_1 = 1. \qquad [4.146]$$

4.9.2. Two-Layered Medium

From Eq. [4.61], it follows that

$$U_1(r,0) = \frac{\rho_1 I}{2\pi}\int_0^\infty (1 + 2A_m) J_0(mr)\mathrm{d}m. \qquad [4.147]$$

Here,

$$A_m = \frac{K_{12} \exp(-2mh_1)}{1 - K_{12} \exp(-2mh_1)}.$$

It is clear that

$$1 + 2A_m = \frac{1 + K_{12} \exp(-2mh_1)}{1 - K_{12} \exp(-2mh_1)} = \frac{(\rho_2 + \rho_1) + (\rho_2 - \rho_1) \exp(-2mh_1)}{(\rho_2 + \rho_1) - (\rho_2 - \rho_1) \exp(-2mh_1)}$$

or

$$1 + 2A_m = \frac{(\rho_2/\rho_1) \coth mh_1 + 1}{(\rho_2/\rho_1) + \coth mh_1} = \coth\left(mh_1 + \coth^{-1} \frac{\rho_2}{\rho_1}\right) \qquad [4.148]$$

since

$$\frac{\rho_2}{\rho_1} = \coth \coth^{-1} \frac{\rho_2}{\rho_1}.$$

Thus, the potential on the earth's surface can be represented as

$$U_1(r, 0) = \frac{\rho_1 I}{2\pi} \int_0^{\infty} R_2 J_0(mr) \mathrm{d}m,$$

where

$$R_2 = \coth\left(mh_1 + \coth^{-1} \frac{\rho_2}{\rho_1}\right). \qquad [4.149]$$

Note, that if $h_1 \to \infty$ or $\rho_2 \to \rho_1$, we have $R_2 \to R_1$.

4.9.3. Three-Layered Medium

In accordance with two last equations of the set [4.132], we can write

$$\frac{C_m \exp[m(h_1 + h_2)] + D_m \exp[-m(h_1 + h_2)]}{C_m \exp[m(h_1 + h_2)] - D_m \exp[-m(h_1 + h_2)]} = -\frac{\rho_3}{\rho_2}. \qquad [4.150]$$

Since

$$\frac{D_m}{C_m} = \exp\left(2 \frac{1}{2} \ln \frac{D_m}{C_m}\right),$$

Eq. [4.150] can be written as

$$\coth\left[m(h_1+h_2)-\frac{1}{2}\ln\frac{D_m}{C_m}\right]=-\frac{\rho_3}{\rho_2}$$

or

$$\frac{1}{2}\ln\frac{D_m}{C_m}=m(h_1+h_2)+\coth^{-1}\frac{\rho_3}{\rho_2}. \qquad [4.151]$$

It is essential that unknowns D_m and C_m are expressed in terms of the parameters of a geoelectric section. From the first two equations of the set [4.132], we have

$$\frac{\exp(-mh_1)+A_m[\exp(mh_1)+\exp(-mh_1)]}{-\exp(-mh_1)+A_m[\exp(mh_1)-\exp(-mh_1)]}=\frac{\rho_2}{\rho_1}\coth\left[mh_1-\frac{1}{2}\ln\frac{C_m}{D_m}\right].$$

Taking into account Eq. [4.151] and the fact that $\coth x$ is an odd function of x, we obtain

$$\frac{\exp(-mh_1)+A_m[\exp(mh_1)+\exp(-mh_1)]}{-\exp(-mh_1)+A_m[\exp(mh_1)-\exp(-mh_1)]}=-\frac{\rho_2}{\rho_1}\coth\left[mh_2+\coth^{-1}\frac{\rho_3}{\rho_2}\right]$$

and

$$\exp(-mh_1)\left(1-\frac{\rho_2}{\rho_1}\coth x\right)=$$
$$A_m\left\{-\frac{\rho_2}{\rho_1}\coth x\cdot[\exp(mh_1)-\exp(-mh_1)]-[\exp(mh_1)+\exp(-mh_1)]\right\}. \qquad [4.152]$$

Here, $x=mh_2+\coth^{-1}(\rho_3/\rho_2)$. This gives

$$A_m=-\frac{[1-(\rho_2/\rho_1)\coth x]\cdot\exp(-mh_1)}{(\rho_2/\rho_1)\coth x\cdot[\exp(mh_1)-\exp(-mh_1)]+[\exp(mh_1)+\exp(-mh_1)]}$$

and

$$1+2A_m=\frac{\exp(-mh_1)-\exp(mh_1)+(\rho_2/\rho_1)\coth x\cdot[\exp(-mh_1)+\exp(mh_1)]}{(\rho_2/\rho_1)[\exp(mh_1)-\exp(-mh_1)]\cdot\coth x+[\exp(mh_1)+\exp(-mh_1)]}$$

or

$$1+2A_m=\frac{1+(\rho_2/\rho_1)\coth x\,\coth mh_1}{(\rho_2/\rho_1)\coth x+\coth mh_1}.$$

Letting

$$\coth u=\frac{\rho_2}{\rho_1}\coth x, \qquad [4.153]$$

we have

$$1+2A_m=R_3=\coth(mh_1+u)$$

or, taking into account Eq. [4.153],

$$R_3=\coth\left\{mh_1+\coth^{-1}\frac{\rho_2}{\rho_1}\coth\left[mh_2+\coth^{-1}\frac{\rho_3}{\rho_2}\right]\right\}. \qquad [4.154]$$

For instance, if $\rho_3=\rho_2$ or $h_2\to\infty$, we again arrive at the two-layered case. Comparison of Eqs. [4.146], [4.149], and [4.154] allows us to see that there is a relation between functions R_n and R_{n-1} for a medium with n and $(n-1)$ layers, which can be written as

$$\begin{aligned}R_n&=\coth\left[mh_1+\coth^{-1}\frac{\rho_2}{\rho_1}R_{n-1}\right],\\R_{n-1}&=\coth\left[mh_2+\coth^{-1}\frac{\rho_3}{\rho_2}R_{n-2}\right],\\R_{n-2}&=\coth\left[mh_3+\coth^{-1}\frac{\rho_4}{\rho_3}R_{n-3}\right],\end{aligned} \qquad [4.155]$$

and so on. It is essential that $R_1=1$. By definition, we have at points on the earth's surface

$$U(r,0)=\frac{\rho_1 I}{2\pi}\int_0^\infty R_nJ_0(mr)\mathrm{d}m=\frac{\rho_1 I}{2\pi}\left[\frac{1}{r}+\int_0^\infty(R_n-1)J_0(mr)\mathrm{d}m\right]$$

and

$$E_\mathrm{r}(r,0)=\frac{\rho_1 I}{2\pi}\left[\frac{1}{r^2}+\int_0^\infty m(R_n-1)J_1(mr)\mathrm{d}m\right]. \qquad [4.156]$$

For instance, in the case of the symmetrical four-electrode array, when $MN \ll AB$ we have

$$\rho_a = \rho_1 \left[1 + r^2 \int_0^\infty m(R_n - 1) J_1(mr) \mathrm{d}m \right]. \qquad [4.157]$$

4.10. ELECTRIC FIELD WHEN THE CURRENT ELECTRODE IS AT THE BOTTOM OF THE UPPER LAYER

Electric methods have been used to map earth structures beneath the ocean by placing electrodes near the bottom of the sea. In particular, the method has been used recently to detect buried, highly resistive thin layers caused by the presence of oil or gas in hydrocarbon reservoirs. To model the field behavior, let us assume that the current electrode is located near the bottom of an upper layer that has the properties of sea water and consider two models of the subsurface. First, let us discuss the asymptotic behavior of the field from an analysis of the charge distribution. When a current-carrying electrode is located anywhere in the upper medium, it acquires a charge equal to $\varepsilon_0 \rho_1 I$. Also, the total surface change induced at the interface at the bottom of the upper layer is

$$e_S = \varepsilon_0 (\rho_2 - \rho_1) I.$$

Correspondingly, the charge on the earth–air interface is

$$e_0 = e_A + e_S = \varepsilon_0 \rho_2 I.$$

This implies that the total charge in the medium is $2\varepsilon_0 \rho_2 I$; that is, we obtained the same result as if the current electrode were located at the upper boundary. In other words, the amount of charges is independent of the position of the current electrode within the upper layer. Therefore, at large distances, the field behaves as that of an elementary charge:

$$U_1(r, z) \to \frac{\rho_2 I}{2\pi R} \quad \text{if } R \to \infty,$$

and the apparent resistivity tends to that of the lower medium:

$$\rho_a \to \rho_2 \quad \text{if } R \to \infty.$$

As shown earlier, the charge on the surface of a current electrode situated at the interface of two media is

$$e_A = \varepsilon_0 \rho_1 I (1 + K_{12}) = \varepsilon_0 I \frac{2\rho_1 \rho_2}{\rho_1 + \rho_2}. \qquad [4.158]$$

Thus, when the observation point approaches the current electrode, we have

$$U_1(r,z) \to \frac{\rho_1\rho_2 I}{2\pi(\rho_1+\rho_2)R} \quad \text{if } R \to 0. \qquad [4.159]$$

We now find the field in a general case for arbitrary values of r. In the simple two-layer model (Fig. 4.11), expressions for the potential are

$$U_1(r,z) = \frac{\rho_1 I}{4\pi}\left\{\frac{1}{R} + \int_0^\infty [A_m \exp(mz) + B_m \exp(-mz)] J_0(mr)\mathrm{d}m\right\}$$

and [4.160]

$$U_2(r,z) = \frac{\rho_1 I}{4\pi}\int_0^\infty C_m \exp(-mz) J_0(mr)\mathrm{d}m.$$

Here,

$$R = \left[(z-d)^2 + r^2\right]^{1/2}$$

is the distance from the current electrode located on the z-axis at depth d. In this case, we have the representation

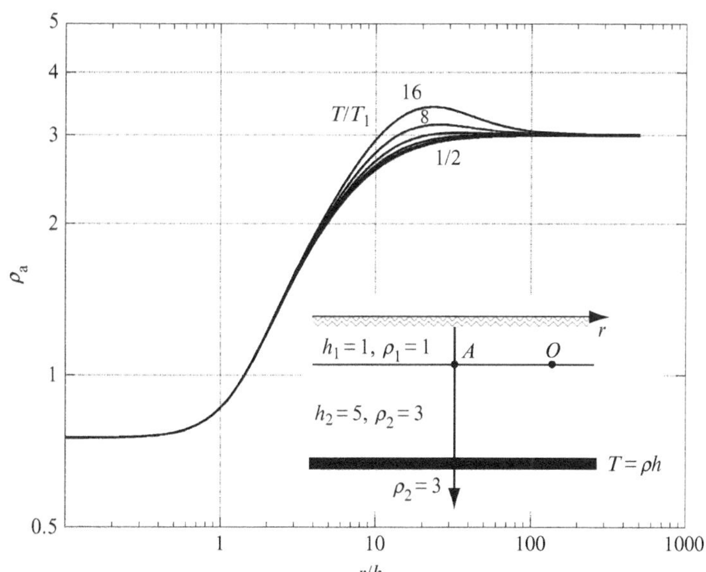

Figure 4.11 Electrodes at the bottom of upper layer (sea) with a thin resistive layer in the lower medium. Apparent resistivity curves for the array *A*0.

$$\frac{1}{R} = \int_0^\infty \exp(-m|z-d|)J_0(mr)\mathrm{d}m$$

or

$$\frac{1}{R} = \int_0^\infty \exp m(z-d)J_0(mr)\mathrm{d}m \quad \text{if } z \le d$$

and [4.161]

$$\frac{1}{R} = \int_0^\infty \exp m(d-z)J_0(mr)\mathrm{d}m \quad \text{if } z \ge d.$$

The boundary condition at the earth–air interface,

$$\frac{\partial U_1}{\partial z} = 0 \quad \text{if } z = 0,$$

gives

$$\exp(-md) + A_m - B_m = 0. \qquad [4.162]$$

Therefore, the potential in the upper layer above the electrode is

$$U_1(r,z) = \frac{\rho_1 I}{4\pi}\int_0^\infty \{[\exp(-md) + A_m][\exp(mz) + \exp(-mz)]\}J_0(mr)\mathrm{d}m$$

and

$$\frac{\partial U_1}{\partial z}(r,z) = \frac{\rho_1 I}{4\pi}\int_0^\infty \{[\exp(-md) + A_m][\exp(mz) - \exp(-mz)]\} mJ_0(mr)\mathrm{d}m \quad \text{if } z \le d. \qquad [4.163]$$

The boundary condition is clearly satisfied when $z=0$. From Eq. [4.161], the potential $U_1(r,z)$ below the electrode near the bottom of the layer ($z=h_1$) is

$$U_1(r,z) = \frac{\rho_1 I}{4\pi}\int_0^\infty \{[B_m - \exp(-md)]\exp(mz) + [B_m + \exp(md)]\exp(-mz)\} J_0(mr)\mathrm{d}m \ \text{if } z \ge d. \qquad [4.164]$$

The boundary conditions at the bottom interface are

$$U_1 = U_2, \quad \gamma_1 \frac{\partial U_1}{\partial z} = \gamma_2 \frac{\partial U_2}{\partial z},$$

and letting $d = h_1$ we obtain the system for the determination of B_m and C_m:

$$B_m[\exp(mh_1) + \exp(-mh_1)] = C_m \exp(-mh_1)$$

and [4.165]

$$\gamma_1\{B_m[\exp(mh_1) - \exp(-mh_1)] - 2\} = -\gamma_2 C_m \exp(-mh_1).$$

This system gives

$$B_m = \frac{2\gamma_1}{(\gamma_2 + \gamma_1)\exp(mh_1) + (\gamma_2 - \gamma_1)\exp(-mh_1)},$$

or

$$B_m = \frac{2}{1 + (\rho_1/\rho_2)} \frac{\exp(-mh_1)}{1 - K_{12}\exp(-2mh_1)}. \quad [4.166]$$

Then, Eq. [4.162] gives

$$A_m = K_{12} \frac{[1 + \exp(-2mh_1)]}{1 - K_{12}\exp(-2mh_1)} \exp(-mh_1). \quad [4.167]$$

Consider two positions of the receiver electrode: $z = 0$ and $z = h_1$:

Case 1 $z = 0$.

If the current source is located at the bottom of the upper layer, then the potential at the earth–air interface can be represented as (Eq. [4.163])

$$U_1(r, 0) = \frac{\rho_1 I}{2\pi}(1 + K_{12}) \int_0^\infty \frac{\exp(-mh_1) J_0(mr) dm}{1 - K_{12}\exp(-2mh_1)}. \quad [4.168]$$

For instance, if the conducting medium is a uniform half-space ($\rho_1 = \rho_2$), we have

$$U_1(r, 0) = \frac{\rho_1 I}{2\pi(r^2 + h_1^2)^{1/2}}.$$

This result is obvious since the observation point is located at the boundary with a nonconducting medium, but the current electrode is located at the distance h_1. If we assume that h_1 tends to zero, that is, the upper layer disappears, then Eq. [4.168] corresponds again to a uniform half-space but with resistivity ρ_2. In fact, in this case, Eq. [4.168] gives

$$U_1(r,0)=\frac{\rho_1 I}{2\pi}\frac{1+K_{12}}{1-K_{12}}\frac{1}{r}=\frac{\rho_2 I}{2\pi r},$$

since both electrodes are located at the same interface.

Case 2 $z=h_1$.

If both electrodes are situated at the bottom of the upper layer, it follows from Eq. [4.164] that the potential is

$$U_1(r,h_1)=\frac{\rho_1 I}{2\pi}\int_0^\infty B_m \cosh mh_1 J_0(mr)\mathrm{d}m$$

or

$$U_1(r,h_1)=\frac{\rho_1 I}{2\pi}\frac{\rho_2}{\rho_1+\rho_2}\int_0^\infty \frac{1+\exp(-2mh_1)}{1-K_{12}\exp(-2mh_1)}J_0(mr)\mathrm{d}m.$$

Using

$$\frac{1}{r}=\int_0^\infty J_0(mr)\mathrm{d}m,$$

we obtain

$$U_1(r,h_1)=\frac{\rho_1 I}{2\pi}\frac{\rho_2}{\rho_1+\rho_2}\left[\frac{1}{r}+(1+K_{12})\int_0^\infty \frac{\exp(-2mh)}{1-K_{12}\exp(-2mh_1)}J_0(mr)\mathrm{d}m\right]. \qquad [4.169]$$

It is instructive to derive asymptotic formulas for the potential from Eq. [4.169]. First, when the receiver approaches the current electrode, $r\to 0$, the integral tends to a finite value while the primary field increases without limit, and we have

$$U_1(r,h_1)\to\frac{\rho_1 I}{4\pi r}(1+K_{12}) \quad \text{if } r\ll h_1,$$

which coincides with Eq. [4.159] when $z=h_1$. When $r\gg h_1$, the integral is defined by small values of m and gives

$$U_1(r,h_1)\to\frac{\rho_1 I}{2\pi}\frac{\rho_2}{\rho_1+\rho_2}\frac{1}{r}\left[1+\frac{1+K_{12}}{1-K_{12}}\right]=\frac{\rho_1 I}{2\pi}\frac{\rho_2}{\rho_1+\rho_2}\frac{2}{(1-K_{12})}\frac{1}{r},$$

or

$$U_1(r,h_1) \to \frac{\rho_2 I}{2\pi r} \quad \text{if } r \gg h_1.$$

As we already know, this result is obvious, because at large distances from the current electrode, the potential coincides with that of an elementary charge equal to the total charge in a medium ($2\varepsilon_0\rho_2 I$). The radial component E_r at the bottom of the upper layer, is obtained by differentiating Eq. [4.169]:

$$E_r(r,h_1)=\frac{\rho_1 I}{4\pi}(1+K_{12})\left[\frac{1}{r^2}+(1+K_{12})\int_0^\infty \frac{\exp(-2mh_1)}{1-K_{12}\exp(-2mh_1)} mJ_1(mr)\mathrm{d}m\right]. \quad [4.170]$$

4.11. ELECTRIC FIELD AT THE BOTTOM OF THE UPPER LAYER IN THE PRESENCE OF THE PLANE *T*

Now, we consider a more complicated model when an underlying medium contains a relatively resistive and thin layer which can be replaced by a horizontal plane with the transversal resistance T (Fig. 4.11). Then, taking into account Eq. [4.164], the potential in the upper layer is

$$U_1(r,z)=\frac{2\rho_1 I}{4\pi}\int_0^\infty [B_m\cosh(mz)-\sinh m(z-d)]J_0(mr)\mathrm{d}m \quad [4.171]$$
$$\text{if } d \le z \le h_1.$$

Also,

$$U_2(r,z)=\frac{\rho_1 I}{4\pi}\int_0^\infty [C_m\exp(mz)+D_m\exp(-mz)]J_0(mr)\mathrm{d}m \quad \text{if } h_1 \le z \le h_2$$

and

$$U_3(r,z)=\frac{\rho_1 I}{4\pi}\int_0^\infty F_m\exp(-mz)J_0(mr)\mathrm{d}m \quad \text{if } z \ge h_2.$$

[4.172]

Here, h_2 is the distance between the upper boundary, $z=0$, and the plane T. As we know, at the bottom of the upper layer

$$U_1(r,h_1)=U_2(r,h_1) \quad \text{and} \quad \gamma_1\frac{\partial U_1(r,h_1)}{\partial z}=\gamma_2\frac{\partial U_2(r,h_1)}{\partial z}, \quad [4.173]$$

while at the plane T (Eq. [4.114]),

$$\frac{\partial U_2(r,h_2)}{\partial z}=\frac{\partial U_3(r,h_2)}{\partial z}, \quad U_3(r,h_2)-U_2(r,h_2)=T\gamma_2\frac{\partial U_3(r,h_2)}{\partial z}. \qquad [4.174]$$

Substituting Eqs. [4.172] into Eqs. [4.173] and [4.174] and letting $d=h_1$, we obtain the system of equations with respect to B_m, C_m, D_m, and F_m:

$$\begin{aligned}
&B_m[\exp(mh_1)+\exp(-mh_1)]=C_m\exp(mh_1)+D_m\exp(-mh_1),\\
&\gamma_1\{B_m[\exp(mh_1)-\exp(-mh_1)]-2\}=\gamma_2[C_m\exp(mh_1)-D_m\exp(-mh_1)],\\
&C_m\exp(mh_2)-D_m\exp(-mh_2)=-F_m\exp(-mh_2),\\
&F_m\exp(-mh_2)-C_m\exp(mh_2)-D_m\exp(-mh_2)=-T\gamma_2 mF_m\exp(-mh_2).
\end{aligned} \qquad [4.175]$$

Eliminating F_m from the last two equations, we obtain the relation between C_m and D_m:

$$C_m=\frac{m\gamma_2 T}{2+m\gamma_2 T}\exp(-2mh_2)D_m. \qquad [4.176]$$

Substitution of the latter into the first two equations of the set [4.175] gives

$$B_m=\frac{1}{\sinh mh_1-(\gamma_2/\gamma_1)P\cosh mh_1}, \qquad [4.177]$$

where

$$P=\frac{m\gamma_2 T\exp(-2md_1)-2-m\gamma_2 T}{m\gamma_2 T\exp(-2md_1)+2+m\gamma_2 T} \qquad [4.178]$$

and

$$d_1=h_2-h_1$$

is the distance from the plate T to the bottom of the upper layer. Consider the particular case, for instance, if the plane T is absent; that is $T=0$, then, $P=-1$ and

$$U_1(r,h_1)=\frac{\rho_1 I}{2\pi}\int_0^\infty \frac{\cosh mh_1}{\sinh mh_1+(\gamma_2/\gamma_1)\cosh mh_1}J_0(mr)dm$$

or

$$U_1(r,h_1)=\frac{\rho_1 I}{2\pi}\frac{\rho_2}{(\rho_1+\rho_2)}\int_0^\infty \frac{[1+\exp(-2mh_1)]}{1-K_{12}\exp(-2mh_1)}J_0(mr)dm, \qquad [4.179]$$

which coincides with Eq. [4.169]. The same result follows when we assume that $d_1 \to \infty$. As usual, it is convenient to represent the potential as a sum of the primary and secondary fields:

$$U_1(r, h_1) = \frac{\rho_1 I}{2\pi}\left[\frac{\rho_2}{\rho_1+\rho_2}\frac{1}{r} + \int_0^\infty \left(B_m \cosh mh_1 - \frac{\rho_2}{\rho_1+\rho_2}\right) J_0(mr)dm\right]. \quad [4.180]$$

It is obvious that with a decrease of the distance when $r < d$, the potential tends to that of the current electrode situated at the interface between media with resistivity ρ_1 and ρ_2. On the contrary, at large distances from the electrode, the integral is defined by small values of m and we have

$$P \to -1, \quad B_m \to \frac{\gamma_1}{\gamma_2}$$

and Eq. [4.171] gives

$$U_1(r, h_1) \to \frac{\rho_2 I}{2\pi r}, \quad [4.181]$$

and it is natural, since the total charge in a medium is equal to $2\varepsilon_0 \rho_2 I$. This is related to the fact that on both sides of the plane T, there is an equal amount of charges which differ by sign. Inasmuch as in these limiting cases the influence of the thin resistive layer is negligible, there is an intermediate range of separations between the current and receiver electrodes when its contribution is maximal. Introducing as before a variable $x = mh_1$, we have

$$P = \frac{x\alpha(T/T_1)\exp(-2\beta x) - 2 - x\alpha(T/T_1)}{x\alpha(T/T_1) + 2 + x\alpha(T/T_1)}. \quad [4.182]$$

Here,

$$\alpha = \frac{\gamma_2}{\gamma_1}, \quad \beta = \frac{d_1}{h_1}, \quad T_1 = \rho_1 h_1, \quad [4.183]$$

and

$$B_m = \frac{1}{\sinh x - \alpha P \cosh x}. \quad [4.184]$$

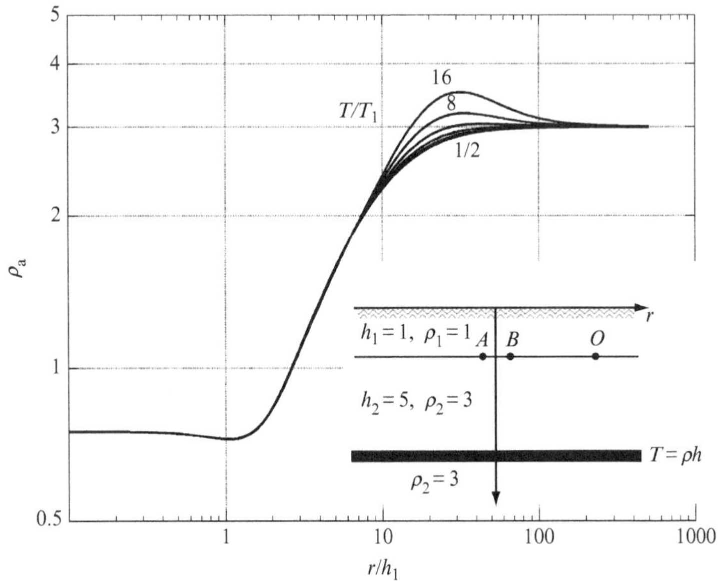

Figure 4.12 Dipole–dipole electrode array at the bottom of the upper layer (on sea bottom) in the presence of a thin resistive sheet.

By definition, the radial component of the electric field at the bottom of the upper layer is

$$E_{1r}(r, h_1) = \frac{\rho_1 I}{2\pi}\left\{\frac{\rho_2}{\rho_1+\rho_2}\frac{1}{r^2} + \int_0^\infty m\left[B_m \cosh mh_1 - \frac{\rho_2}{\rho_1+\rho_2}\right]J_1(mr)dm\right\} \quad [4.185]$$

and, correspondingly, measuring the electric field, we have for the apparent resistivity

$$\frac{\rho_a}{\rho_1} = \left\{\frac{\rho_2}{\rho_1+\rho_2} + r^2\int_0^\infty m\left[B_m - \frac{\rho_2}{\rho_1+\rho_2}\right]J_1(mr)dm\right\}. \quad [4.186]$$

Similarly, we can introduce the apparent resistivity for the axial dipole array. The behavior of the apparent resistivity for the axial dipole–dipole arrays is shown in Fig. 4.12.

4.12. POTENTIAL AND ELECTRIC FIELD IN AN ANISOTROPIC MEDIUM

Until now, we assumed that the medium is an isotropic one; that is, the resistivity is independent of the direction. At the same time, the process of sedimentation suggests that any elementary volume of sedimentary

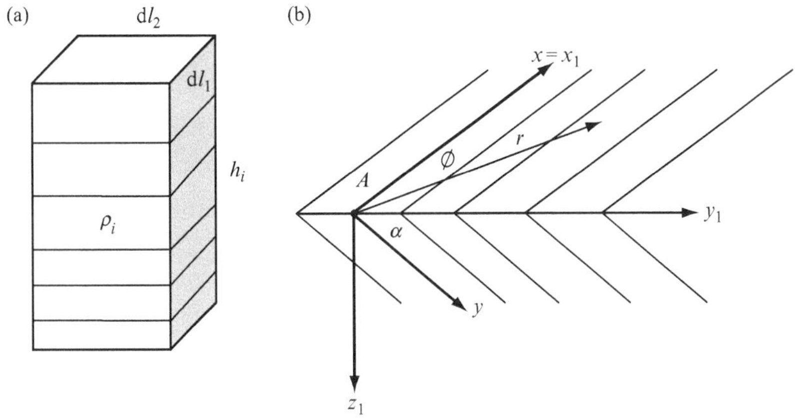

Figure 4.13 (A) An elementary volume in a system of thin layers with thickness h_i and resistivity ρ_i and (B) potential on the earth's surface.

formation may consist of a system of very thin layers (Fig. 4.13A) and it is natural to expect that the resistance along and perpendicular to the boundaries between them is different. This implies that the thickness h_i of each thin layer is many orders greater than the atomic distances, whereas it is much smaller than the geometrical parameters of the arrays and the separation between the observation points. Also, an isotropic medium inside these layers is characterized by the resistivity ρ_i. Our first goal is to perform a transition from a system of extremely thin layers to an anisotropic medium assuming that all boundaries are parallel to each other and that the area of the top and bottom of the volume is $dS = dl_1 dl_2$.

4.12.1. Longitudinal and Transversal Resistivity of an Anisotropic Medium

First, suppose that the vector of the current density is parallel to the boundaries between layers and the corresponding lateral surfaces have a constant potential (Fig. 4.13A). Therefore, we can apply the concept of resistance to such an elementary volume where all layers are connected in parallel. For instance, if the vector $\mathbf{j}$ is parallel to $\mathbf{dl}_2$, the resistance of this volume R_t is defined as

$$\frac{1}{R_t} = \frac{1}{R_1} + \frac{1}{R_2} + \cdots + \frac{1}{R_i} + \cdots + \frac{1}{R_{n-1}} + \frac{1}{R_n}. \qquad [4.187]$$

By definition,

$$R_i = \frac{\rho_i dl_2}{h_i dl_1}, \quad R_t = \frac{\rho_t dl_2}{dl_1 \sum_{i=1}^{n} h_i}. \quad [4.188]$$

Here, R_t is called the longitudinal resistance of the medium which causes the same resistance as the system of thin layers provided that the vector of current density is parallel to interfaces, while ρ_t is its longitudinal resistivity. Substitution of Eqs. [4.188] into Eq. [4.187] gives

$$\frac{\sum_{i=1}^{n} h_i}{\rho_t} = \left[\frac{h_1}{\rho_1} + \frac{h_2}{\rho_2} + \cdots + \frac{h_i}{\rho_i} + \cdots + \frac{h_{n-1}}{\rho_{n-1}} + \frac{h_n}{\rho_n}\right] \quad [4.189]$$

or

$$S = \sum_{i=1}^{n} S_i \quad [4.190]$$

is called conductance of the system of layers and it is equal to the sum of the conductances of each layer. It is clear that

$$\rho_t = \frac{\sum_{i=1}^{n} h_i}{S}. \quad [4.191]$$

The same result will be obtained if the vector of the current density is parallel to the side dl_1. Now, we assume that the vector of current density is perpendicular to the boundaries between thin layers. Then, the latter are connected in series and the total resistance of this system is

$$R_n = \rho_n \frac{\sum_{i=1}^{n} h_i}{dl_1 \cdot dl_2} = \frac{1}{dl_1 \cdot dl_2}[\rho_1 h_1 + \rho_2 h_2 + \cdots + \rho_n h_n] \quad [4.192]$$

or

$$R_n = \frac{\sum_{i=1}^{n} T_i}{dl_1 dl_2} = \frac{T}{dl_1 dl_2}, \quad [4.193]$$

where T is the transversal resistance of the system of thin layers and it is equal to the sum of transversal resistances of each layer. By definition,

$$\rho_n = \frac{T}{\sum_{i=1}^{n} h_i} \quad [4.194]$$

is the transversal resistivity of a uniform medium when the current is perpendicular to a system of layers and both of them have the same resistance. Thus, we replaced a layered medium by a uniform one which is characterized by two values of resistivity: ρ_t and ρ_n. In other words, this layered medium is equivalent to a uniform anisotropic medium. At each of its points, the resistivity is characterized by two different values corresponding to two perpendicular directions so that one of them is along the boundaries of the original model. The ratio

$$\lambda = \left(\frac{\rho_n}{\rho_t}\right)^{1/2} \tag{4.195}$$

defines the degree of anisotropy of the medium and it is usually called the coefficient of anisotropy. Now, we demonstrate that the latter exceeds unity; that is $\rho_n > \rho_t$. In fact, consider the ratio

$$\frac{\rho_n}{\rho_t} = \frac{\left(\sum_{i=1}^{n} T_i\right)\cdot\left(\sum_{i=1}^{n} S_i\right)}{\left(\sum_{i=1}^{n} h_i\right)^2} = \frac{1}{h^2}\left[(h_1\rho_1 + h_2\rho_2 + \cdots + h_n\rho_n)\left(\frac{h_1}{\rho_1} + \frac{h_2}{\rho_2} + \cdots + \frac{h_n}{\rho_n}\right)\right].$$

Here, h is the total thickness of layers of an elementary volume. Performing multiplication at the right-hand side of this equality, we obtain

$$\begin{aligned}
\frac{\rho_n}{\rho_t} = {} & \frac{1}{h^2}\left(h_1^2 + h_1 h_2\frac{\rho_2}{\rho_1} + h_1 h_3\frac{\rho_3}{\rho_1} + \cdots + h_1 h_n\frac{\rho_n}{\rho_1} + h_1 h_2\frac{\rho_1}{\rho_2} + h_2^2 + \right. \\
& \left. h_2 h_3\frac{\rho_2}{\rho_3} + \cdots + h_2 h_n\frac{\rho_2}{\rho_n} + \cdots + h_1 h_n\frac{\rho_1}{\rho_n} + h_2 h_n\frac{\rho_2}{\rho_n} + \cdots + h_n^2\right) \\
= {} & \frac{1}{h^2}\left[h_1^2 + h_2^2 + \cdots + h_n^2 + h_1 h_2\left(\frac{\rho_2}{\rho_1} + \frac{\rho_1}{\rho_2}\right) + h_1 h_3\left(\frac{\rho_3}{\rho_1} + \frac{\rho_1}{\rho_3}\right) + \cdots + \right. \\
& \left. h_i h_k\left(\frac{\rho_i}{\rho_k} + \frac{\rho_k}{\rho_i}\right) + \cdots + h_{n-1}h_n\left(\frac{\rho_n}{\rho_{n-1}} + \frac{\rho_n}{\rho_{n-1}}\right)\right]
\end{aligned} \tag{4.196}$$

since

$$\frac{\rho_k}{\rho_i} + \frac{\rho_i}{\rho_k} = \frac{\rho_k^2 + \rho_i^2}{\rho_i\rho_k} = \frac{2\rho_i\rho_k + \rho_k^2 - 2\rho_i\rho_k + \rho_i^2}{\rho_i\rho_k} = 2 + \frac{(\rho_k - \rho_i)^2}{\rho_k\rho_i}.$$

Therefore, coefficients in front of products $h_i h_k$ in the square brackets are greater than 2 and, correspondingly, the numerator at the right-hand side of Eq. [4.196] exceeds h^2; that is, $\rho_n > \rho_t$. This means that the resistance of a layered medium is greater when the vector of current density is normal to the interfaces. Values of the coefficient of anisotropy are usually close to unity and seldom exceed 2. Next, we will study some interesting features of the field behavior in anisotropic medium, and with this purpose in mind, consider the simplest case of a uniform medium.

4.12.2. The Potential and Electric Field in a Uniform Anisotropic Medium

As in the case of the isotropic medium, we will proceed from the same equations of the field:

$$\text{curl}\,\mathbf{E} = 0, \qquad \text{div}\,\mathbf{j} = 0.$$

However, in an anisotropic medium, the relation between the vector of current density and electric field is more complicated since the conductivity is a tensor. In the Cartesian system of coordinates, Ohm's law can be written as

$$j_x = \gamma_x E_x, \quad j_y = \gamma_y E_y, \quad j_z = \gamma_z E_z. \qquad [4.197]$$

Here, γ_x, γ_y, and γ_z are the conductivities along the coordinate axes. Let us choose the direction of these axes in such a way that the z-axis is perpendicular to a system of thin layers, but two other axes are located in the plane tangential to the interfaces. Then, we can write for resistivity

$$\rho_x = \rho_y = \rho_t \quad \text{and} \quad \rho_z = \rho_n. \qquad [4.198]$$

Taking into account Eqs. [4.197] and [4.198], the equation for the current density ($\text{div}\,\mathbf{j}=0$) has the form

$$\frac{1}{\rho_t}\left(\frac{\partial E_x}{\partial x} + \frac{\partial E_y}{\partial y}\right) + \frac{1}{\rho_n}\frac{\partial E_z}{\partial z} = 0. \qquad [4.199]$$

Bearing in mind that $\mathbf{E} = -\,\text{grad}\,U$, Eq. [4.199] becomes

$$\frac{1}{\rho_t}\left(\frac{\partial^2 U}{\partial x^2} + \frac{\partial^2 U}{\partial y^2}\right) + \frac{1}{\rho_n}\frac{\partial^2 U}{\partial z^2} = 0. \qquad [4.200]$$

Certainly, it is not Laplace's equation, but introducing new variables it is possible to obtain this equation. Performing the following transformation of coordinates

$$\xi = (\rho_t)^{1/2} x, \quad \eta = (\rho_t)^{1/2} y, \quad \zeta = (\rho_n)^{1/2} z, \qquad [4.201]$$

we have

$$\frac{\partial^2 U}{\partial x^2} = \rho_t \frac{\partial^2 U}{\partial \xi^2}, \quad \frac{\partial^2 U}{\partial y^2} = \rho_t \frac{\partial^2 U}{\partial \eta^2}, \quad \frac{\partial^2 U}{\partial z^2} = \rho_n \frac{\partial^2 U}{\partial \zeta^2},$$

and in place of Eq. [4.200], we arrive at the Laplace's equation:

$$\frac{\partial^2 U}{\partial \xi^2} + \frac{\partial^2 U}{\partial \eta^2} + \frac{\partial^2 U}{\partial \zeta^2} = 0. \qquad [4.202]$$

Assuming as usual that the current electrode can be described as a point source, we choose from among an infinite number of solutions of Eq. [4.202] the simplest one with a singularity which corresponds to this charge:

$$U(\xi, \eta, \zeta) = \frac{C}{(\xi^2 + \eta^2 + \zeta^2)^{1/2}} = \frac{C}{[\rho_t(x^2 + y^2) + \rho_n z^2]^{1/2}}. \qquad [4.203]$$

Here, C is unknown. In an isotropic medium, equipotential surfaces of the field caused by the electrode charge are spherical, but in our case, they are surfaces of the spheroid of rotation and their equation is

$$\rho_t(x^2 + y^2) + \rho_n z^2 = \text{constant} \quad \text{or} \quad \frac{x^2 + y^2}{\rho_n} + \frac{z^2}{\rho_t} = \text{constant}. \qquad [4.204]$$

By definition, at each point, the electric field is perpendicular to these spheroidal surfaces. Bearing in mind that the direction of the Cartesian axes is related with an anisotropy, we see that the minor axis of ellipsoids is perpendicular to boundaries between thin layers and these axes are proportional $\rho_t^{1/2}$ and $\rho_n^{1/2}$, respectively. It is clear that the ratio of the semi-axes of all equipotential surfaces is the same and is equal to the square of the coefficient of anisotropy:

$$\frac{\rho_n}{\rho_t} = \lambda^2. \qquad [4.205]$$

4.12.3. Charge of the Current Electrode

First, we find the constant C in Eq. [4.203], and with this purpose in mind, consider a behavior of the vector of current density. From Ohm's law, we have

$$j_x = -\frac{1}{\rho_t}\frac{\partial U}{\partial x}, \quad j_y = -\frac{1}{\rho_t}\frac{\partial U}{\partial y}, \quad j_z = -\frac{1}{\rho_n}\frac{\partial U}{\partial z}$$

and performing differentiation using Eq. [4.203], we have

$$j_x = \frac{Cx}{[\rho_t(x^2+y^2)+\rho_n z^2]^{3/2}},$$
$$j_y = \frac{Cy}{[\rho_t(x^2+y^2)+\rho_n z^2]^{3/2}}, \qquad [4.206]$$
$$j_z = \frac{Cz}{[\rho_t(x^2+y^2)+\rho_n z^2]^{3/2}}.$$

These formulas show a very interesting behavior of the field $\mathbf{j}(x, y, z)$, namely,

$$j_x : x = j_y : y = j_z : z,$$

where x, y, z are the components of the radius vector $\mathbf{R}$. Therefore, vectors $\mathbf{j}$ and $\mathbf{R}$ are parallel to each other. This means that in the spherical system of coordinates with the origin at the electrode, the vector of the current density has only a radial component in spite of the fact that in general, the field $\mathbf{E}$ has a different direction. As follows from Eqs. [4.206], the magnitude of the current density is

$$j = C\frac{(x^2+y^2+z^2)^{1/2}}{[\rho_t(x^2+y^2)+\rho_n z^2]^{3/2}}. \qquad [4.207]$$

Now, we are prepared to determine the unknown C, since the current I through the electrode is given. It is convenient to introduce the spherical system of coordinates with the same origin, so that the z-axis is directed perpendicular to interfaces of the thin layers, and consider a spherical surface with any radius R:

$$R = (x^2+y^2+z^2)^{1/2}.$$

In as much as

$$x = R\sin\theta\cos\phi, \quad y = R\sin\theta\sin\phi, \quad z = R\cos\theta,$$

Eq. [4.207] becomes

$$j_R = \frac{1}{R^2}\frac{1}{[\rho_t\sin^2\theta + \rho_n\cos^2\theta]^{3/2}}.$$

Note that, unlike an isotropic medium, the current density changes on a spherical surface surrounding the current electrode. For instance, it has maximal magnitude parallel to the layers and minimum in the perpendicular direction. Performing integration over the spherical surface and taking into account that

$$dS = R^2 \sin\theta \, d\theta d\phi,$$

we obtain for the total current flowing through this surface

$$I = C\int_0^{2\pi} d\phi \int_0^{\pi} \frac{\sin\theta \, d\theta}{[\rho_t \sin^2\theta + \rho_n \cos^2\theta]^{3/2}}$$

or

$$I = 2\pi C\int_0^{\pi} \frac{\sin\theta \, d\theta}{[\rho_t \sin^2\theta + \rho_n \cos^2\theta]^{3/2}}. \qquad [4.208]$$

Performing integration, we obtain

$$I = \frac{4\pi C}{\rho_t(\rho_n)^{1/2}} \quad \text{or} \quad C = \frac{I\rho_t(\rho_n)^{1/2}}{4\pi}, \qquad [4.209]$$

and finally, we obtain an expression for the potential:

$$U(x, y, z) = \frac{I\rho_t(\rho_n)^{1/2}}{4\pi} \frac{1}{[\rho_t(x^2 + y^2) + \rho_n z^2]^{1/2}} \qquad [4.210]$$

or

$$U(x, y, z) = \frac{I(\rho_t\rho_n)^{1/2}}{4\pi} \frac{1}{(x^2 + y^2 + \lambda^2 z^2)^{1/2}}. \qquad [4.211]$$

The latter suggests that the charge on the surface of the electrode is equal to

$$e_A = \varepsilon_0(\rho_t\rho_n)^{1/2} I. \qquad [4.212]$$

As follows from Eq. [4.211], the components of the electric field are

$$E_x(x, y, z) = \frac{e_A}{4\pi\varepsilon_0} \frac{x}{(x^2 + y^2 + \lambda^2 z^2)^{3/2}},$$

$$E_y(x, y, z) = \frac{e_A}{4\pi\varepsilon_0} \frac{y}{(x^2 + y^2 + \lambda^2 z^2)^{3/2}}, \qquad [4.213]$$

$$E_z(x, y, z) = \frac{e_A}{4\pi\varepsilon_0} \frac{\lambda^2 z}{(x^2 + y^2 + \lambda^2 z^2)^{3/2}}.$$

Correspondingly, for points of the vector lines of the electric field we have

$$\frac{dx}{x} = \frac{dy}{y} = \frac{dz}{\lambda^2 z},$$

and this shows that unlike vector lines of the current density, the lines of the field $\mathbf{E}$ are not straight. Also, performing differentiation of Eqs. [4.213], it is a simple matter to see that at each point of a uniform anisotropic medium div $\mathbf{E} = 0$; that is, the volume density of charges is zero. Earlier, we pointed out that equipotential surfaces coincide with the surfaces of the ellipsoid of rotation, and for a solution of the boundary-value problem, it is natural to suppose that the electrode has the same shape. Application of electrodes of a different shape results in the distortion of the shape of the equipotential surfaces in the vicinity of the current source but with an increase in the distance, they are again described by Eq. [4.211].

4.12.4. The Field on the Surface of a Uniform Anisotropic Half-Space

If the current electrode is located on the earth's surface, then the influence of this surface, as in the case of the isotropic medium, can be easily taken into account by increasing the charge of the electrode twice. Then, we have

$$U(x, y, z) = \frac{\rho_m I}{2\pi} \frac{1}{\left(x^2 + y^2 + \lambda^2 z^2\right)^{1/2}}. \qquad [4.214]$$

Here, $\rho_m = \left(\rho_t \rho_n\right)^{1/2}$. Note that the direction of the coordinate axes x, y, z is related to anisotropy of a medium, and it does not characterize the position of an observation point. In order to describe the behavior of the field, it is convenient to introduce the system of coordinates x_1, y_1, z_1 with the origin on the earth's surface, shown in Fig. 4.13B. Coordinate lines x and x_1 coincide in both systems and

$$x = x_1, \quad y = y_1 \cos\alpha + z_1 \sin\alpha, \quad z = -y_1 \sin\alpha + z_1 \cos\alpha. \qquad [4.215]$$

Here, α is the angle characterizing the orientation bedding with respect on the earth's surface. Substituting Eqs. [4.215] into Eq. [4.214] and letting $z_1 = 0$, we have for the potential on the earth's surface

$$U(x_1, y_1) = \frac{\rho_m I}{2\pi} \frac{1}{\left\{x_1^2 + \left[1 + \left(\lambda^2 - 1\right)\sin^2\alpha\right] y_1^2\right\}^{1/2}}. \qquad [4.216]$$

The latter shows that the equation of equipotential lines can be written as

$$x_1^2 + [1 + (\lambda^2 - 1)\sin^2\alpha]y_1^2 = C = \text{constant}$$

or

$$\frac{x_1^2}{a^2} + \frac{y_1^2}{b^2} = 1. \qquad [4.217]$$

Thus, equipotential lines on the earth's surface are ellipses with semi-axes

$$a = C^{1/2} \quad \text{and} \quad b = C^{1/2}[1 + (\lambda^2 - 1)\sin^2\alpha]^{-1/2},$$

and, therefore, the ratio of semi-axes

$$\frac{a}{b} = [1 + (\lambda^2 - 1)\sin^2\alpha]^{1/2} \qquad [4.218]$$

depends on the coefficient of anisotropy and the angle characterizing the direction of interfaces with respect to the earth's surface. For instance, this means that if λ is known, we can obtain information about the orientation of the bedding.

4.12.5. Behavior of Apparent Resistivity and Paradox of Anisotropy

Let us define the position of an observation point on the earth's surface in the polar system of coordinates r and ϕ (Fig. 4.13B). Then, we have

$$x_1 = r\cos\phi \quad \text{and} \quad y_1 = r\sin\phi$$

and the potential on the earth's surface can be written in the form

$$U(r,\phi) = \frac{\rho_m I}{2\pi r}\frac{1}{[1 + (\lambda^2 - 1)\sin^2\alpha \cdot \sin^2\phi]^{1/2}}. \qquad [4.219]$$

Performing differentiation with respect to r, we have for the component E_r

$$E_r(r,\phi) = \frac{\rho_m I}{2\pi r^2}\frac{1}{[1 + (\lambda^2 - 1)\sin^2\alpha \cdot \sin^2\phi]^{1/2}}. \qquad [4.220]$$

As follows from this equation, the electric field is proportional to ρ_m if the bedding is parallel to the earth's surface ($\alpha = 0$):

$$E_r = \frac{\rho_m I}{2\pi r^2}.$$

On the contrary, in the case of a vertical bedding, we have

$$E_r(\phi) = \frac{\rho_m I}{2\pi r^2[\cos^2\phi + \lambda^2 \sin^2\phi)^{1/2}}.$$

To illustrate the behavior of the apparent resistivity, consider the three-electrode array *AMN*.

Then, the apparent resistivity is represented as

$$\rho_a = \frac{\rho_m}{[1 + (\lambda^2 - 1)\sin^2\alpha \cdot \sin\phi]^{1/2}}. \quad [4.221]$$

The polar diagram of the function $\rho_a(\phi)$ is an ellipse and, as before, the ratio of semi-axes is defined by Eq. [4.218]. Let us consider two special cases: $\phi = 0$ and $\phi = \pi/2$. In the first case, the array is directed along bedding and $\rho_a(0) = \rho_m$, but in the second case, the array is perpendicular to bedding, and we have

$$\rho_a\left(\frac{\pi}{2}\right) = \frac{\rho_m}{(\cos^2\alpha + \lambda^2 \sin^2\alpha)^{1/2}} < \rho_m = \rho_a(0). \quad [4.222]$$

This relation is called the paradox of anisotropy because, for resistivity characterizing a medium in these directions, the opposite relation takes place: $\rho_n > \rho_t$. Intuitively, such behavior may be obvious since it is natural to expect that the current density is higher along the bedding. The latter takes place only if the component of the electric field and, therefore, the apparent resistivity, prevail in this direction.

CHAPTER FIVE

Interpretation of Electric Soundings

This chapter describes some of the practical issues that arise in interpreting electric soundings over layered media, including the numerical calculation of apparent resistivity curves and questions related to the inverse problem for electrical measurements with steady currents.

5.1. SYMMETRIC AND DIPOLE ARRAYS OF ELECTRIC SOUNDINGS

5.1.1. Introduction

As described in Chapter 4, two main types of arrays are used for electrical sounding: symmetrical arrays ($AMNB$) and dipole arrays ($ABMN$). The symmetrical array measures the apparent resistivity as a function of the average separation $AB/2$ between current and receiver electrodes (Fig. 5.1A). The central point 0 is the "record point"; that is, we assume that the results characterize a geoelectric section beneath this point. It is convenient to visualize the measurements as apparent resistivity curves on a log–log scale (Fig. 5.1B); to provide a uniform distribution of points on such a curve, the separation $AB/2$ is usually expanded as terms of a geometric progression. This can be achieved in various ways: for example, the entire array can be expanded symmetrically about the central point, or just the outer (current) electrodes can be moved. Since the measured voltage V_{MN} becomes smaller as the distance between the current and receiver electrodes increases, it is customary to repeat the measurements at certain values of $AB/2$ with two different values of the separation MN (Fig. 5.1B). Several kinds of dipole arrays are used, sometimes called the axial, azimuthal, and equatorial arrays. The axial and equatorial dipole arrays are the most common for geoelectric sounding or profiling. Another common arrangement—especially when measurements are made over a two-dimensional grid of points on the surface—keeps the current line AB and the receiver MN line parallel at all times. One convenience of

Methods in Geochemistry and Geophysics, Volume 44
ISSN 0076-6895, DOI: 10.1016/S0076-6895(10)44005-6
© 2010 Elsevier B.V.
All rights reserved.

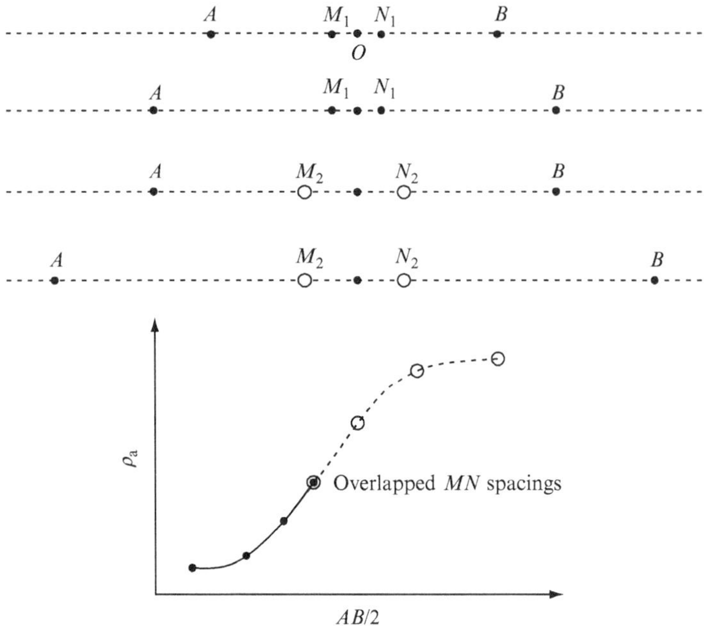

Figure 5.1 (A) Expanding symmetrical array *AMNB*, with overlapped separations of the voltage electrodes. (B) Apparent resistivity curve.

dipole arrays is that it is easy to perform soundings in different directions, leaving the current dipole fixed. Because of this, sounding and profiling with dipole arrays have been used for many years in marine geophysics. Again, with all the dipole arrays, the measurements are often visualized as curves of apparent resistivity as a function of the separation between the current and receiver dipoles, with the central point of the array taken as the record point. Also, the length of the current and receiver lines is usually increased as their separation increases to provide a measurable signal at large separations. When the medium is horizontally layered, apparent resistivity curves obtained with the azimuth dipole arrays coincide with those of the symmetrical (*AMNB*) array. Expressions for the apparent resistivity of dipole arrays over a layered medium can be derived from the equation for the potential at an observation point p created by the current electrode located at the origin 0 [Eq. 4.156]:

$$U(r) = \frac{\rho_1 I}{2\pi} \int_0^\infty R_N(m) J_0(mr) \mathrm{d}m, \qquad [5.1]$$

where R_N is the "kernel function" representing the response of the N-layer medium.

Now, suppose that in place of a single current electrode at point 0 we have two electrodes with charges equal in magnitude but opposite in sign. Then, the potential at point p is

$$U^{\mathrm{d}}(p) = U(r_A) - U(r_B). \qquad [5.2]$$

Here, r_A and r_B are the distances between the point p and electrodes A and B, respectively. Letting AB be the length of the current dipole, the potential of the dipole oriented along line l can be represented approximately as

$$U^{\mathrm{d}}(p) \approx -\frac{\partial U}{\partial l} l = -\frac{\partial U}{\partial r}\frac{\partial r}{\partial l} l = -AB\frac{\partial U}{\partial r}\cos\theta. \qquad [5.3]$$

Using Eq. [5.1], we, thus, have for the potential of the electric dipole on the earth's surface:

$$U^{\mathrm{d}}(r,\theta) = \frac{\rho_1 IAB}{2\pi}\cos\theta\int_0^\infty mR_n J_1(mr)\mathrm{d}m. \qquad [5.4]$$

The radial and azimuthal components of the electric field of the dipole are

$$E_r = -\frac{\partial U^{\mathrm{d}}}{\partial r} \text{ and } E_\theta = -\frac{1}{r}\frac{\partial U^{\mathrm{d}}}{\partial \theta},$$

and Eq. [5.4] gives

$$E_r(r,\theta) = -\frac{\rho_1 IAB}{2\pi}\cos\theta\int_0^\infty m^2 R_n J'_1(mr)\mathrm{d}m$$

and [5.5]

$$E_\theta(r,\theta) = \frac{\rho_1 IAB}{2\pi r}\sin\theta\int_0^\infty mR_n J_1(mr)\mathrm{d}m,$$

where $J'_1(mr)$ is the derivative of the Bessel function.

Now, we are ready to introduce the apparent resistivity for each dipole array.

5.1.2. Axial Dipole Array ($\theta = 0$)

It follows from Eqs. [5.5] that

$$E_r(r,0) = -\frac{\rho_1 IAB}{2\pi}\int_0^\infty m^2 R_n J'_1(mr)\mathrm{d}m,$$

whereas the electric field at the surface of a uniform half-space is

$$E_r^0 = \frac{\rho_1 IAB}{2\pi r^3}.$$

Thus, the apparent resistivity is given by

$$\frac{\rho_a}{\rho_1} = -r^3 \int_0^\infty m^2 R_n J'_1(mr)dm. \qquad [5.6]$$

5.1.3. Azimuthal Dipole Arrays

The second equation of the set Eq. [5.5] gives the azimuthal component of the electric field in a layered medium, whereas the azimuthal component of the electric field of a uniform half-space is

$$E_\theta^0 = \frac{\rho_1 IAB \sin\theta}{4\pi r^3}.$$

Thus,

$$\frac{\rho_a}{\rho_1} = 2r^2 \int_0^\infty mR_n J_1(mr)dm. \qquad [5.7]$$

This expression gives the apparent resistivity for the equatorial dipole array when $\theta = \pi/2$. As pointed out earlier, the apparent resistivity curves for these arrays coincide with those for the symmetrical array *AMNB*.

5.2. SOLUTION OF FORWARD AND INVERSE PROBLEMS IN ELECTRICAL METHODS

5.2.1. Numerical Integration of Integrals with Bessel Functions

Interpretation of electric soundings ultimately relies on a comparison of experimental data (measurements) with theoretical models. Solution of the forward problem is an important part of this procedure. Bearing in mind that geoelectric sections are characterized by an extremely broad range of parameters, such as the thicknesses and resistivities of layers, it was clear from the beginning that methods were needed for accurate and rapid calculation of integrals describing the potential and electric field at the earth's surface, as well as at the sea or river bottom. The forward problem for horizontally layered media was solved more than 80 years ago by Sabba Stefanescu and, as shown in Chapter 4, the field is represented in the form of integrals containing Bessel functions. Until the middle of the last century, the main tools for numerical integration were calculators. In principle, they allowed one to perform numerical integration, but due to the oscillatory character of Bessel functions in the integrand, this task required a lot of time to provide sufficient accuracy. It is proper to discuss this question in some detail because it strongly influenced the development of practical methods in geophysics. Let us write a typical integral for the potential as

$$\int_0^\infty F(m)J_0(mr)\mathrm{d}m. \tag{5.8}$$

Here, $F(m)$ is a nonoscillating function that changes relatively slowly and tends to zero at large values of m. The Bessel function behaves differently. As was shown earlier, for large values of its argument ($mr > 1$), it is an oscillating function,

$$J_0(x) \approx \left(\frac{2}{\pi x}\right)^{1/2} \cos\left(x - \frac{\pi}{4}\right),$$

with slowly decreasing magnitude and period equal to

$$\Delta m = \frac{2\pi}{r}. \tag{5.9}$$

As the separation between electrodes (r) increases, the period becomes smaller and the integrand changes sign more often. Since the function $F(m)$ varies slowly, the partial integrals over successive periods at large r have nearly the same magnitudes but opposite signs, leading to severe cancelation of terms. To obtain an accurate numerical result, it is necessary to calculate the integrand to high precision. Before the availability of modern computers, this was a formidable numerical problem and various methods were developed to replace the integral by a series. This development was an important contribution to the theory of electric sounding.

5.2.2. Representation of the Field as a Series

To illustrate this method, we first consider the simple case of a two-layer medium. As shown in Chapter 4, the expression for the potential in the upper layer is

$$U_1(r,z) = \frac{\rho_1 I}{2\pi}\left\{\frac{1}{R} + \int_0^\infty C_1[\exp(-mz) + \exp(mz)]J_0(mr)\mathrm{d}m\right\}, \tag{5.10}$$

where $R = (r^2 + z^2)^{1/2}$, and $C_1 = \dfrac{K_{12}\exp(-2mh_1)}{1 - K_{12}\exp(-2mh_1)}$.

Expanding the fraction

$$\frac{1}{1 - K_{12}\exp(-2mh_1)}$$

in a series, we have

$$\frac{1}{1 - K_{12}\exp(-2mh_1)} = 1 + K_{12}\exp(-2mh_1) + K_{12}^2\exp(-4mh_1) + K_{12}^3\exp(-6mh_1) + \cdots$$

Thus,

$$C_1 = \sum_1^\infty K_{12}^n \exp(-2mnh_1). \quad [5.11]$$

Substituting Eq. [5.11] into Eq. [5.10] and changing the order of integration and summation gives

$$U_1(r,z) = \frac{\rho_1 I}{2\pi}\left\{\frac{1}{R} + \sum_{n=1}^\infty K_{12}^n \left[\int_0^\infty \exp m(-2nh_1 - z) J_0(mr)\mathrm{d}m + \int_0^\infty \exp m(-2nh_1 + z) J_0(mr)\mathrm{d}m\right]\right\}.$$

The individual terms can now be integrated analytically using the Lipschitz integral introduced in Chapter 4:

$$\int_0^\infty \exp(-mz) J_0(mr)\mathrm{d}m = \frac{1}{(r^2+z^2)^{1/2}}, \quad \text{if } z \geq 0.$$

Applying this to every term of the sum, we obtain

$$U(r,z) = \frac{\rho_1 I}{2\pi}\left\{\frac{1}{(r^2+z^2)^{1/2}} + \sum_{n=1}^\infty \frac{K_{12}^n}{\left[r^2 + (2nh_1 + z)^2\right]^{1/2}} + \sum_{n=1}^\infty \frac{K_{12}^n}{\left[r^2 + (2nh_1 - z)^2\right]^{1/2}}\right\}.$$

Finally, letting $z = 0$ gives an expression for the potential on the earth's surface:

$$U_1(r,0) = \frac{\rho_1 I}{2\pi}\left\{\frac{1}{r} + 2\sum_{n=1}^\infty \frac{K_{12}^n}{\left[r^2 + (2nh_1)^2\right]^{1/2}}\right\}. \quad [5.12]$$

Summation of the series in Eq. [5.12] is often much simpler than numerical integration of a rapidly oscillating function. In particular, if $\rho_2 > \rho_1$, all terms are positive and the series converges quickly once $2nh_1 > r$. Differentiating with respect to r gives an expression for the radial component of the electric field:

$$E_r(r,0) = \frac{\rho_1 I}{2\pi}\left\{\frac{1}{r^2} + 2\sum_{n=1}^{\infty}\frac{K_{12}^n r}{\left[r^2 + (2nh_1)^2\right]^{3/2}}\right\}. \qquad [5.13]$$

The apparent resistivity for the symmetrical array is, therefore,

$$\rho_a = \rho_1\left\{1 + 2\sum_{n=1}^{\infty}\frac{K_{12}^n (r/h_1)^3}{\left[(r/h_1)^2 + 4n^2\right]^{3/2}}\right\}. \qquad [5.14]$$

This formula was often used at the beginning to compute apparent resistivity curves for two-layer media. The corresponding asymptotic formulas also follow simply. For small values of r, we clearly have

$$\rho_a \to \rho_1, \quad \text{if } r \to 0,$$

whereas in the opposite case ($r \to \infty$), the series can be summed analytically,

$$\rho_a \to \rho_1\left(1 + 2\sum_{n=1}^{\infty} K_{12}^n\right) = \rho_1\left(1 + 2\frac{K_{12}}{1 - K_{12}}\right) = \rho_2,$$

showing that the apparent resistivity tends to the resistivity of the lower half-space. The same approach can be used for a medium with more layers. For instance, the coefficient A_m describing the potential in the upper layer of a three-layer medium can be written as

$$A_m = \frac{K_{12}g^{p_1} + K_{23}g^{p_2}}{1 - K_{12}g^{p_1} - K_{23}g^{p_2} + K_{12}K_{23}g^{(p_2 - p_1)}}. \qquad [5.15]$$

Here,

$$K_{12} = \frac{\rho_2 - \rho_1}{\rho_2 + \rho_1}, \quad K_{23} = \frac{\rho_3 - \rho_2}{\rho_3 + \rho_2}, \quad g = \exp(-mh),$$

and

$$h_1 = p_1\frac{h}{2}, \quad h_2 = p_2\frac{h}{2},$$

where p_1, p_2 are integers; that is, h_1 and h_2 are commensurate (multiples of a single length). The coefficient A_m can be expanded in a series,

$$A_m = \sum_{n=1}^{\infty} q_n g^n = \sum_{n=1}^{\infty} q_n \exp(-mhn). \qquad [5.16]$$

To match Eq. [5.15], the unknown coefficients q_n have to satisfy the equation

$$K_{12}g^{p_1} + K_{23}g^{p_2} = \left[1 - K_{12}g^{p_1} - K_{23}g^{p_2} + K_{12}K_{23}g^{(p_2-p_1)}\right] \cdot \sum_{n=1}^{\infty} q_n g^n. \qquad [5.17]$$

Expanding the right-hand side and equating the coefficients of like powers of g on both sides, we obtain a system of equations for q_n. This procedure is greatly simplified, because there are recursive relations between coefficients with different indexes. Substituting Eq. [5.16] into expression of the potential, we obtain

$$U_1(r,z) = \frac{\rho_1 I}{2\pi}\left\{\frac{1}{(r^2+z^2)^{1/2}} + \sum_{n=1}^{\infty} q_n J_0(mr)\exp(-mhn)[\exp(-mz) + \exp(mz)]\mathrm{d}m\right\}$$

$$= \frac{\rho_1 I}{2\pi}\left\{\frac{1}{(r^2+z^2)^{1/2}} + \sum_{n=1}^{\infty} q_n \int_0^{\infty} J_0(mr)\exp[-m(nh+z)]\mathrm{d}m + \sum_{n=1}^{\infty} q_n \int_0^{\infty} J_0(mr)\exp[-m(nh-z)]\mathrm{d}m\right\}$$

or

$$U_1(r,z) = \frac{\rho_1 I}{2\pi}\left\{\frac{1}{(r^2+z^2)^{1/2}} + \sum_{n=1}^{\infty} \frac{q_n}{\left[r^2+(nh+z)^2\right]^{1/2}} + \sum_{n=1}^{\infty} \frac{q_n}{\left[r^2+(nh-z)^2\right]^{1/2}}\right\}.$$

In particular, on the earth's surface, we have

$$U_1(r,0) = \frac{\rho_1 I}{2\pi}\left\{\frac{1}{r} + 2\sum_{n=1}^{\infty} \frac{q_n}{\left[r^2+(nh)^2\right]^{1/2}}\right\}. \qquad [5.18]$$

The similarity with Eq. [5.12] is obvious. Differentiating with respect to r gives a series for the electric field,

$$E_r(r,0) = \frac{\rho_1 I}{2\pi}\left\{\frac{1}{r^2} + 2\sum_{n=1}^{\infty}\frac{q_n r}{\left[r^2 + (nh)^2\right]^{3/2}}\right\}. \quad [5.19]$$

The apparent resistivity for the symmetric array is, therefore,

$$\rho_a = \rho_1\left\{1 + 2\sum_{n=1}^{\infty}\frac{q_n (r/h)^3}{\left[(r/h)^2 + n^2\right]^{3/2}}\right\}, \quad [5.20]$$

where $r = AB/2$ (for the symmetric array). The relationship between the series for the apparent resistivity and the parameters of a three-layer medium is more complicated, because each coefficient q_n depends on coefficients K_{12} and K_{23}, as well as on the ratio $\nu_1 = h_2/h_1$. During the early years of hand calculation, interpretation was mainly based apparent resistivity curves for two-layer media and for the following three-layer cases:

$$\rho_3 : 0, \frac{\rho_2^2}{\rho_1}, \rho_1, \infty,$$

$$\mu_1 = \frac{\rho_2}{\rho_1} = 39,\ 19,\ 9,\ 4,\ 7/3,\ 3/2,\ 2/3,\ 3/7,\ 1/4,\ 1/9,\ 1/19,\ 1/39,$$

$$\nu_1 = \frac{h_2}{h_1} = 1/9,\ 1/5,\ 1/3,\ 1/2,\ 1,\ 2,\ 3,\ 5,\ 9,\ 24.$$

Equation [5.18] was obtained from the integral representation of the field derived by Sabba Stefanescu. The same result was obtained slightly earlier by J.N. Hummel using the method of mirror reflection. To illustrate this approach, consider a two-layer medium (Fig. 5.2) and suppose that the current electrode with charge $e_A = 2\varepsilon_0\rho_1 I$ is located on the earth's surface. We will satisfy the boundary conditions at each surface by constructing a series of mirror images of the source charge and the accompanying induced charges. In Chapter 2, we demonstrated that the boundary conditions at a single plane interface can be satisfied by placing a fictitious ("image") charge at a point which is the mirror reflection across the interface of the location of the current electrode. In the case of two plane interfaces—here, the earth–air interface and the internal interface between the two layers—a series of image sources is obtained as follows. Assume that the electrode is placed at the origin. First, to satisfy the condition at the boundary $z = h_1$, we place a fictitious

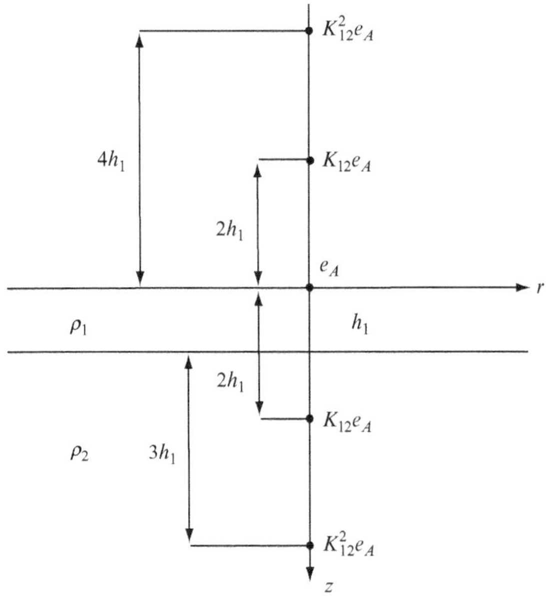

Figure 5.2 Illustration of the method of mirror reflections.

charge $K_{12}e_A$ at the point that is the mirror reflection of the origin with respect to this interface, that is, at the point $z=2h_1$. To satisfy the boundary condition at the earth–air interface, the field of this first mirror image charge must be countered by a second fictitious charge (of equal value) a distance $2h_1$ above the earth's surface (at $z=2h_1$). The second image charge must now be mirrored in the interface at $z=h_1$ by charge of magnitude $K_{12}^2 e_A$ at $z=3h_1$. Continuing this process, we obtain a system of fictitious charges located at the vertical axis at points:

$$z_n = 2nh_1,$$

for all integers n. Note that the magnitudes of the mirror charges at the same relative distances above or below the earth's surface are the same, because the reflection coefficient across the earth–air interface is equal to unity. But, the magnitude increases by one power of K_{12} at successive distances away from the interface. The potential on the earth's surface caused by these charges is

$$U(r,0) = \frac{\rho_1 I}{2\pi}\left[\frac{1}{r} + 2\sum_{n=1}^{\infty}\frac{K_{12}^n}{(r^2+4n^2)^{1/2}}\right],$$

which coincides with Eq. [5.12]. If the lower medium is more resistive, all terms of the series are positive, but the series is alternating when $\rho_2 < \rho_1$. Proceeding from this equation, Hummel obtained expressions for the electric field and apparent resistivity and then extended this approach to a three-layer medium.

5.2.3. Interpretation of Electric Sounding by Master Curves

For many years, the interpretation of field curves of apparent resistivity was based on catalogs of theoretical curves for two- and three-layer media. A famous example is the set of Schlumberger master curves, which can still be found today. Examples of such curves were given in Chapter 4. The interpretation simply consisted of the comparison of the experimental and theoretical curves. On a log–log scale, the equations for the curves of two- and three-layer media are

$$\log\rho_a = \log\rho_1 + \log f(q_n, \log r), \quad [5.21]$$

and

$$\log\rho_a = \log\rho_1 + \log f[q_n, (\log r - \log h_1)]. \quad [5.22]$$

With a two-layer medium, there is only one set of the master curves (see, e.g., Fig. 4.6). The shape of these curves depends on the module μ; the parameters h_1 and ρ_1 control the location of a given curve along the horizontal or vertical axis. Taking into account these facts, we can illustrate the process of interpretation. The experimental curve is first drawn on a transparent paper on the same numerical log–log scale as the master curves, but of course with the axes being the actual separation r and the measured apparent resistivity ρ_a. The field curve is then translated over the master curves to find the best match with one of the master curves. When it happens, the point where the origin (1,1) of the master curves falls on the experimental curve indicates the values of h_1 and ρ_1—that is, the resistivity and thickness of the upper layer. The value of ρ_2 can then be obtained from the right asymptote of the master curve or its index value μ. In this process, the similarity of shape of the experimental curve with one of the master curve provides a natural check on whether the experimental data are consistent with the two-layer model. If the resistivity ρ_2 greatly exceeds ρ_1, then there is broad range of electrode separations over which the experimental curve behaves as a straight line (zone S, as described in Chapter 4). The value of the longitudinal conductance

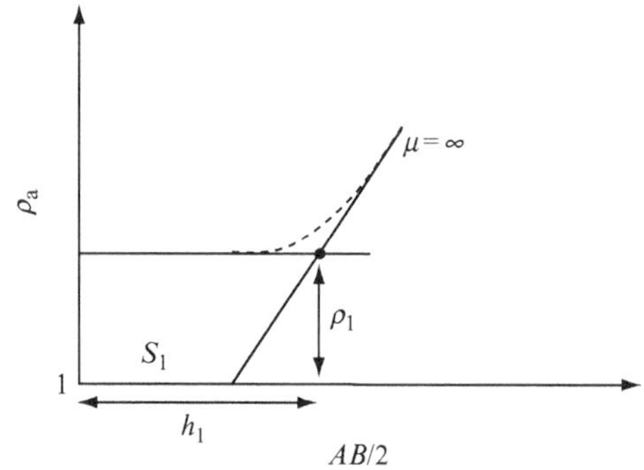

Figure 5.3 Determination of parameters of the upper layer when $\rho_2 \gg \rho_1$.

$S_1 = h_1/\rho_1$ of the upper layer can, in this case, be obtained directly without the use of master curves. In fact, the straight line in zone S has a slope 45°, and its intersection with the horizontal axis gives the thickness of the top layer h_1. The resistivity of the top layer ρ_1 is given by the asymptote of the curve for small separations (Fig. 5.3). On a log–log scale, the equations of these two lines are

$$\log\rho_a = \log\frac{AB}{2} - \log S \text{ and } \log\rho_a = \log\rho_1.$$

Thus, abscissa of the point of intersection is defined from the equality

$$\log\rho_1 = \log\frac{AB}{2} - \log h_1 + \log\rho_1 \text{ or } h_1 = \frac{AB}{2}.$$

The interpretation of models with three and more layers was carried out in a similar way: by hand comparison of theoretical and experimental curves, often using "auxiliary curves" for different types of geoelectric sections. Successful application of the sounding method over many years revealed two main problems with this method of interpretation: the first was the nonavailability of master curves to cover all cases encountered in the field, and the second, more serious, related to the principle of equivalence.

5.2.4. The Principle of Equivalence

The principle of equivalence can be illustrated by two simple cases. Suppose that the intermediate layer in a three-layer medium is relatively thin and its resistivity satisfies the conditions

$$\rho_1 \gg \rho_2 \ll \rho_3 \text{ and } h_2 < h_1. \qquad [5.23]$$

It can easily be shown that, in this case, the apparent resistivity curves practically coincide for different values of ρ_2 and h_2, provided that the ratio $S_2 = h_2/\rho_2$ remains constant. Given that field measurements are only accurate with some tolerance or error, the field curve can be matched with a range of different master curves satisfying the above conditions. In other words, the different geoelectric sections are equivalent to each other. This phenomenon is sometimes called "equivalence by S" (or equivalence by conductance). In this case, only the longitudinal conductance of the intermediate layer can be determined, not its thickness and resistivity separately. In contrast, if the second layer is thin but has much higher resistivity than the surrounding layers ($\rho_1 \ll \rho_2 \gg \rho_3$), then "equivalence by T" is observed, where $T_2 = \rho_2 h_2$ is the transverse resistance of the intermediate layer. In this case, geoelectric sections with the same value of T_2 produce nearly the same curves. Of course, in a section with multiple layers, similar equivalences can be observed with any intermediate layer. The principle of equivalence demonstrates the practical limitations of electric soundings and shows that in many cases, additional information is needed to find some of the parameters. As discussed later, this principle does not contradict the fact that, in principle, a solution of the inverse problem of electric sounding over a horizontally layered medium is unique. Nevertheless, it is worth noting that a knowledge of the different types of equivalence can be helpful in solving the inverse problem for electric sounding, because it allows one to identify the "stable parameters" of a geolectric section, that is, only those parameters that are not highly sensitive to the precision of the measurements.

5.2.5. Apparent Resistivity Curves and Triangle of Anisotropy

To overcome the first problem—the availability of only relatively few master curves for media with three and more layers—geophysicists developed an approximate method that used known master curves to construct apparent resistivity curves for new models with an accuracy of 5–10%. This was an important development that enabled for many years the broad application of soundings to problems of petroleum, groundwater, mining, and engineering geology. This empirical method was the result of a careful study of the two- and three-layer curves along with the use of the concepts of anisotropy for a horizontally layered medium. This was a natural approach because, as we have seen in the previous chapter, one way that anisotropy can arise is by the replacement of a stack of many thin layers by a single equivalent anisotropic layer. Similarly, we will now describe a procedure that reduces several layers of finite thickness to a smaller number of layers. As in the case of a uniform anisotropic medium, we introduce the following quantities:

1. The total thickness of all N layers:

$$H = h_1 + h_2 + h_3 + \cdots + h_N.$$

2. The total longitudinal conductance of the layers:

$$S = \sum_{n=1}^{N} S_i.$$

3. The mean longitudinal resistivity ρ_l of the uniform layer that has the same conductance S as a system of layers

$$\rho_l = \frac{H}{S}. \qquad [5.24]$$

4. The total transverse resistance of layers:

$$T = \sum_{n=1}^{N} T_i.$$

5. The mean transverse resistivity of uniform layer that has the same transversal resistance as a system of layers:

$$\rho_n = \frac{T}{H}. \qquad [5.25]$$

6. The mean geometric resistivity, which is used sometimes as a parameter of a layered medium:

$$\rho_m = (\rho_n \rho_l)^{1/2}. \qquad [5.26]$$

These quantities can be calculated from the above formulas, or graphically, using the "triangle of anisotropy" (Fig. 5.4). Taking the logarithm of Eq. [5.24] gives

$$\log \rho_l = \log H - \log S,$$

which on a log–log scale is a straight line with unit slope. This line defines the locus of points (ρ_l,H) whose mean longitudinal resistivity and total thickness give the same value of S. Similarly, Eq. [5.25] can be written on a log–log scale as

$$\log \rho_n = -\log H + \log T.$$

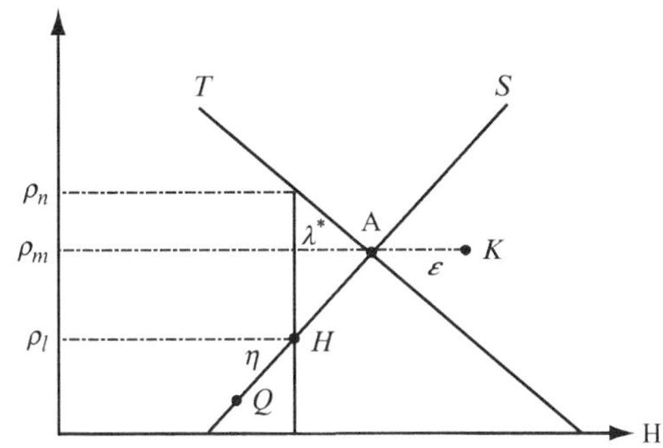

Figure 5.4 Triangle of anisotropy.

This straight line with slope -1 represents the locus of points (ρ_n, T) whose mean transversal resistivity and total thickness give the same value for the total transverse resistance T. To these two lines, we add a vertical line parallel to the resistivity axis and shifted by a distance H from this axis. The three points of intersection of these straight lines form the triangle of anisotropy (Fig. 5.4). The apex of the triangle, formed by intersection of lines H and S, is denoted by H and it is usually called the Hummel point. It is clear that its ordinate is equal to ρ_1 The intersection of lines S and T gives the point A and its ordinate is equal to ρ_m (Eq. 5.26). The ordinate of the intersection of lines T and H is ρ_n. It is easy to show that the height of the triangle is equal to the coefficient of macro-anisotropy:

$$\lambda^* = \left(\frac{\rho_n}{\rho_1}\right)^{1/2} = \frac{(TS)^{1/2}}{H}.$$

As was pointed out, master curves for manual interpretation of apparent resistivity were generally plotted for two- and three-layer media. From physical considerations, it is natural to expect that the initial portion (short electrode separations) of any apparent resistivity curve for an arbitrary model should coincide with the curve for a two-layer model whose parameters match those of the first two layers in the actual model. Analysis of the theoretical curves for a three-layer medium showed that, beyond a certain electrode separation, the apparent resistivity curve also resembles that of two-layer model. The bottom layer (underlying half-space) of this equivalent two-layer model naturally has the resistivity ρ_3 of the underlying half-space in the original model, but the top layer is obtained by replacing the first two layers of the original model by a single equivalent layer. The rules of replacement were found empirically and strongly influenced the successful application of electric soundings. For models of type $H(\rho_1 > \rho_2 < \rho_3)$, the thickness and resistivity of the equivalent layer were found to be

$$h_e = h_1 + h_2, \quad \rho_e = \frac{h_1 + h_2}{S_1 + S_2}, \tag{5.27}$$

which define the coordinates of the point H in the triangle of anisotropy. For models of type $A(\rho_1 < \rho_2 < \rho_3)$, the parameters of the equivalent layer are found to be

$$h_e = [(S_1 + S_2)(T_1 + T_2)]^{1/2} = \lambda^*(h_1 + h_2)$$

and

$$\rho_e = \left(\frac{T_1 + T_2}{S_1 + S_2}\right)^{1/2} \tag{5.28}$$

which define the point A on the triangle. For models of type $Q(\rho_1 > \rho_2 > \rho_3)$, the following equivalent parameters define the coordinates of the point Q,

$$h_e = \frac{1}{\eta}(h_1 + h_2), \quad \rho_e = \frac{1}{\eta}\frac{h_1 + h_2}{S_1 + S_2}, \tag{5.29}$$

where η is a parameter to be determined and represents the distance of point Q from the point H along the line S. Finally, for sections of type $K(\rho_1 < \rho_2 > \rho_3)$, the following equations define the location of point K:

$$h_e = \varepsilon[(S_1 + S_2)(H_1 + H_2)]^{1/2}, \quad \rho_e = \left(\frac{T_1 + T_2}{S_1 + S_2}\right)^{1/2}, \tag{5.30}$$

where ε is a parameter to be determined and represents the distance of K from the point A along the horizontal axis (Fig. 5.4). Values of the parameters ε, η are found with the help of special auxiliary curves. Construction of an approximate curve of apparent resistivity for a three-layer model consists simply of plotting on the log–log scale two-layer curves with parameters h_1, ρ_1 and h_e, ρ_e. Afterwards, these two-layer curves were connected with the help of two known three-layer curves. Similarly, the apparent resistivity curves for four and more layers are obtained. Now, with computers and algorithms available to compute apparent resistivity curves almost instantly even for complicated models, this method looks strange and cumbersome. But as mentioned above, it was for many years a powerful tool enabling many different applications of electric sounding.

5.2.6. Comparison of the Theoretical and Experimental Kernel Functions

Next, we briefly describe a method of interpretation of electric soundings based on construction and analysis of the "kernel function" $R(m)$ in place of the apparent resistivity. This approach starts from the recognition that the formula for the apparent resistivity of a layered medium takes the form of a Fourier–Bessel or Hankel transform, with the corresponding inverse transform. For illustration, suppose that the forward and inverse transforms for the apparent resistivity curves can be written as

$$\rho_a^*(r) = \int_0^\infty m R_1^*(m) J_1(mr) \mathrm{d}m$$

and [5.31]

$$R_1^* = \int_0^\infty r \rho_a^*(r) J_1(mr) \mathrm{d}r$$

where the second equation follows directly from the theory of Hankel transforms which resembles closely the theory of Fourier transforms. A similar pair of equations can be written with the zero-order Bessel function $J_0(mr)$. Comparing with Eqs. (4.157) and (4.155) in Chapter 4, we make the following identifications:

$$\rho_a^*(r) = \frac{\rho_a(r) - \rho_1}{\rho_1 r^2}, \; R_1^*(m) = R_1(m) - 1,$$

and

$$R_1(m) = \rho_1 \left\{ \coth\left[mh_1 + \coth^{-1}\left(\frac{\rho_2}{\rho_1}\right)\left[\coth(mh_2 + \cdots + \coth^{-1})\frac{\rho_n}{\rho_{n-1}}\right]\right\} \quad [5.32]$$

Note that the recursion [5.32] derived in Chapter 4 is stable numerically if the resistivity increases monotonically with depth. If it decreases with depth, or varies up and down through the geoelectric section, it is always possible to obtain a numerically stable recursion using the identity,

$$\coth\left(mh_l + \coth^{-1}\frac{\rho_{l+1}}{\rho_l}\right) = \tanh\left(mh_l + \tanh^{-1}\frac{\rho_{l+1}}{\rho_l}\right).$$

In particular, in the case of uniform half-space and a two-layered medium, we have

$$R_1 = \rho_1, \quad \text{and } R_1(m) = \rho_1 \coth\left[mh_1 + \coth^{-1}\frac{\rho_2}{\rho_1}\right],$$

$$\rho_2 > \rho_1.$$

The function $R_1(m)$ depends, of course, on the variable of integration m and on parameters of the geoelectric section. For instance, we have

$$R_1 \to \rho_N, \quad \text{if } m \to 0 \quad \text{and} \quad R_1 \to \rho_1, \quad \text{if } m \to \infty.$$

Here, ρ_n is the resistivity of underlying half-space. We see from Eq. [5.31] that the kernel function $R_1(m)$ and the apparent resistivity $\rho_a(r)$ are complementary descriptions of a geoelectric section. In addition, we know from the theory of Hankel transforms that the relationship between the two functions is unique; knowledge of the kernel function for all m uniquely defines the apparent resistivity through the first equation in the set Eq. [5.31], while knowledge of the apparent resistivity for all r uniquely defines the kernel function through the second equation in the set. Calculation of the kernel function for a given geoelectric section is, of course, simpler than calculation of the integral that gives the apparent resistivity—because the oscillating behavior of Bessel function requires that the integrand be computed for many values of m to achieve an accurate result. Since both functions $R_1(m)$ and $\rho_a(r)$ contain the same information about a geoelectric section, it was suggested early in the development of electric methods to base interpretation solely on the properties of the kernel function $R_1(m)$, by reconstructing this function from the measured values of the apparent resistivity using the second equation of the set [5.31]. Notice, however, that this approach simply shifts the problem of accurately calculating the Hankel transform to the measurement domain, requiring accurate values of the apparent resistivity for a wide range of electrode separations r. In the 1970s, a breakthrough occurred when it was discovered that the integrals in Eq. [5.31] could be computed quickly and with high accuracy by a simple numerical integration method based on the theory of the linear digital filters. This approach, which will be described in a later section, allowed interpretation to proceed by comparison of the theoretical and experimental curves of $R_1(m)$, where the experimental curve is reconstructed from measured values of the apparent resistivity using the second equation [5.31]. We first describe a classical way of computing the integral for the apparent resistivity based on a numerical method first developed by Filon in the 1920s.

5.2.7. Numerical Integration of the Apparent Resistivity, Eq. [5.31], by Fillon's Method

The first equation of the set [5.31] can be written simply as

$$L = \int_0^\infty F(m) J_1(mr) \mathrm{d}m. \quad [5.33]$$

Its integrand is the product of two functions, one of which, the kernel function $F(m)$, changes relatively slowly, while the other, the Bessel function $J_1(mr)$, oscillates more and more rapidly as the separation r increases. This means that even within a small interval Δm, the integrand can change its sign several times. Consider a small interval of integration, defined by three points,

$$m_i < m_{i+1} < m_{i+2}.$$

A standard approach, which leads to the well-known Simpson's rule, is to fit a quadratic polynomial to the values of the integrand at the three points,

$$F(m) J_1(mr) = a_0 + a_1 m + a_2 m^2, \quad \text{for } m = m_i, m_{i+1}, m_{i+2},$$

and to integrate the polynomial exactly. However, if the integrand changes sign more than twice within the interval, it will not be well approximated by a quadratic polynomial, which has at most two zeros. Of course, it is possible to reduce the number of sign changes by making the interval smaller. But as r becomes large, this requires more and more integration points, which can increase both the time of calculation and the error because the final result can end up being the difference of many positive and negative values. To avoid these difficulties with oscillating integrals, Filon suggested a modification of Simpson's method for integrands involving the product of two functions, only one of which oscillates. In Filon's method, only the nonoscillating function, here $F(m)$, is approximated by a quadratic polynomial. Applying this method to the integral in Eq. [5.33] for any interval, Δm gives

$$\int_{m_i}^{m_{i+2}} (a_0 + a_1 m + a_2 m^2) J_1(mr) \mathrm{d}m = a_0 \int_{m_i}^{m_{i+2}} J_1(mr) \mathrm{d}m + \int_{m_i}^{m_{i+2}} m J_1(mr) \mathrm{d}m + a_2 \int_{m_i}^{m_{i+2}} m^2 J_1(mr) \mathrm{d}m. \quad [5.34]$$

The integrals are now standard integrals of Bessel's functions, depending only on r and on the grid of points m_i used to approximate the kernel function $F(m)$. Since this function varies smoothly, the grid can be fixed in advance independent of the parameters of a given geoelectric section. Therefore, the integral on the right-hand side can be calculated once for a given set of separations. The coefficients are determined for each interval Δm by interpolating the kernel function:

$$\begin{aligned} a_0 + a_1 m_i + a_2 m_i^2 &= F(m_i), \\ a_0 + a_1 m_{i+1} + a_2 m_{i+1}^2 &= F(m_{i+1}), \\ a_0 + a_1 m_{i+2} + a_2 m_{i+2}^2 &= F(m_{i+2}). \end{aligned}$$

Before the availability of digital computers, Filon's method was a major advance in the computation of oscillatory integrals, including Fourier integrals. In applying this method to the computation of apparent resistivity curves, the integral is best split into two pieces. The first interval of integration is taken to be small, for example,

$$0 \le m \le m_0 = x_0/r,$$

where x_0 is the first zero of the Bessel function. Within this interval, the integrand does not oscillate, and standard procedures (such as Simpson's rule) provide a result with sufficient accuracy. Filon's method is then applied over the remainder of the interval,

$$m_0 \le m \le M,$$

where the integrand oscillates. Here, M is taken to be sufficiently large so that integrand becomes tiny and the rest of the integration out to infinity can be neglected. Typically, the spacing of the grid of points can be allowed to increase geometrically, according to a rule of the form,

$$m_{i+1} = 2^{1/n} m_i,$$

where n defines an accuracy of calculations. This choice can usually be justified because the contribution to the integral from intervals at large m generally decreases rapidly (but should always be checked in careful numerical work). Application of Filon's method more than 50 years ago enabled a large expansion of the set of master curves of apparent resistivity used in surface and marine geophysics. The same approach found wide application in the theory of frequency and transient electromagnetic soundings and in induction logging.

5.3. FUNCTIONS WITH FINITE SPECTRUM AND INTEGRATION OF EQ. [5.31]

Filon's method and other integration rules for computing apparent resistivity curves were largely replaced in the 1970s by a different approach that greatly reduces the computational time by exploiting analytical properties of Fourier–Bessel integrals of smooth functions.

5.3.1. Fourier Integral and Fourier Series

We first recall some fundamental relations of spectral analysis. Almost any function $s(x)$ can be represented through a Fourier integral of the form,

$$s(x) = \frac{1}{2\pi}\int_{-\infty}^{\infty} S(\omega)\exp(\mathrm{i}\omega x)\mathrm{d}\omega. \qquad [5.35]$$

Here, x is a variable that, in particular applications, can be time, distance, etc. $S(\omega)$ is generally a complex function called the "spectrum" of $s(x)$ and characterizes the amplitude and phase of each oscillation at the frequency ω. By definition, ω is the angular frequency of oscillation

$$\omega = 2\pi f = \frac{2\pi}{T}, \qquad [5.36]$$

where T is the period of each oscillation. According to Eq. [5.35], the function $s(x)$ is represented as an infinite sum of sinusoidal oscillations $\exp(\mathrm{i}\omega x)$ with infinitely small amplitudes $S(\omega)\mathrm{d}\omega$. There is a corresponding Fourier integral inverse to Eq. [5.35],

$$S(\omega) = \int_{-\infty}^{\infty} s(x)\exp(-\mathrm{i}\omega x)\mathrm{d}x, \qquad [5.37]$$

which allows one to calculate the spectrum for (nearly) any function. We also recall the fact that a function $s(x)$ defined only within a finite interval $-x_c < x < x_c$ can be expanded in an (infinite) Fourier series,

$$s(x) = \sum_{k=-\infty}^{\infty} C_k \exp\left(\pi k \frac{x}{x_c}\right), \qquad [5.38]$$

where

$$C_k = \frac{1}{2x_c}\int_{-x_c}^{x_c} s(x)\exp\left(-\pi k \frac{x}{x_c}\right)\mathrm{d}x. \qquad [5.39]$$

In this case, the function $s(x)$ is represented as a sum of oscillations with finite amplitude and discrete frequencies, which are multiples of the fundamental frequency $2\pi/2x_c = \pi/x_c$. Note that the sum [5.38] may contain a constant ($k=0$). After this short introduction, we now turn to the special case of functions whose spectrum is limited: that is, the spectrum contains no frequencies beyond some upper limit ω_c. Such functions are called "band limited" and possess some remarkable properties.

5.3.2. Representation of Band-Limited Functions

Band-limited functions play an important role in the theory of transmission of information in electromagnetic or other signals, a subject originally called communication theory and now generally known as information theory. These functions posses a remarkable feature discovered more than 80 years ago by Harry Nyquist and developed into a systematic theory by Claude Shannon in the late 1940s. A general function $s(x)$ is defined by its values at all points x: in principle, a continuously infinite set of values. A band-limited function, however, can be defined by its discrete samples $s(x_i)$ at a collection of individual points x_i (to be defined precisely below). Geometrically, this means that if we have only the samples $s(x_i)$, then the full graph of the function for all x can still be drawn through these points uniquely. That certain functions can be represented in this way is, of course, not surprising. For example, if we know that a given function is linear, then obviously, the values of the function at two points allow us to determine the function everywhere. Similarly, a quadratic function is defined by its value at three points, etc. With a band-limited function, however, we make no assumptions about the specific shape of the function, and rely only on the fact that its spectrum is zero beyond some upper frequency. Intuitively, we expect that the absence of high frequencies in the spectrum imposes strong limitations on the way the two nearby values of the function can be connected. The precise meaning of the so-called "sampling theorem" for band-limited functions is to show that the complete function can be reconstructed from samples of the function taken at points sufficiently close together. Now, we are ready to prove this property.

5.3.3. Sampling Theorem

In 1928, Harry Nyquist published a paper regarding the transmission of information in telegraph signals that showed the following: Any function

$s(t)$ whose spectrum is limited to the frequency range $0 \leq f \leq f_c$ can be transmitted with arbitrary accuracy using only its discrete values taken at intervals $\Delta t = 1/2f_c$. This is the original form of the "sampling theorem" that describes the remarkable property of band-limited functions. To prove this theorem, note first of all that for a band-limited function, Eq. [5.35] can be written as

$$s(t) = \frac{1}{2\pi}\int_{-\omega_c}^{\omega_c} S(\omega)\exp(\mathrm{i}\omega t)\mathrm{d}\omega, \qquad [5.40]$$

since by definition $S(\omega)=0$ if $\omega > |\omega_c|$. Here, ω is angular frequency, given by $\omega=2\pi/f$. Next, recall that since the function $S(\omega)$ is given within the finite interval $(-\omega_c, \omega_c)$, it can be expanded in a Fourier series according to Eq. [5.38], making an obvious identification $(x, x_c) \rightarrow (\omega, \omega_c)$,

$$S(\omega) = \sum_{k=-\infty}^{+\infty} D_k \exp\left(\mathrm{i}\pi k\frac{\omega}{\omega_c}\right), \qquad [5.41]$$

where

$$D_k = \frac{1}{2\omega_c}\int_{-\omega_c}^{\omega_c} S(\omega)\exp\left(-\mathrm{i}\pi k\frac{\omega}{\omega_c}\right)\mathrm{d}\omega. \qquad [5.42]$$

Substituting Eq. [5.41] into Eq. [5.40] and changing the order of integration and summation gives

$$s(t) = \frac{1}{2\pi}\sum_{k=-\infty}^{k=\infty} D_k\int_{-\omega_c}^{\omega_c} \exp[\mathrm{i}\omega(t+(k\pi/\omega_c))]\mathrm{d}\omega. \qquad [5.43]$$

The integral can be evaluated exactly:

$$\int_{-\omega_c}^{\omega_c} \exp[\mathrm{i}\omega(t+(k\pi/\omega_c))]\mathrm{d}\omega = \frac{1}{\mathrm{i}\omega(t+(k\pi/\omega_c))}\exp[\mathrm{i}\omega(t+(k\pi/\omega_c))],$$

So, we have

$$s(t) = \frac{1}{\pi}\sum_{k=-\infty}^{\infty} D_k\frac{\sin\omega_c(t+(k\pi/\omega_c))}{\omega_c(t+(k\pi/\omega_c))}. \qquad [5.44]$$

We can determine the unknown coefficients D_k by comparing Eqs. [5.40] and [5.42]. The integrals in these two equations are clearly identical when

$$t = \frac{-k\pi}{\omega_c}.$$

Thus,

$$\pi s\left(t = \frac{-k\pi}{\omega_c}\right) = \omega_c D_k \text{ or } D_k = \frac{\pi}{\omega_c} s\left(\frac{-k\pi}{\omega_c}\right). \qquad [5.45]$$

Substituting the latter into Eq. [5.44], we obtain

$$s(t) = \frac{1}{\omega_c} \sum_{k=-\infty}^{\infty} s\left(\frac{-k\pi}{\omega_c}\right) \frac{\sin\ \omega_c(t + (k\pi/\omega_c))}{t + (k\pi/\omega_c)}.$$

The right-hand side is a weighted summation of discrete samples of the function f at points,

$$t_k = \frac{k\pi}{\omega_c} = k\Delta t,$$

where $\Delta t = \pi/\omega_c = 1/2f_c$.

Since the summation is over all positive and negative k, we can make the substitution $k \rightarrow -k$, and write simply

$$s(t) = \sum s(k\Delta t) \frac{\sin\ \omega_c(t - k\Delta t)}{\omega_c(t - \Delta t)}.$$

In the general case for a function of the variable x,

$$s(x) = \sum s(k\Delta x) \frac{\sin\ \omega_c(x - k\Delta x)}{\omega_c(x - k\Delta x)}. \qquad [5.46]$$

Thus, a band-limited function can be represented by the series [5.46], where the coefficients are samples of the function taken at intervals,

$$\Delta x = \frac{\pi}{\omega_c} = \frac{1}{2f_c}. \qquad [5.47]$$

The function $s(x)$ can, thus, be calculated with arbitrary accuracy at any point x if its values are known at points $k\Delta x = k\pi/\omega_c$. To represent the function exactly using its samples requires an infinite sum. But in most practical cases, the function will differ from zero appreciably only within some interval $x_1 < x < x_2$; in addition, the contribution from distant samples falls off as inverse distance because of the weighting function. Then, in place of Eq. [5.46], we have the approximation

$$s(x) \approx \sum_{k_1}^{k_2} s(k\Delta x) \frac{\sin \omega_c (x - k\Delta x)}{\omega_c (x - k\Delta x)}. \qquad [5.48]$$

It follows from Eq. [5.47] that the "cutoff" frequency f_c determines the sampling interval.

When f_c is small (low frequency content), the distance between neighboring samples is large and the number of points needed to represent the function can be relatively small. As the cutoff frequency increases, the sampling density must also increase.

5.3.4. Application of Eq. [5.48] for Numerical Integration

We now apply the sampling theorem to integrals of the form,

$$I = \int_{-\infty}^{\infty} F(m) J_1(mr) \mathrm{d}m. \qquad [5.49]$$

Suppose that we know values of the function $F(m)$ at points

$$k\Delta m = k\frac{\pi}{\omega_c}.$$

Substituting the representation [5.48] into Eq. [5.49] and changing the order of summation and integration gives

$$I = \sum_{k_1}^{k_2} F(k\Delta m) \int_{-\infty}^{\infty} \frac{\sin \omega_c (m - k\Delta m)}{\omega_c (m - k\Delta m)} J_1(mr) \mathrm{d}m$$

$$\text{or} \quad I = \sum_{k_1}^{k_2} a_k(r) F(k\Delta m), \qquad [5.50]$$

where

$$a_k(r) = \int_{-\infty}^{\aleph} \frac{\sin \omega_c (m - k\Delta m)}{\omega_c (m - k\Delta m)} J_1(mr) \mathrm{d}m. \qquad [5.51]$$

The coefficients are functions of the parameter r—in our case, the separation between current and receiver electrodes—and are independent of the parameters of the geoelectric section (which are contained in the function F). Correspondingly, they can be calculated once for different parameters of the layered medium. In addition, let us make two comments. (1) To apply Eq. [5.50], we have to have an expression for the field or apparent resistivity in the form of Eq. [5.49], with the limits of integration from $-\infty$ to ∞. The field is actually given by the integral

$$\int_0^{\infty} mF(m)J_1(mr)\mathrm{d}m. \quad [5.52]$$

To change the limits of integration, we introduce new variable $m=\exp(u)$. Then,

$$\mathrm{d}m = \exp(u)\mathrm{d}u$$

and Eq. [5.52] becomes

$$\int_{-\infty}^{\infty} F(\exp u)\exp(2u)J_1[\exp(u)r]\mathrm{d}u.$$

2. Numerical study of the Fourier spectrum of functions $F(m)$; that is

$$\int_0^{\infty} F(m)\exp(\mathrm{i}2\pi fm)\mathrm{d}m$$

for different parameters of a geoelectric section shows that the contribution of relatively high harmonics ($f>f_c$) can be neglected, and correspondingly these functions can be treated practically as band-limited functions.

5.4. CONCEPT OF UNIQUENESS AND THE SOLUTION OF THE INVERSE PROBLEM

5.4.1. Main Steps of Interpretation

The electric field measured at an observation point is caused by charges arising at interfaces between regions of different conductivity; their contributions to the measured field depend on the separation between electrodes r and on locations of these interfaces. For instance, if the separation is relatively small, then the main influence is caused by charges situated at interfaces located sufficiently close to the observation point. As the separation increases, the contribution of charges at more remote boundaries becomes stronger. In other words, the influence of charges at different interfaces changes with the distance from the current electrode to an observation point, while at each point, the measured field contains some information about all geoelectric parameters of a medium. Taking into account this simple but fundamental fact, let us formulate the main steps of interpretation of electric soundings for the case of the horizontally layered medium.

1. At the beginning, we will make some assumptions about a number of layers, their resistivity, and thickness. Such a step is usually called the first guess or the first approximation. It is mainly based on known geological information, or possibly on data from electric logging measurements. Also, the behavior of the apparent resistivity curves themselves approximately defines the number of parameters and their numerical values. Of course, the difference between the first approximation and actual values of geoelectric parameters can vary significantly depending on the quality of additional information.
2. The second step of interpretation consists of calculating the apparent resistivity curve as a function of separation, using the first approximation, and comparing the measured and calculated values of ρ_a. A reasonable coincidence of these values at each value of separation may indicate that the chosen parameters of the model of the medium are close to the real ones.
3. If there is a difference between the measured and calculated apparent resistivity, all parameters of the first approximation, or some of them, are changed in such a way that a better fit to curves of the apparent resistivity is achieved. Thus, we obtain a second approximation for parameters of the section. Of course, in those cases where even the new set of parameters does not provide a satisfactory match of curves of the apparent resistivity, this process of calculation has to be continued. As we see from this procedure, every step of interpretation requires the application of Eq. [5.31], that is, a solution of the forward problem.

Later, we will add one more element, caused by the fact that the electric field containing information about the geolectric section (useful signal) is never known exactly. Inasmuch as every step of interpretation is reasonably well defined, we may arrive at the impression that the solution of the inverse problem is straightforward and does not have any complications. Unfortunately, this is not true in general, except in cases of purely theoretical interest. In order to illustrate some of the difficulties in the solution of the inverse problem for electric methods, it is useful to start from an ideal situation.

5.4.2. Uniqueness and its Application

Suppose that both the calculated and measured electric fields are known with infinitely high accuracy. This means that any digit after the decimal point is known exactly regardless of how small its contribution is. Of course, in reality, it is impossible to know the value of the field without some error. Any voltmeter provides a value up to a finite number of digits, and any computational scheme for calculating the field will have only a finite precision. Nevertheless, it is useful to discuss the ideal case for understanding the principles of interpretation of the electric soundings. Thus, assume that we know the fields exactly and perform all steps of interpretation described above. Suppose that sequentially repeating the solution of the forward problem at each step and comparing the calculated and measured fields or apparent resistivity, we obtain a set of parameters such that the difference between these fields is infinitely small. Then, the following question arises: Does this mean that by providing a perfect fit between the measured and calculated fields, it is possible to determine with infinitely small error the arbitrary distribution of resistivity of a medium that created the given field? In general, the answer is negative and there are cases when the solution of the inverse is not unique; that is, media with different distributions of resistivity may create exactly the same field along a profile or a system of profiles. In other words, in general, but not always, different geoelectric sections can provide an exact match to the measured field. At the same time, if we look more carefully at this subject, then it becomes clear that the phenomenon of nonuniqueness is hardly obvious. In fact, in accordance with Coulomb's law, the electric field at any point of a piecewise uniform medium is

$$E(p) = \frac{1}{4\pi\varepsilon_0}\sum\int_{S_i}\frac{\sigma(q)L_{qp}}{L_{qp}^3}\,dS, \qquad [5.53]$$

and a change of a position of interfaces S_i, as well as a resistivity of layers, should result in a change of the electric field. However, from nonuniqueness, it follows that different geoelectric sections can create exactly the same field even if we know the field without error. In other words, it is impossible to detect the difference between fields generated by charges of such sections. It is difficult to get rid of the impression that nonuniqueness is an amazing, unexpected fact which is more natural to be treated as a paradox rather than an obvious consequence of the behavior of the electric fields. In this light, let us imagine for a moment that nonuniqueness is always present. Then, it is clear that in such a case, the interpretation of electric soundings would be impossible. In fact, having determined parameters of a section that generates a given field, we also have to assume that due to nonuniqueness, there are always other geoelectric sections which create the same field. Certainly, we can say that such an ambiguity would be a disaster for the application of the electric methods. Fortunately, this whole subject of uniqueness, to some extent, is irrelevant in practice, because we never know the fields exactly, and for this reason, it would be natural to avoid the discussion of this topic. At the same time, it is worthwhile to clarify some aspects of uniqueness in solving the inverse problem. First of all, as the theory shows, our treatment of nonuniqueness as a paradox is usually correct. More than 70 years ago, R. Langer proved that a solution of the inverse problem for the electric sounding in a horizontally layered medium is unique, provided that we know exactly the field caused by a current source at each point of some element of the earth's surface. This suggests that, by applying analytical continuation, the field is known at each point of the earth's surface. Then, in accordance with this theorem, we can distinguish geolectric sections regardless of how small the difference is between the parameters of each layer. In particular, both resistivity and thickness of each layer can be determined even if this layer is very thin and almost invisible. In this light, let us recall that such effects as equivalence by S or T are observed in real conditions when the field is known with an error and these effects are absent when the field is known exactly. Later, V. Druskin provided a solution of the inverse problem in a much more general case. Suppose that the field is known at any point of a given portion of the Earth's surface for any position of the current electrode at this area and a medium is piecewise uniform where each region with a constant resistivity has a smooth boundary. Then, he showed that a solution of the inverse problem is unique. It is obvious that this model includes a wide range of geolectric sections; besides, it is quite possible that there are other classes of models for which the solution of the inverse problem is also unique. It seems that this information is already sufficient to think that nonuniqueness is not obvious, but looks like a paradox. In a fact, if anisotropic media are excluded, the solution of the inverse problem for electric resistivity measurements is unique, provided that a complete set of ideal measurements is available.

Now, we continue to discuss a solution of the inverse problem. Assuming that the measured and calculated fields caused by charges at interfaces are known exactly, it is a simple matter to outline the main steps of interpretation, provided that we deal with a class of models for which uniqueness holds. As was pointed out earlier, it is a straightforward task, and the main steps of interpretation were already formulated and they are as follows: (1) Proceeding from the observed field and making use of additional information, we approximately define the parameters of a geoelectric section (first guess). (2) Substituting values of these parameters into an expression for the apparent resistivity (Eq. 5.31), we solve the forward problem. (3) The process of comparison of the measured and calculated fields or apparent resistivity allows us to determine how the parameters of the first guess have to be changed in order to decrease the difference between the measured and calculated values. By solving the forward problem with the new parameters, we again compare the fields and this process can continue until the accuracy of determination of the parameters satisfies our requirements. It is essential that in solving the inverse problem, we can, in principle, reduce the error in evaluating the parameters of a geoelectric section to zero. Note that if the first guess contains parameters that do not characterize a section and they are introduced by error, this procedure of solution of the inverse problem allows us to detect and eliminate them. Of course, there are classes of geoelectric sections for which the solution of the inverse problem in the electric sounding is not unique. The same is true if sections of different classes are considered. This fact is not surprising, and is also observed in all geophysical methods. For this reason, it is not simple to understand why the geophysical literature often emphasizes the fact that a solution of the inverse problem in the so-called "potential methods" (gravitational, magnetic, and electric methods) is not unique without reference to other methods. In this light, it may be proper to notice that a concept of potential is used in all geophysical methods, and so, a division on the potential and nonpotential methods is hardly proper. At the same time, theorems of uniqueness are proven for certain well-specified class of models of a medium, and if they belong to different classes, they may create exactly the same fields. For instance, in the case of gravitational method, it is always possible to find two completely different distributions of masses which create exactly the same gravitational field on the earth's surface. It is also appropriate to point out that the theorem of uniqueness for electrical measurements is proved for a much more broad class of models of a medium than in the gravitational and magnetic methods. As was pointed out earlier, the uniqueness of a solution of an inverse problem has purely mathematical interest because it implies that the calculated and measured fields are known exactly. Since, in reality, this does not hold, this subject will be put aside, even though for an interpretation of the electric sounding, it is important to know whether a solution of the inverse problem for a given class of models is unique.

5.4.3. Solution of Inverse Problem for Soundings and Influence of Noise

Now, we are ready to discuss some aspects of interpretation for real conditions when the electric field caused by charges at horizontal interfaces is measured with some error:that is, the numbers that describe the field are accurate to only a few decimal places. This is the fundamental difference from the previous case where we discussed uniqueness and assumed that this field is known exactly. The presence of error is caused mainly by two factors: (1) Measuring devices can only operate to a precision that depends on the design of the instrument as well as external factors such as variation of temperature. (2) The measured field is always subjected to the influence of geological noise caused by lateral changes of resistivity, such as topography and a presence of different inhomogeneities beneath the earth's surface. Thus, the measured field is a sum:

$$E = E_U + E_N. \qquad [5.54]$$

Here, E_U is the useful signal caused by charges in the horizontal layer medium and E_N is the noise. The total field E can be measured with a relatively high accuracy but the error of determination of the useful signal,

$$E_U = E - E_N, \qquad [5.55]$$

depends upon the noise. Note that reduction of noise is one of the most important elements of interpretation of any method, including of course, electric soundings, since the determination of parameters of a horizontally layered medium must be based on comparison of a calculated field (the solution of the forward problem) not with the measured field but rather with the useful signal that contains information about the medium. Bearing in mind that at any separation between the current and receiver electrodes we never know the value of the noise signal but rather its range of variation, it is appropriate to speak also of an interval of variation of the useful signal. To emphasize this fact, consider the curves of the apparent resistivity in Fig. 5.6. Because of the influence of noise, we only know that the value of E_U is located somewhere inside the interval (bar). Its boundary is obtained from Eq. [5.55] assuming a certain level of noise. If, at an observation point, the measured field, and that due to noise, have different signs, then, we have $E_U > E$. In contrast, when E and E_N have the same sign, the value of the useful signal is smaller than the measured field. Notice that the boundary of a bar, where the useful signal is located, is defined approximately and based on some additional

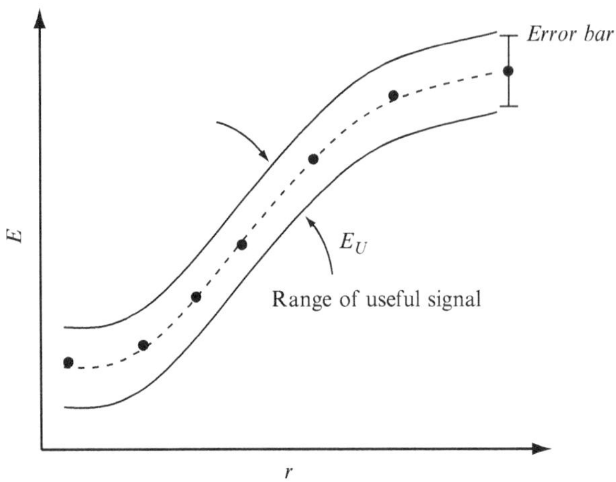

Figure 5.5 Measured field (dotted line) and range of useful signal.

information about the accuracy of measurements and the influence of geological noise. From this consideration, it is clear that the accuracy of model calculation (forward problem) can, in practice, be about the same as that of the measured field. Because of this, we should never expect to get an exact agreement between the measured and modeled fields. In other words, an attempt to fit measured data with a field calculated to a higher precision than the range of useful signal, has no sound meaning. Recognizing that the useful signal is known with an error that can often exceed several percent, let us consider again the process of interpretation. First, as was pointed out earlier, the measured field can be represented at every point as a sum of fields caused by surface charges located at different interfaces of the geolectric section, and their contributions depend essentially on the electrode separation r and the locations of the interfaces with respect to the earth's surface. For instance, we can imagine such changes of resistivity and thickness of some layers that the useful signal would still remain somewhere inside the range of useful signal (error bar). In other words, because of the presence of noise, there can be an unlimited number of different geolectric sections that generate practically the same useful signal, even when the inverse problem for a particular model (such as a layered medium) has a unique answer for ideal measurements. Since the measured field is an integrated effect caused by all surface charges, possible changes in some parts of the geoelectric section can have a significant effect on the field at a given point whereas similar changes in other parts of the section will result in much smaller changes. For this reason, in performing an interpretation, it is natural to distinguish at least two groups of parameters describing a geolectric section:

1. Parameters that have a strong effect on the field; that is, relatively small changes of their values, produce a change of the useful signal that can be detected.
2. Parameters that have detectable influence on the field only if their values are greatly changed. This simply means that they cannot be defined from the useful signal measured with some error.

Therefore, we can say that an interpretation or a solution of the inverse problem consists mainly of determining the parameters in group 1, that is, the resistivities and thicknesses of layers that have a significant effect on the measurements, even though they do not completely characterize the geolectric section. It is clear that this so-called "stable" group of parameters describes a section that differs to some extent from the actual one, but both of them have common parameters. For instance, these can be the resistivity and thickness of relatively thick layers, along with the longitudinal conductance S or the transversal resistance T of sufficiently thin layers. Certainly, the most important factor that defines the essence of interpretation is that the useful signal is known with some error and, because of this, the error of evaluation of certain parameters can be extremely large. In other words, these parameters cannot be practically determined. Such inverse problems are called ill-posed problems. In general, the inverse problem of electric sounding is ill-posed. To illustrate this fact, imagine that we change the value of a single parameter of the geoelectric section by the amount Δp_i. Let $\Delta E_i(r)$ be change in the calculated field (the useful signal) at different electrode separations caused by the change in the ith parameter. We can write the following relation between the change of the useful signal and Δp_i:

$$\Delta E_i(r) = k_i(r)\Delta p_i \text{ or } \Delta p_i = k_1^{-1}(r)\Delta E_i(r). \quad [5.56]$$

Here, $k_i(r)$ is the coefficient of proportionality between the change in the ith parameter of a geoelectric section and the change in the useful signal $\Delta E_i(r)$ at the electrode separation r. The factor $k_i(r)$ is usually called a "sensitivity coefficient" and will be a different function for different parameters. If it is large, then, a small change of parameter p_i causes a large change of useful signal. This means that in performing a solution of the inverse problem, we can determine the value of this parameter with relatively high accuracy. In contrast, when k_i is relatively small, a large change of parameter p_i will make only a small change of useful signal ΔE_i. It follows from the second equation of the set [5.56] that, as the sensitivity coefficient k_i gets smaller and smaller, our ability to determine the ith parameter deteriorates, and it may not be possible to determine a given parameter to any reasonable precision. In principle, a lower limit cannot be established for some coefficients k_i. This is the most important feature of an ill-posed problem. In other words, even an unlimited change of some parameter—for instance, the resistivity of an intermediate layer in a geoelectric section—does not produce a noticeable variation of the field exceeding an interval, and as a result of this, it is impossible to define these parameters. Thus, in any practical inverse problem, there are almost always two groups of parameters: stable parameters and unstable parameters. The goal of interpretation and inversion is to separate them and determine the stable ones. The latter characterize a new model of a real geolectric section and the coefficients k_i for its parameters are relatively large. In this case, we arrive at a well-posed problem, in which some parameters remain unknown, because they are not well defined by the measurements. This situation may seem unsatisfactory, but in fact, reflects a reality of the solution of nearly all inverse problems in geophysics, in particular, the inverse problems of electric methods. The transition from an ill-posed problem to well-posed one is called the regularization of the inverse problem and is of great practical interest. It is obvious that interpretation of electric soundings is useful if the parameters of a medium that can be resolved in a well-posed inversion are determined within a narrow enough range of values to be useful for some practical application. Often, selection of this group of parameters can be done automatically by making use of the standard algorithms for solving inverse problem, employing Eq. [5.56]. Of course, with an increase in the number of model parameters, the approximation of a real section can in principle be better. However, the error with which some of these parameters are determined also increases. As in the theoretical case when the field is known exactly, we can expect that the interpretation of the electric sounding is greatly facilitated by the presence of additional information about a section derived from geology and other geophysical methods. Now, let us formulate the main steps of a solution of the inverse problem taking into account the fact that the useful signal is subjected to the influence of error. These steps are as follows:

(1) Making use of preliminary information about parameters of the geolectric section, we formulate the first guess as is done in the abstract case when a signal is known exactly.

(2) The second step is a solution of the forward problem, Eq. [5.31].

(3) As result of this calculation, we obtain a set of values of the electric field which can be graphically represented as a curve $E_U(r)$ (Fig. 5.6B). Suppose that this curve is situated beyond the interval. Then, by changing the parameters of the section, we again use Eq. [5.31] and obtain a curve of the electric field or apparent resistivity which is closer to the interval of the useful signal. This process is continued until a curve of the calculated field is located inside the interval (bar). So far, the steps of interpretation are identical to those when we considered the case of uniqueness. Now, we will observe a fundamental difference.

(4) When values of a calculated field are located inside the observation interval (bar), further improvement of matching between the measured and calculated fields does not have any well-defined meaning because we do not know where inside the interval the useful signal is located. Therefore, we stop the process of fitting of fields and start a new procedure which also requires a solution of the forward problem; that again emphasizes the importance of the development of very efficient algorithms for calculation of the field. In the last stage of matching, we obtain a set of parameters,

$$p_1, p_2, p_3, \ldots, p_n,$$

that places the calculated useful signal inside the interval. Our goal is to determine the range of change of each parameter so that the useful signal remains inside the interval. This procedure is usually repeated several times and it is accompanied by a solution of the forward problem. Of course, every step causes a movement of the curve of the useful signal and, so long as its position is inside the interval, it is equivalent to the previous one. As result of these steps, we obtain for each parameter its range

$$p_1^{\min} < p_1 < p_1^{\max}, \ldots, p_n^{\min} < p_n < p_n^{\max}.$$

As was pointed out earlier, within these ranges, the degree of "matching" with the measured useful signal is the same for any set of parameters. Certainly, knowledge of variations of these parameters is the most important step in solving the inverse problem, because this allows us to separate the stable from unstable ones, and correspondingly, perform a transformation from the ill-posed to a well-posed problem. Thus, the interpretation gives us a set of stable parameters of a geoelectric section which causes the useful signal and the range of its change. From a review of the solution of the inverse problem, it is clear that with a decrease of the interval width, the range of each possible parameter of the section decreases too and the number of unstable parameters may become smaller. At the same time, stable parameters can be determined with higher confidence. For this reason, reduction of different types of noise—first of all the geological noise—is a very important subject. We outlined the main features of interpretation for the class of sections where uniqueness takes place. It turns out that, if models can be described by a finite number of parameters, the solution of inversion will be stable and one can get inequalities for errors. In regard to interpretation within the class of models with no uniqueness, the situation is completely different. We can say that a solution of the inverse problem in such cases is hardly possible (speaking strictly, it is senseless). In fact, even in the absence of noise, we always have an infinite number of models which create the same measured field. At the same time, it may be possible to determine some generalized characteristics of a section. Thus, uniqueness theorems, which appear to be mainly of academic interest, are actually very important in solving practical inverse problems.

CHAPTER SIX

Methods of Electrical Profiling and Mapping

This chapter briefly describes the use of steady currents to study lateral changes of resistivity, an application called electrical profiling or mapping. Unlike the vertical soundings described in previous chapters, which aim to provide values of electrical resistivity with depth at a specific location, profiling methods have traditionally been designed to provide only a qualitative map of how the apparent resistivity varies over some regions. Modern equipment of shallow geophysics, however, has made it possible to collect data in ways that combine profiling and sounding, and thereby, may allow a quantitative study of electrical resistivity, both laterally and vertically in the earth. As in the rest of this book, we will focus mainly on the physical principles that determine the response of electrical profiling to the resistivity distribution in the earth, as well as on the key model parameters in different applications.

6.1. ELECTRIC PROFILING

To illustrate electrical profiling, we consider several examples of geoelectric sections and arrays.

6.1.1. Example 1: Vertical Contact

We start by studying in more detail a model considered earlier of a vertical interface that intersects the earth's surface (Fig. 6.1A). The model can be taken as the simplified representation of a geological fault, juxtaposing different uniform regions. We will formulate the boundary-value problem by first assuming that the current electrode e_A is placed slightly below the earth's surface at depth h, located in a medium with resistivity ρ_1. The charge on the electrode is then $e_A = \varepsilon_0 \rho_1 I$. We also place the origin of coordinates x,y,z at the electrode, with the x-axis directed perpendicular to the vertical contact surface and the z-axis perpendicular to the earth's surface and positive downward. As usual, we assume that air is an ideal

Methods in Geochemistry and Geophysics, Volume 44
ISSN 0076-6895, DOI: 10.1016/S0076-6895(10)44006-8
© 2010 Elsevier B.V.
All rights reserved.

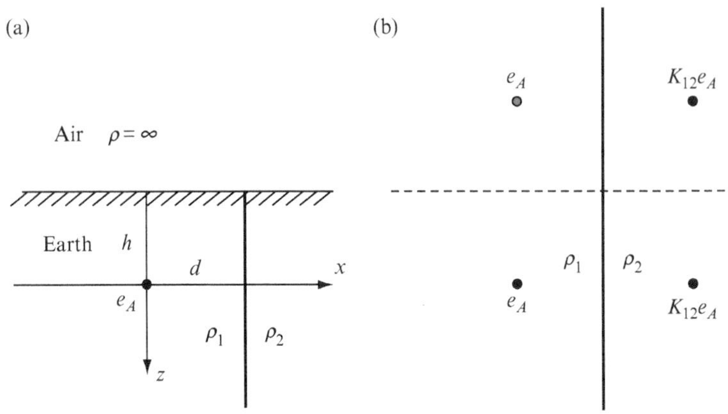

Figure 6.1 (A) Vertical boundary intersecting the earth–air interface. (B) Equivalent model with distribution of image charges obtained by the method of mirror reflection.

insulator (zero conductivity). Let U_1 and U_2 be the electric potential in the two regions separated by the contact and R the distance from an observation point to the origin. The boundary-value problem for the field has the following form:

1. At regular points, the potential obeys Laplace's equation:

$$\nabla^2 U_1 = \nabla^2 U_2 = 0;$$

2. Close to the electrode in region 1, the potential approaches the field of an electrode in the uniform medium:

$$U_1 \to U_0 = \frac{\rho_1 I}{4\pi R}, \quad \text{as}\, R \to 0;$$

3. There is zero current flow across the earth–air interface:

$$\frac{\partial U_1}{\partial z} = \frac{\partial U_2}{\partial z} = 0, \quad \text{if}\, z = -h;$$

4. The potential and current are continuous across the vertical contact:

$$U_1 = U_2 \quad \text{and} \quad \gamma_1 \frac{\partial U_1}{\partial x} = \gamma_2 \frac{\partial U_2}{\partial x}, \quad \text{if}\, x = d;$$

5. The potential goes to zero at infinity:

$$U_1 \to 0 \quad \text{and} \quad U_2 \to 0, \quad \text{if}\, R \to \infty.$$

To solve this boundary-value problem, we will use the method of mirror reflection and replace the actual model with a fictitious model in which the charge and medium are reflected across the earth–air interface (Fig. 6.1B). From the results of Chapter 3, we conclude that both the source charge e_A at $z = -h$ and its mirror image at $z = -2h$ induce surface charge at the vertical interface. As we know, the fields of these induced charges can be represented by the fields of image charges of magnitude $K_{12}e_A$, located symmetrically with respect to the vertical plane $x = d$ (Fig. 6.1B). It is easy to show that this arrangement of four point charges satisfies all the boundary conditions of the problem, including the condition of zero current flow across the earth–air interface (condition 2 above) and the continuity of current across the vertical contact (condition 4). Inasmuch as we are interested in the case when the current electrode is located on the earth's surface, we take the limit $h \to 0$, which simply amounts to doubling the electrode charge:

$$e_A = 2\varepsilon_0 \rho_1 I.$$

Note that the method of mirror reflection does not allow us to find the field in the nonconducting region above the earth's surface. Proceeding from equations derived in Chapter 3, we have for the potential when the current electrode is located in the first medium,

$$U_1(x) = \frac{\rho_1 I}{2\pi}\left[\frac{1}{|x|} + \frac{K_{12}}{2d - x}\right], \quad U_2(x) = \frac{\rho_1 I}{2\pi}(1 + K_{12})\frac{1}{x}, \qquad [6.2]$$

where

$$K_{12} = \frac{\rho_2 - \rho_1}{\rho_2 + \rho_1}.$$

If the current electrode is located in the second medium, we obtain by analogy

$$U_1(x) = \frac{\rho_2 I}{2\pi}(1 + K_{21})\frac{1}{|x|}, \quad U_2(x) = \frac{\rho_2 I}{2\pi}\left[\frac{1}{|x|} + \frac{K_{21}}{2d + x}\right]. \qquad [6.3]$$

We use these equations to study profiling over a vertical interface with several different arrays.

Case 1. Two-Electrode Array AM

Consider first a four-electrode array ($AMNB$), with electrodes B and N located far away relative to the distance x between the A and M electrodes, so that we can treat it as a two-electrode array. As shown earlier, the apparent resistivity of such an array is

$$\rho_a = 2\pi|x|\frac{U(M)}{I}. \quad [6.4]$$

There are three cases to consider: (1) both electrodes are in the first medium; (2) current electrode is in the first medium with the potential electrode in the second medium; and (3) both electrodes are in the second medium. The corresponding apparent resistivities are

$$\rho_a^{(1)} = \rho_1\left[1 + \frac{K_{12}|x|}{2d - x}\right], \quad [6.5]$$

$$\rho_a^{(2)} = \rho_1(1 + K_{12}),$$

$$\rho_a^{(3)} = \rho_2\left[1 + \frac{K_{21}x}{2d + x}\right]. \quad [6.6]$$

Consider the behavior of the potential and apparent resistivity when this array moves from left to right over the contact. First, suppose that $\rho_2 > \rho_1$, which means that positive charge arises at the contact and at the earth–air interface. Assume that this array approaches the contact with the receiver electrode M in front. At very large distances from the contact, $d \gg |x|$, its influence is negligible and

$$\rho_a \to \rho_1. \quad [6.7]$$

Since all charges in the model are positive, the presence of the contact must increase the potential. On approaching the contact, the potential, along with the apparent resistivity, therefore increases. When the electrode M touches the contact $x = d$, the first equation of the set [6.5] gives

$$\rho_a^{(1)} = \rho_1(1 + K_{12}), \quad \text{if } x = d. \quad [6.8]$$

The second equation of [6.5] shows that the value of the apparent resistivity does not change when the potential electrode enters the second medium; that is, we again observe continuity of the potential across an interface. Also interesting is that the potential and apparent resistivity are constant when the electrodes are located on opposite sides of the contact. Qualitatively, this behavior can be understood as follows. As the electrode A approaches the contact, the induced surface charge becomes more concentrated near the x-axis, but at the same time, the receiver electrode moves away from the induced charge, and these two effects compensate each other completely. When the current electrode A crosses this boundary, the charge induced on the interface becomes negative, but the potential remains continuous. In fact, we have from Eqs. [6.5] and [6.6],

$$\rho_a^{(1)} = \rho_1(1 + K_{12}) = \frac{2\rho_1\rho_2}{\rho_1 + \rho_2} = \rho_2(1 - K_{12}) = \rho_a^{(3)}.$$

This result is not surprising since the potential is a continuous function when the current electrode crosses the interface, even though the sign of the surface charge changes. As the electrodes move away from the interface into the second medium, the influence of the induced charge diminishes, and the apparent resistivity tends to the resistivity of the second medium, $\rho_a \to \rho_2$ (Fig. 6.2A). Now, suppose that the second medium is more conductive than the first, $\rho_2 < \rho_1$. In this case, negative charge appears at the interface when the current electrode is in the first medium, and the potential decreases on approaching the interface. When the electrode M touches the interface, the apparent resistivity is still defined by Eq. [6.8], but now K_{12} is negative. There is again an interval of constant potential and constant ρ_a. When the current electrode intersects the contact, the sign of the surface charge changes and, as the distance from the contact increases, the apparent resistivity gradually decreases and approaches ρ_2 (Fig. 6.2B).

Case 2. Three-Electrode Array AMN

Consider now a three-electrode array with the distance between the voltage electrodes MN much less than the distance to the current electrode A:

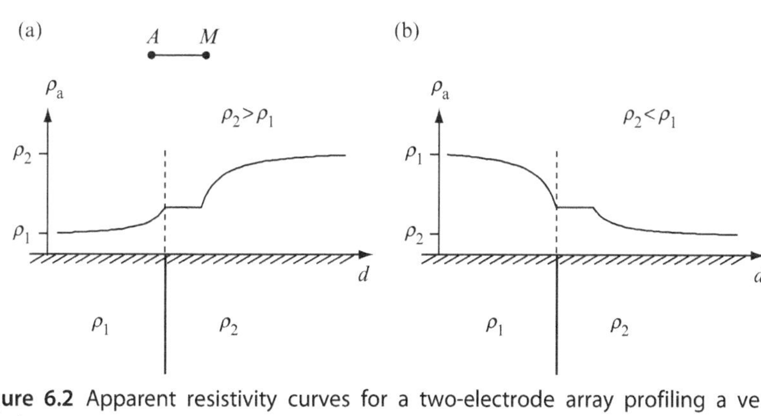

Figure 6.2 Apparent resistivity curves for a two-electrode array profiling a vertical interface.

that is, $MN << AM$. (The second current electrode is assumed to be at infinity.) The electric field between the receiver electrodes is practically constant, so

$$V_{MN} = E_x MN.$$

This allows us to consider an equivalent array AO that measures the electric field E_x at the midpoint of the receiver electrodes. We first derive expressions for the electric field when the electrode O is in front of the current electrode. Using

$$E_x = -\frac{\partial U}{\partial x},$$

and the expressions for the potential [Eq. 6.2], we have for the apparent resistivity in the three cases described above,

$$\rho_a^{(1)} = \rho_1\left[1 - \frac{K_{12}x^2}{(2d-x)^2}\right], \qquad [6.9]$$

$$\rho_a^{(2)} = \rho_1(1 + K_{12}),$$

$$\rho_a^{(3)} = \rho_2\left[1 + K_{21}\frac{x^2}{(2d+x)^2}\right]. \qquad [6.10]$$

First, consider the function ρ_a when $\rho_2 > \rho_1$. As follows from the first equation of the set [6.9], in approaching a contact, the apparent resistivity decreases, and when the receiver is about to touch the contact, we have

$$\rho_a^{(1)} = \rho_1(1 - K_{12}). \qquad [6.11]$$

This behavior is understandable because the charge of the electrode and the induced charge at the contact are both positive and the observation point O is located between them, so the two electric fields oppose each other. When the receiver O crosses the contact, the second equation of the set [6.9] gives

$$\rho_a^{(2)} = \rho_1(1 + K_{12}). \qquad [6.12]$$

Thus, the apparent resistivity is discontinuous at the boundary in the ratio

$$\frac{\rho_a^{(2)}}{\rho_a^{(1)}} = \frac{1 + K_{12}}{1 - K_{12}} = \frac{\rho_2}{\rho_1} > 1. \qquad [6.13]$$

This result is expected because the normal component of the current density j_x is continuous, and, therefore, the component E_x has a discontinuity equal to the ratio of the resistivities. This behavior of the field is useful for detecting the position of a contact. The apparent resistivity stays constant when the array straddles the contact, and remains continuous when the source electrode crosses the interface, even though the induced charge changes sign. In particular, when the source electrode is in medium 2, but touching the contact, the apparent resistivity is given by

$$\rho_a^{(3)} = \rho_2(1 + K_{21}) = \rho_2(1 - K_{12}),$$

which coincides with Eq. [6.12]. Again, as the electrode array moves away from the contact in medium 2, the apparent resistivity approaches ρ_2 (Fig. 6.3A). Next, suppose that the second medium is more conductive $(\rho_2 < \rho_1)$. In this case, negative charge is induced at interfaces when the current electrode is in the first medium, and the apparent resistivity increases when approaching the contact, because the x-component of the electric fields of the electrode and the induced charge reinforce each other. When the receiver O crosses the contact, these components oppose each other, and the apparent resistivity abruptly decreases and remains constant at the value

$$\rho_a^{(2)} = \rho_2(1 - K_{12}) > \rho_2 \quad (\text{since}\, K_{12} < 0).$$

As the electrode array moves away from the contact in medium 2, the apparent resistivity gradually decreases and approaches ρ_2. Now, consider the apparent resistivity curve when the receiver electrode O is behind the current electrode and $\rho_2 > \rho_1$. Recall that, by our convention,

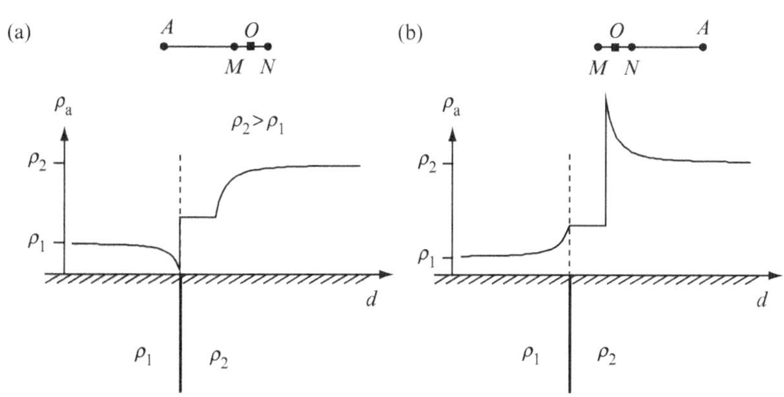

Figure 6.3 Apparent resistivity curves for the arrays *AO* and *OA* over a vertical interface.

the position x of the receiver O is always negative when the receiver trails the source electrode. In the first medium, we have

$$\rho_a^{(1)} = \rho_1 \left[1 + K_{12} \frac{x^2}{(2d - x)^2} \right]. \qquad [6.14]$$

In this case, both charges are positive, and in approaching the contact, the apparent resistivity increases. When the source electrode touches the contact $(d = 0)$, we have

$$\rho_a^{(2)} = \rho_1 (1 + K_{12}) > \rho_1. \qquad [6.15]$$

As before, the apparent resistivity is continuous when the source electrode A crosses the contact, even though the induced charge changes sign. The apparent resistivity stays constant when the array straddles the contact, then jumps discontinuously when the receiver crosses the interface. When both the source and the receiver are in medium 2, the apparent resistivity is given by

$$\rho_a^{(3)} = \rho_2 \left[1 + K_{12} \frac{x^2}{(2d + x)^2} \right]. \qquad [6.16]$$

In particular, at the contact, we have

$$\rho_a^{(3)} = \rho_2 (1 + K_{12}) > \rho_2.$$

As the array moves far into medium 2, the apparent resistivity again asymptotically approaches ρ_2 (Fig. 6.3B). We have considered apparent resistivity curves over the vertical contact with two- and three-electrode arrays. Using the principle of superposition, it is a simple matter to obtain the function ρ_a for four-electrode arrays, such as symmetrical and dipole arrays. Finally, let us make one obvious comment. If the contact does not extend to the earth's surface but is buried beneath an upper layer, then there is no discontinuity in the electric field, and the apparent resistivity measured with a three-electrode array is continuous.

6.1.2. Example 2: Profiling Over a Resistive Layer with Symmetrical Array *AMNB*

Profiling with a symmetrical *AMNB* array of fixed separation (Fig. 6.4) is generally effective with simple geoelectric sections. A good example is when a single object of investigation is surrounded by a relatively uniform medium and buried under a surface layer ("overburden") that does not

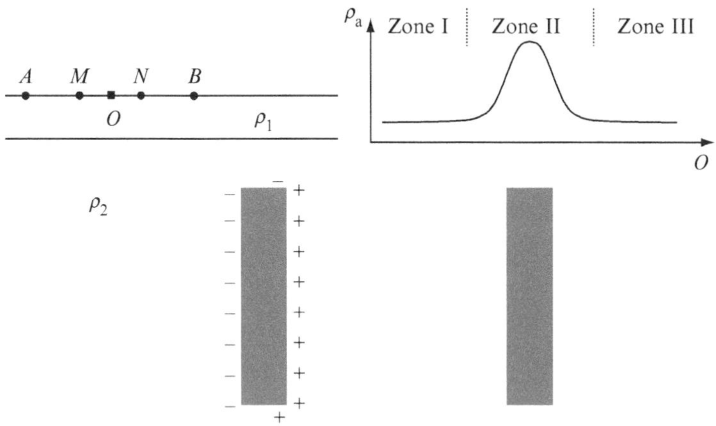

Figure 6.4 (A) Symmetrical array. (B) Profiling over the resistive layer.

vary much in its thickness or resistivity. Consider the case shown in Fig. 6.4, where a tabular resistive body is embedded in a homogenous medium and covered by a layer of constant thickness. The symmetrical array can be represented as a sum of the two-electrode arrays: *AMN* and *MNB*. As usual, assume that positive current goes into the medium through electrode *A* and, therefore, its charge is positive. Correspondingly, current enters electrode *B*, which has negative charge. Consider the apparent resistivity curve as the array moves along the *x*-axis, that is, from the left to right. At the beginning, the vertical layer is far away and the apparent resistivity practically coincides with that for a two-layer medium, consisting of the overburden and underlying half-space at the given electrode separation. Near the resistive body, the field of the charges induced on the resistive body becomes significant. The total field measured by the receiver line *MN* is equal to the sum

$$E_x = E_0 + E_s, \tag{6.17}$$

where E_0 and E_s are the *x*-components of the primary and secondary fields, respectively. In the interval *I*, the current electrode *B* is located closer to the vertical body and its influence on the charge distribution is stronger (Fig. 6.4B). In this case, both terms in Eq. [6.17] have the same sign and, correspondingly, the apparent resistivity increases as the array approaches the body. For the array *AMN*, the field components E_0 and E_s have opposite signs; but its contribution to the response is smaller because of its greater distance from the body. When the full array *AMNB* is above the layer (interval II), the responses of both three-electrode arrays are positive and reinforce each other, giving a maximum of the apparent resistivity curve. The roles of A and B are reversed as the array moves to the right of the body, and the apparent resistivity curve returns to the background level in interval III.

6.1.3. Example 3: Profiling with a Symmetric Array at Two Electrode Separations

In rapid profiling with symmetrical arrays, it is useful to repeat the measurements at two different separations of the current electrodes. For example, consider the behavior of the apparent resistivity with the symmetric array *AMNB* in the presence of the two geoelectric sections shown in Fig. 6.5. Case (a) corresponds to a basin or "syncline" of resistive sediment bulging downward into more conductive basement, whereas case (b) corresponds an arch or "anticline" of resistive basement bulging upward into a conductive top layer.

Applying the principles described earlier, it is not hard to see that the apparent resistivity will increase over the anticline, because the more resistive bottom layer gets closer to the surface. But the apparent resistivity will also increase over the syncline, because in this case, the more resistive top layer gets thicker. Use of two current electrode separations for each position of the potential electrodes, *AMNB* and *A'MNB'*, where $AB > A'B'$, can help resolve the ambiguity. The array *A'MNB'* will have a smaller depth of investigation at each location and, therefore, be more sensitive

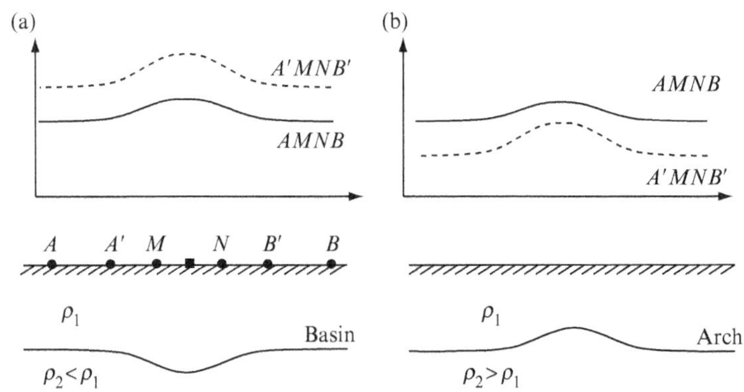

Figure 6.5 Electrical profiling with a symmetrical array at two electrode separations. Apparent resistivity curves over a resistive basin and a resistive arch show similar patterns at one separation, but can be distinguished by changing the depth of investigation.

to the upper layer. Therefore, in case (a), its apparent resistivity curve $\rho_a(A'MNB')$ will be located below the curve of the deeper-looking array, $\rho_a(AMNB)$; whereas in case (b), the opposite will occur.

6.1.4. Example 4: Profiling with the Array $AMNB$–C_∞

Figure 6.6A shows a special type of profiling array designed to detect thin vertical conductors located close to the surface and surrounded by a more resistive medium. This is often the configuration of metallic ore bodies formed near the surface by circulation of hydrothermal fluids through vertical fractures in the surrounding rock. The full array consists of two separate four-electrode arrays, *AMNC* and *BMNC*, with the common current electrode *C* placed far away from the line of profiling in the direction perpendicular to the profile. Values of apparent resistivity are measured separately for each array at every observation point along the profile. To understand how the method works, consider first ρ_a measured with the three-electrode array *AMN* (which is a good approximation to the full array *AMNC* when *C* is far from the profile line). When the *AMN* array is located far to the left of the conductor, negative charge appears on the conductor's left flank, with positive charge on the opposite side. In this configuration, this is because current flows from the more resistive surrounding rock into the less resistive body through its left flank, and the contrast coefficient K_{12} is negative. Current flows out of the body through its right flank. Because the negative charge is closer to the receiver *MN*, the secondary and primary fields reinforce each other: the electric

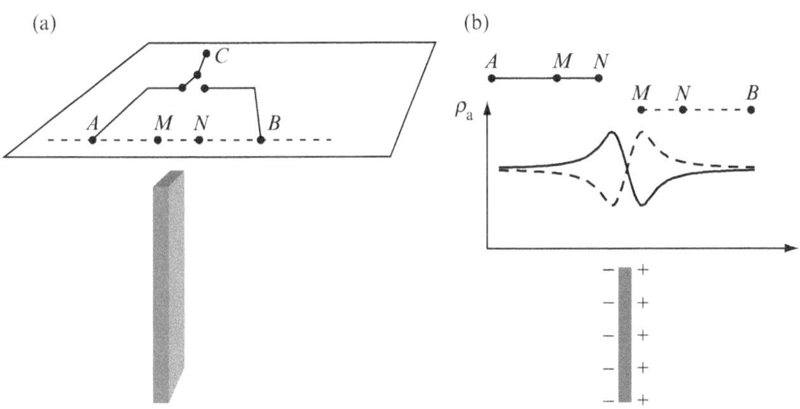

Figure 6.6 (A) Combined profiling array with a second current electrode perpendicular to the profile. (B) Schematic profiles over a thin conductor with each array.

field of the positive charge on the electrode and the negative charge on the left flank of the conductor, both have a positive x-component at the location of the receiver. Therefore, the apparent resistivity initially increases on approaching the body from the left. Closer to the conductor, however, the relative contribution of the positive charge becomes stronger and acts to decrease the horizontal component of the secondary electric field. Thus, a maximum of ρ_a is observed somewhere to the left side of the body. At some point, the horizontal components of the secondary field caused by the induced negative and positive charges cancel each other, and the apparent resistivity becomes equal to the resistivity of the surrounding medium, ρ. As the array moves over the body, the influence of positive charge increases and the value of ρ_a becomes even smaller. Further along the profile movement of the array, the horizontal component of the secondary field becomes positive, and the apparent resistivity curve reaches a minimum value. Then, it returns again to the resistivity of the surrounding medium as the array moves far to the right of the body. From the symmetry of the configuration, the apparent resistivity curve for the array MNB will be a mirror reflection of the curve for AMN, and the two curves should ideally intersect above the vertical conductor. This type of survey is useful for detecting a thin conductor, especially in areas of relatively complicated geology and topography.

6.1.5. Example 5: Method of Middle Gradient

The profiling technique called the "method of middle gradient" consists of a series of current and receiver lines located on the earth's surface as shown in Fig. 6.7. The current piece consists of two parallel lines A and B connected to each other by line C, which has the external current source. Electric contact with the ground is established by the electrodes placed at regular intervals along the two lines A and B. Suppose that current flows into ground from the line A. Then, positive charge arises on each electrode of this line in the amount $e_A = 2\varepsilon_0 \rho I$, where I is current flowing through the electrode and ρ is resistivity of the ground surrounding the electrode. Negative charge of the same magnitude arises at the electrodes of the negative current line B, if the resistivity of the ground near these electrodes is the same as that around the line A. (Note that at the surface of the insulated wire C, there are also electric charges of both signs, but due to the electrostatic induction, these external charges create no electric field beneath the earth's surface.) Thus, we have a system of linear charges

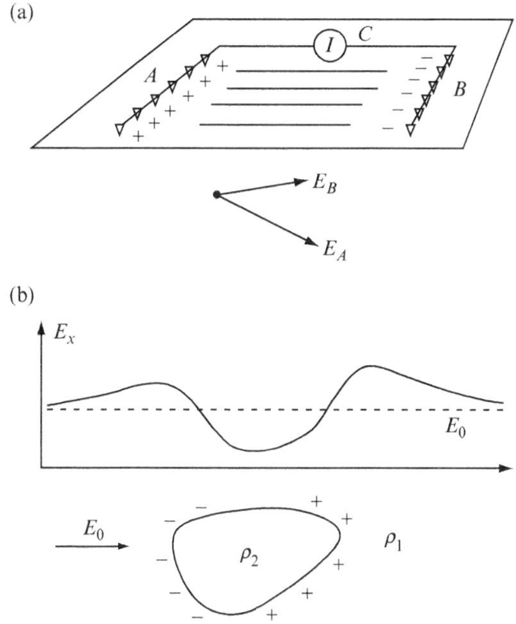

Figure 6.7 (A) Method of middle gradient and central profiles. (B) Schematic behavior of the horizontal component of the field in the presence of a conducting body.

of different signs located along two lines A and B, which create the electric field in the earth. The direction of the fields is shown in Fig. 6.7A. It is essential that the sign of horizontal component of the field E_x caused by both sets of charges is the same. The field of an array of linear charge of the same sign decreases at a rate that is almost inversely proportional to the distance from an observation point to the line (provided that this distance is much smaller than the length of the line). It is obvious, then, that near the middle part of the area between current lines A and B the tangential component of the electric field E_0 is almost constant. Moreover, this field practically remains the same at points beneath the earth's surface if their distance to the earth's surface is smaller than that of the current lines. Thus, with the help of such a system, it is possible to create within some volume a very simple field, which is usually called the normal or primary field. This simplifies the detection of conductors located beneath the earth's surface. Thus, over a uniform half-space, measuring the voltage along lines of observation between the two current lines would show a constant normal field (Fig. 6.7). Next, suppose that there is a conductor (ore body) beneath the earth's surface. Since the inhomogeneity is more conductive, negative and positive charges will appear, respectively, on the left (A) and right sides (B) of the conductor. These charges are sources of the secondary field E_s. The total horizontal component of the electric field at each point and, in particular, between receiver electrodes M and N, is $E_x = E_0 + E_s$. The secondary field contains information about parameters of the inhomogeneity, such as its position, size, and conductivity. Measurements of the voltage along parallel survey lines are usually represented in the form of a system of curves $E_x(x)$; a typical curve is shown schematically in Fig. 6.7B. The behavior of such curves can be easily explained by reference to the position of the induced charges. At points of part I of the profile, negative charges on a conductor create a horizontal component of the same sign as the normal field, whereas positive charges create a horizontal component of the opposite sign. Since the sum of the positive and negative charges is zero, and the negative charges are located closer, we conclude that at this part of the survey line, the field increases, $E_x > E_0$. Near the middle of the profile (part II), located above the conductor, both positive and negative charges create a horizontal component of negative sign for the secondary field, giving $E_x < E_0$. Finally, within part III of the profile, the horizontal components of the normal and secondary fields have the same sign. Since the positive charge is located closer to observation points than the negative charge, the total field again increases. A full set of such curves covering the area between the two current lines provides a qualitative map of the structure, and can also enable quantitative estimates to be made of some parameters. In particular, the distance between the maxima and their magnitudes can often be used to characterize the size of a buried conductor. The separation between observation points and survey lines is chosen mainly from available geological information.

6.1.6. Example 6: Profiling and Sounding, Electrical Resistivity "Tomography"

A modern survey technique for shallow geophysics made possible by advances in electronics is to lay out an array of electrodes on a dense regular grid covering the survey area. A computer can control the electrode array injecting current between any two electrodes and measuring the electrical potential at all the other electrodes (with respect to a common reference), or more simply between pairs of electrodes. The measurement sequence can then be repeated with two different electrodes serving as the current electrodes, etc. The complete set of data can then be plotted in various formats to give a qualitative picture of the subsurface geology or analyzed quantitatively by inversion software that systematically adjusts the parameters of a model to fit the data. This type of data collection and analysis is sometimes called "electrical resistivity tomography."

6.2. THE CHARGED-BODY OR MISE-À LA-MASSE METHOD

One of the oldest and most effective methods for electrical mapping is called the "charged-body" or the "mise-à la-masse" method, in which a current electrode is placed into direct contact with a large conductor, usually a metallic ore body. The next sections briefly describe the applications of this method.

6.2.1. Isometric Ore Body

Assume that the current electrode A is connected directly to a large buried conductor, such as an ore body, while the other electrode B is removed far enough away to neglect its influence. The current from the electrode A flows through the ore body into the surrounding rock, causing positive electric charge to appear on its surface (Fig. 6.8). These charges are sources of the electric field inside and outside the body. If the body has much higher conductivity than the surrounding medium, these surface charges will create a nearly equipotential surface over the body, which is the most

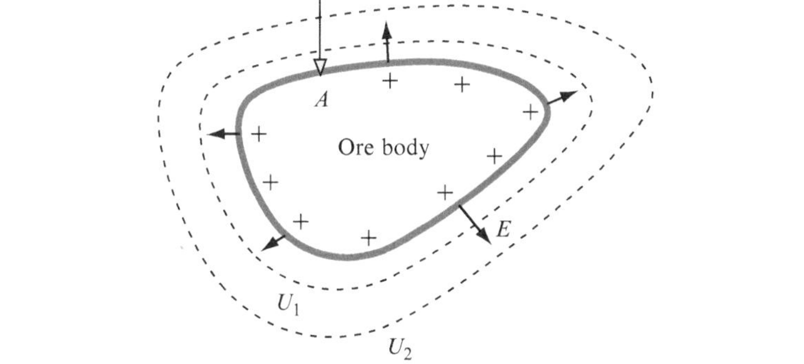

Figure 6.8 Charged-body (mise-à la-masse) method in which electric current is injected directly into a buried conductor, with schematic electric field and equipotential lines.

important feature of the method. (The same conclusion can be reached by applying Ohm's law to the current flow within the highly conducting body.) Since the surface of the body is almost equipotential, the vector lines of the electric field will be perpendicular to the body and the surrounding equipotential surfaces will also have practically the same shape as that of the conductor. Of course, far away from the body, the equipotential surfaces will tend to a spherical shape. In effect, the entire conducting body plays the role of the current electrode in this method. It is also obvious that the intersection of the earth's surface with the equipotential ones forms closed lines of equal potential (equipotential lines) that approximately map the projection of the body's shape onto the surface. Measuring the voltage on a grid on the earth's surface thus gives information about the dimensions and shape of the conductor (Fig. 6.8). There are two main approaches allowing one to study the field on the earth's surface. In the first, the receiver electrode N is placed in some point located relatively far away from the electrode A, while the second receiver electrode M roams over the surface above the body. Measuring the voltage difference between electrode M and the fixed reference N is equivalent to measuring the potential over the body (which is in fact defined only to within an arbitrary constant). Correspondingly, these data allow us to plot a map of equipotential lines on the earth's surface (Fig. 6.8).

A second approach (Fig. 6.9) is based on measuring voltage differences along a profile or a set of profiles. The behavior of the voltage can be easily explained using the principles described earlier.

6.2.2. Elongated Conducting Body

Geological structures are often elongated in one direction, usually called the "strike" direction. If an ore body is highly elongated, it may be necessary to take into account its finite resistivity, which will eventually cause the potential to drop along the current flow. At the same time, the fact that the conductor is elongated, means that positive charges at its surface

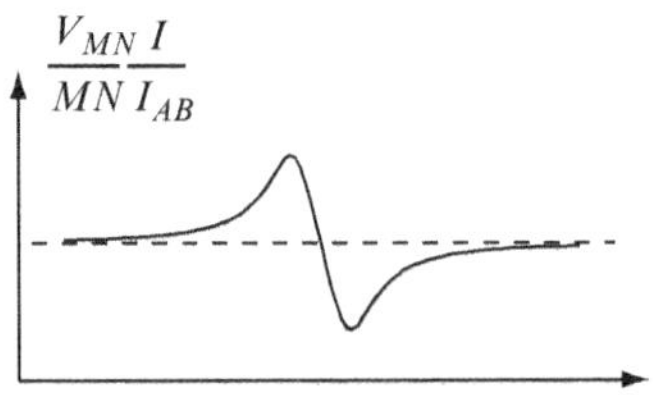

Figure 6.9 Profile of a two-electrode array over a conductor with the charged-body method.

act to direct the flow mainly along the body, so that leakage into a more resistive medium is minimized. (The current wants to follow the path of least resistance.) This suggests that the density of charge in the charged-body method decreases relatively slowly along the axis of elongation of a conductor, and outside the electric field, is mainly directed perpendicular to this direction. Measurements along profiles perpendicular to the strike direction generally have a characteristic shape (Fig. 6.10) that is easy to explain from the distribution of charge. With measurements V_{MN} made between closely spaced electrodes, there is a point, usually directed over the body, where the voltage difference is equal to zero. A system of such profiles will often define the strike of the conductor and even its shape. Often, instead of measuring the voltage between receiver lines, a coil is used to measure the vertical and horizontal components of the magnetic field on the earth's surface caused by the low-frequency current flowing

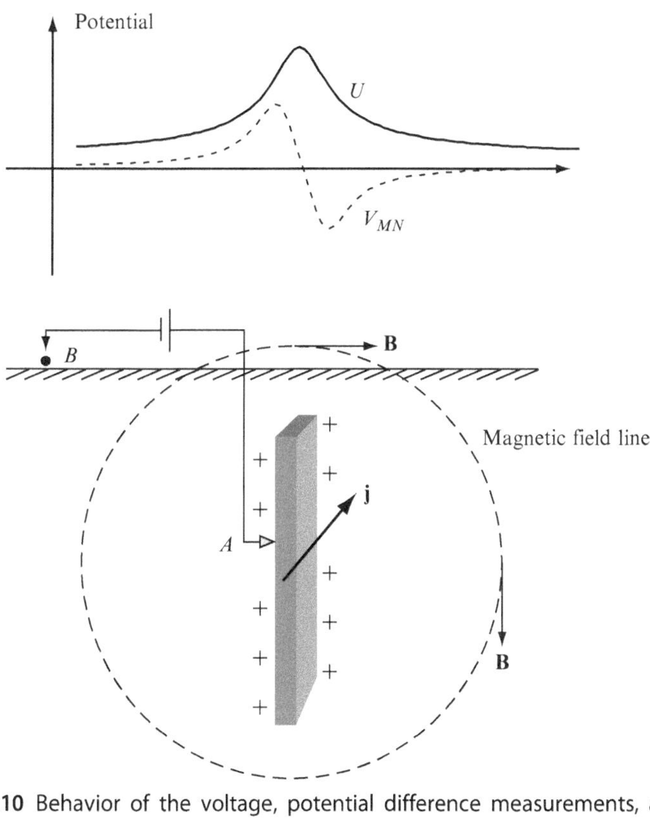

Figure 6.10 Behavior of the voltage, potential difference measurements, and magnetic field over an elongated conducting body in the charged-body method.

through the body. If the body is highly elongated, the magnetic field resembles that from a linear current filament and has a particular simple behavior: the components B_x and B_z have profiles of the same shape as the potential and electric field, respectively: that is, B_x peaks over the body, whereas B_z changes sign (Fig. 6.10).

6.2.3. Correlation of Conducting Layers Between Boreholes

The charged-body method can be used to correlate geoelectric sections in neighboring boreholes, or to outline the contours of a conducting body crossed by one of the holes. Suppose that each borehole crosses two conducting layers at different depths (Fig. 6.11). Our goal is to correlate the layers labeled 1 and 2 in the first borehole with layers 3 and 4 in the second borehole. Note that in the presence of complex geology, including faults and overturned structures, any type of correlation is possible, including a "reverse" correlation in which layers 1 and 4, and 2 and 3 match up, or no correlation at all. With this in mind, we place electrode A in contact with layer 1 and measure the potential in the second borehole. If the intervals 1 and 3 represent the same geological layer, then the maximum

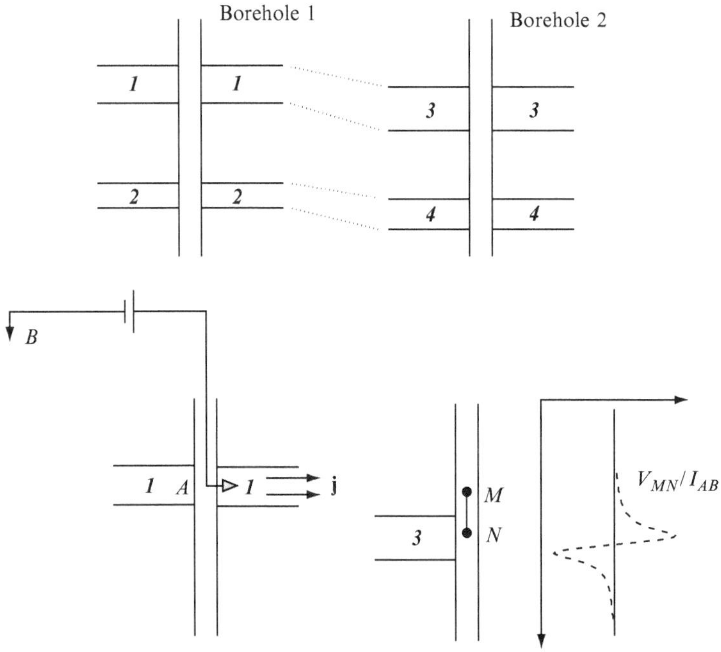

Figure 6.11 Correlations between layers with the charged-body method.

of potential in the second borehole will be observed near layer 3, because positive charge will appear at the top and bottom of the layer, as current injected in the first borehole flows along the layer and leaks out into the surrounding medium. If there is no pronounced maximum of the potential in the second borehole near layer 3, it is likely that these conducting zones are not connected. Similarly, placing the current electrode A in layer 2 can determine whether layers 2 and 4 are connected. The results of interpretation become more reliable if the measurements are repeated with the roles of the two boreholes reversed: that is, with the current electrode in layers 3 and 4 and potential measurements made in the first borehole. In general, the profile of electric potential in the measurement borehole—the positions of maxima and minima of the potential and any sign changes—gives useful information about the geometry of the body between the holes.

6.2.4. Determination of the Velocity and Direction of the Groundwater Flow

A modification of the charged-body method can be used to determine the direction and velocity of underground water flow. This application works as follows. The current electrode A is placed inside a porous bag of salt (NaCl) and lowered into the borehole to the depth of the water-bearing formation. The second current electrode B is located on the earth's surface far from the borehole (Fig. 6.12A). At the start of the experiment, the charge on the electrode A dominates the electric field, whose

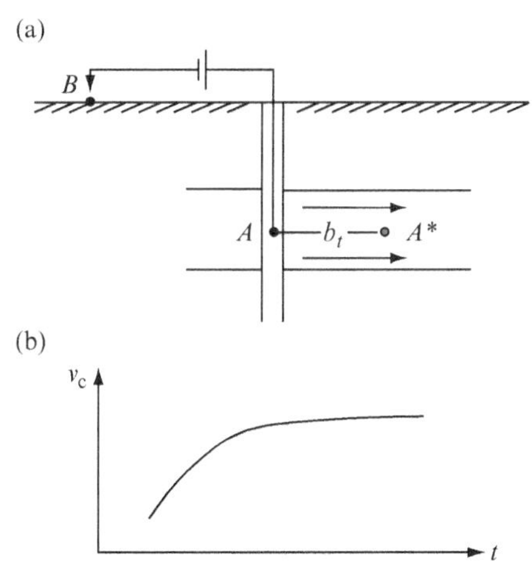

Figure 6.12 (A) Method of charged body for determination of the direction and velocity of water flow. (B) Dependence of the velocity of the center of the equipotential line on time.

equipotential surfaces are nearly spherical (if the resistivities near the borehole do not vary much). The corresponding equipotential lines measured on the surface are approximately circles centered on the borehole axis. As the salt dissolves in the water layer, the region with higher concentration of salt—and, therefore, higher conductivity—expands away from the borehole in the direction of the water flow. Surface measurements of the equipotential lines at different times (usually intervals of several hours or even days) will show elongation of the equipotential lines in the direction of the flow. An approximate method can be derived for calculating the velocity of the flow. Let us assume that with time, the center of the zone with high concentration of the salt can be treated as the point electrode A^*. This model has two current electrodes A and A^*, with charges

$$e_A = \varepsilon_0 \rho \alpha I \quad \text{and} \quad e_{A^*} = \varepsilon_0 \rho (1 - \alpha) I. \qquad [6.18]$$

The value of α varies between 0 and 1, and characterizes the distribution of current between the electrodes. If we assume that the potential is caused by only these charges, which are located at a distance b_t from each other at time t, then at any given time, the potential on the earth's surface is

$$U(r) = \frac{\rho I}{4\pi} \alpha \frac{1}{(r^2 + h^2)^{1/2}} + \frac{\rho I}{4\pi} (1 - \alpha) \frac{1}{\left[(r - b_t)^2 + h^2\right]^{1/2}}, \qquad [6.19]$$

where h is the depth of the electrodes and ρ is the resistivity of the surrounding medium.

Note that this model assumes that the depth of water flow does not vary laterally. The maximum of the potential as a function of radial distance from the borehole is where the derivative of the potential with respect to r equals zero:

$$\alpha \frac{r}{(r^2 + h^2)^{3/2}} + (1 - \alpha) \frac{(r - b)}{\left[(r - b_t)^2 + h^2\right]^{3/2}} = 0. \qquad [6.20]$$

The solution of this equation depends on α. Assuming that current is equally split between A and A^* so that $\alpha = 1/2$, we obtain $r_{\max} = b_t/2$. The displacement of the maximum of potential (the center of the equipotential curves) is therefore one-half of the displacement of the center of the fictitious electrode A^*, which measures the average displacement of the mass of dissolved salt. In other words, the velocity of displacement of this center is twice smaller than the velocity of the water flow. If all current goes through the electrode $A^*(\alpha = 0)$, then the coordinate of the center of equipotential line is equal to b. In this case, the velocity of this point coincides with that of the water flow. These considerations can be used to put bounds on the velocity. Note that the velocity of the center of equipotentials gradually approaches that of the water flux (Fig. 6.12), and is related to the fact that the development of the salt halo from the borehole requires some time.

6.3. SELF-POTENTIAL METHOD

We have so far described electric methods involving steady currents that are injected into the earth with an external source. There are also steady currents that occur naturally in the earth as a consequence of electromotive forces of different origins, including currents driven by the natural underground flow of conducting water through porous rocks. In surface electrical methods, these forces are mainly related to the effects of electrochemical and electrokinetic origin which generate a steady electric field.

6.3.1. Self-Potential of Electrochemical Origin in Mineral Exploration

Suppose that an ore body with electronic conductivity is surrounded by sedimentary rock whose conductivity is dominated by ions in its water-filled pores, and that part of the body extends above the water table. The surface of the ore body below the water table is a boundary between media with different mechanisms of conductivity (electronic and ionic), and a natural electrochemical reaction takes place generating an external non-Coulomb electromotive force in the form of a double layer of charge: net positive charge arises on the external surface of the ore body and is balanced by negative charge of equal magnitude on its internal surface (Fig. 6.13). Above the water table, the process of oxidation plays an important role and reverses the double layer: net negative charge appears on the external surface, balanced by positive charge on the internal surface. The ore body, thus, acts as a natural battery. The strength of the double layer below the water table depends on the mineral content of the ore body and on the concentration of ions in the surrounding fluid-filled rocks. The difference of potential across the layer can reach several hundred millivolts. Two conditions must be met to generate a steady electric field outside the body. The first is a variation of the dipole density along the double layer (a uniform double layer does not create an external

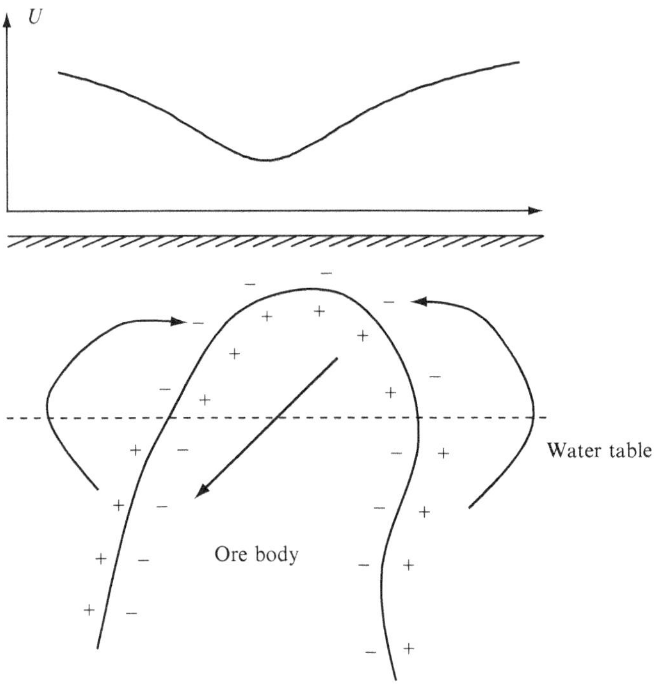

Figure 6.13 Distribution of charge and self-potential around an ore body that extends above the water table.

electric field). This first condition is the reason that massive sulfide ore bodies, which usually contain different types of minerals, cause a strong self-potential field. The second condition for generating a steady external field is a process that continuously renews the double layer by removing stray ions created by electrochemical processes near the contact that act to neutralize the surface charge. Often, the flow of the groundwater performs this function by carrying oxygen to the region. The self-potential field can be studied either by measuring the potential at observation points with respect to a fixed reference electrode, or by measuring potential differences between a moving pair of electrodes. In the first case, the reference electrode N is usually located at a large distance from the survey grid occupied by the second electrode M. This method gives a direct map of the self-potential field. Inasmuch as the dipole moment of the double layer in the upper part of the ore body is directed downward, the potential usually has a minimum directly above the body (Fig. 6.13A). A map of equipotential lines on the earth's surface will have a similar depression. The second approach is based on measuring of the voltage difference between observation points along survey lines. In this case, the difference of potentials between two points 1 and 2 can be written as

$$\nabla U_1 = \nabla U_{MN} + e_1 - e_2,$$

where ∇U_{MN} is the voltage between electrodes M and N caused by the self-potential field, and e_1 and e_2 are the electrode potentials. Their effect can be eliminated by interchanging the electrodes and repeating the measurement, giving

$$\nabla V_2 = \nabla U_{MN} + e_2 - e_1.$$

The average of the two measurements is the desired quantity. (With an artificial current source, the influence of electrode potentials can be similarly "averaged out" by repeating the measurement after reversing the direction of the current.) Over a massive ore body, the potential can be relatively large, reaching several hundred millivolts. When the mineralization is disseminated through the rock, each clump of mineralized rock acquires a double layer, but the net effect is usually small because of random orientations.

6.3.2. Self-potential of Electrokinetic Origin

A self-potential field also arises from underground electrokinetic processes that accompany filtration of groundwater through pores of geological formations. This phenomenon is used to study groundwater and civil engineering problems. The mechanism for generating electrokinetic potentials is complicated but can be summarized as follows. At the interface between solid rock and pore fluids in underground rocks, there is often a net absorption of dissolved ions of one charge by minerals lining the pores. For example, clay minerals in the pore walls of sedimentary rocks generally absorb negative ions from the pore fluids. This process changes the electrochemical equilibrium of the solution (which acquires an excess of positive ions), but the total charge in any macroscopic elementary volume is still zero. In fact, the net positive ions in solution are balanced by net negative charge on the pore walls; that is, the macroscopic charge density vanishes,

$$\delta = 0, \qquad [6.21]$$

and despite the redistribution of ions, there is no macroscopic electric field. The double layer that arises in the pore space is, however, asymmetric in the sense that the negative ions are more or less rigidly attached to the solid rock while the positive ions are free to move in the pore fluid. In reality, the positive ions accumulate in a thin diffuse layer of fluid lining the pore walls. Next, suppose that there is mechanical movement of the groundwater. Some of the positive charges in the diffuse layer can be dragged along by the flow, which is equivalent to a net electric current of mechanical origin. The presence of this "convection current" itself is not sufficient to generate a macroscopic electric field, since the macroscopic charge density may still vanish everywhere. Of course, there will always be a magnetic field associated with the convection current. To understand how a steady electric potential—in this case, called a "streaming potential"—can be generated by electrokinetic processes, we must find places where electric charge can arise from the convection current. To solve this problem, we first consider the forces acting on each elementary volume of moving water. In general, there are three forces: the Coulomb force caused by charges, the mechanical force due to a change of pressure, and finally, the force of resistance of the porous medium. When a sum of these three forces is equal to zero, we observe a movement of ions with a constant velocity. The relation between the current density and the first two forces for an elementary volume has the form

$$\mathbf{j} = \gamma\mathbf{E} - L\nabla p. \qquad [6.22]$$

Here, $\mathbf{E}$ is the Coulomb electric field, p is the fluid pressure, γ is the electrical conductivity, and L is a "coupling coefficient" between mechanical and electrical effects. L generally increases with increasing permeability of the porous medium. The negative sign in front of the second term arises because the motion of the fluid is directed toward smaller values of pressure. The following notations are usually used: $\mathbf{j} = \mathbf{j}_t$ is the total current density, $\mathbf{j}_{\text{cond}} = \gamma\mathbf{E}$ is the conduction current density, and $\mathbf{j}_{\text{conv}} = -L\nabla p$ is the convection current density. Correspondingly, the total current density is the sum:

$$\mathbf{j}_t = \mathbf{j}_{\text{cond}} + \mathbf{j}_{\text{conv}}. \qquad [6.23]$$

The total current density in the steady state has one important feature, namely, conservation of charge:

$$\text{div}\,\mathbf{j}_t = 0, \qquad [6.24]$$

that is, its flux through a closed surface surrounding an elementary volume is equal to zero, otherwise, an accumulation of charges would take place generating a time-varying electric field by Coulomb's law. We therefore have

$$\text{div}\mathbf{j}_{\text{cond}} + \text{div}\mathbf{j}_{\text{conv}} = 0.$$

In particular, at points where $\text{div}\mathbf{j}_{\text{conv}} = 0$, we also have $\text{div}\mathbf{j}_{\text{cond}} = 0$. Expanding the individual terms gives

$$\text{div}\mathbf{j}_{\text{cond}} = \text{div}(\gamma\mathbf{E}) = \gamma\ \text{div}\mathbf{E} + \mathbf{E}\cdot\nabla\gamma = \gamma\frac{\delta}{\varepsilon_0} + \mathbf{E}\cdot\nabla\gamma, \qquad [6.25]$$

where δ is the volume density of the electric charge, and

$$\text{div}\mathbf{j}_{\text{conv}} = -\ \text{div}(L\nabla p) = -L\cdot\nabla^2 p - \nabla L\cdot\nabla p. \qquad [6.26]$$

Let us represent Eq. [6.22] in the known form of the generalized Ohm's law:

$$\mathbf{j} = \gamma(\mathbf{E} + \mathbf{E}^{\text{ext}}). \qquad [6.27]$$

Here, $\mathbf{E}$ is the field-caused charges, while $\mathbf{E}^{\text{nc}}$ is an extraneous "non-Coulomb" force. Comparing Eqs. [6.22] and [6.27], it is clear that

$$\mathbf{E}^{\text{nc}} = -\rho L\nabla p, \qquad [6.28]$$

where ρ is a resistivity and $C = \rho L$ is called the "streaming current coefficient." We therefore see that a non-Coulomb force arises at every point where groundwater moves through a resistive porous medium. We are now ready to find the distribution of macroscopic electric charge associated with this motion. Using Eqs. [6.22], [6.24], and [6.25], we obtain

$$0 = \frac{\gamma}{\varepsilon_0}\delta + \mathbf{E}\cdot\nabla\gamma + \text{div}\mathbf{j}_{\text{conv}},$$

or

$$\delta = -\varepsilon_0\rho\mathbf{E}\cdot\text{grad}\gamma - \varepsilon_0\rho\ \text{div}\mathbf{j}_{\text{conv}}. \qquad [6.29]$$

The first term,

$$\delta_1 = -\varepsilon_0\rho\mathbf{E}\cdot\nabla\gamma,$$

shows that charge appears where the electric field is aligned with changes in conductivity, that is, where $\mathbf{E}\cdot\nabla\gamma$ differs from zero. This type of charge was studied in detail earlier. In particular, in the vicinity of a point where a medium is uniform by conductivity δ_1 is equal to zero. The second type of charge is caused by the divergence of the convection current density,

$$\delta_2 = -\varepsilon_0\rho\ \text{div}\mathbf{j}_{\text{conv}}. \qquad [6.30]$$

This charge arises because of the motion of groundwater and is the ultimate driving force of the electric field. For instance, we may observe the appearance of this type of charge (and its electric field) even when $\delta_1 = 0$. But, in the absence of the convection current density and its charge δ_2, charges of the first kind δ_1 do not arise because we are dealing with a natural field with no extraneous sources. In other words, charges with density δ_2 are "primary" sources of the electric field of electrochemical origin. Nevertheless, since the measured electric field is caused by charges of both types, it may happen that the maximum (or minimum) of the potential on the earth's surface is not located over "primary" sources. Next, consider a contact between media with different values of resistivity and parameter L. From continuity of the total current density, we have for its component normal to the interface:

$$j_{t,2n} - j_{t,1n} = \gamma_2 E_{2n} - \gamma_1 E_{1n} + j_{\text{conv},2n} - j_{\text{conv},1n} = 0,$$

or

$$\tfrac{1}{2}[(\gamma_2 - \gamma_1)(E_{2n} + E_{1n}) + (\gamma_2 + \gamma_1)(E_{2n} - E_{1n})] + j_{\text{conv},2n} - j_{\text{conv},1n} = 0. \quad [6.31]$$

Taking into account that

$$E_{2n} - E_{1n} = \frac{\sigma}{\varepsilon_0},$$

we have for the surface charge density σ:

$$\sigma = 2\varepsilon_0 \frac{\gamma_1 - \gamma_2}{\gamma_1 + \gamma_2} E_n^{\text{av}} + \frac{\varepsilon_0}{\gamma^{\text{av}}} (j_{\text{conv},1n} - j_{\text{conv},2n}), \quad [6.32]$$

where γ_1, E_{1n} and γ_2, E_{2n} are conductivity and normal components of the electric field at both sides of the interface, respectively. Also

$$\gamma^{\text{av}} = \frac{\gamma_1 + \gamma_2}{2}, \quad E_n^{\text{av}} = \frac{E_{1n} + E_{2n}}{2},$$

and $j_{\text{conv},1n}$, $j_{\text{conv},2n}$ are the normal components of the convection current density. In accordance with Eq. [6.32], we can define two types of surface charge:

$$\sigma_1 = 2\varepsilon_0 K_{12} E_n^{\text{av}} \quad \text{and} \quad \sigma_2 = \frac{\varepsilon_0}{\gamma^{\text{av}}} (j_{\text{conv},1n} - j_{\text{conv},2n}), \quad [6.33]$$

where

$$K_{12} = \frac{\rho_2 - \rho_1}{\rho_2 + \rho_1}.$$

The first type of surface charge, σ_1, is caused by a change of resistivity in the direction of current flow and has been studied in detail in earlier chapters. The second type, σ_2, appears at points of the interface where the density of convection current has discontinuity. For example, positive charge appears at the interface if $j_{\mathrm{conv},1n} > j_{\mathrm{conv},2n}$, and vice versa. The second equation in the set [6.33] can also be written as

$$\sigma_2 = \frac{\varepsilon_0}{\gamma^{\mathrm{av}}}\left(L_1\frac{\partial p_1}{\partial n} - L_2\frac{\partial p_2}{\partial n}\right), \qquad [6.34]$$

where L_1,p_1 and L_2,p_2 are values of the coupling parameter L and pressure p on the two sides of the interface.

The configuration shown in Fig. 6.14 illustrates a case where the self-potential method can detect leakage of groundwater from a near-surface aquifer. If some of the water flowing through the top layer is diverted into a vertical layer with higher permeability, then negative charge arises at the top of the vertical layer (see Eq. [6.33]), and a negative self-potential anomaly will appear at the surface.

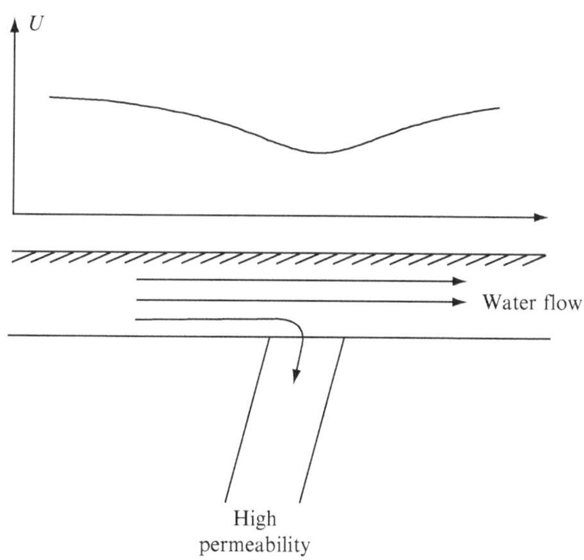

Figure 6.14 Leakage of groundwater flow into a vertical layer of high permeability and associated self-potential anomaly.

PART THREE

Electric Methods of Borehole Geophysics

CHAPTER SEVEN

Electrical Methods of Borehole Geophysics

Electrical methods of borehole geophysics use electrodes arranged along a well to determine the resistivity of the surrounding rocks. Usually, the well itself is filled with fluid, which can be either native formation fluid such as groundwater or the "drilling mud" used in the process of creating the borehole. Conductive borehole fluid provides a conduit for electrical current to flow into the surrounding rocks from electrodes suspended in the well, and thus allows measurements to be recorded nearly continuously as the electrodes are moved up or down the well in a process called "electrical logging."

Logging methods are nearly as old as surface electrical methods. The first systematic log of an oil well was made in 1927 by Henri-Georges Doll, a French engineer working for a company owned by Conrad and Marcel Schlumberger. The experiment was carried out in an oil field near the town of Pechelbronn in eastern France, and was quickly followed by commercial tests in the United States and Venezuela. But many of the advances in the interpretation of borehole electrical measurements that would establish logging as an indispensable step in oil exploration were made during a remarkable period in the early 1930s when a group of Schlumberger engineers worked with geologists, geophysicists, and reservoir engineers from the Soviet oil trusts in the great oil fields of the Caucasus. In the early 1940s, Gus Archie, an electrical engineer working for Shell Oil, determined empirically that the resistivity of porous reservoir rock can be tied quantitatively to its oil content. Since then, "Archie's law" relating electrical resistivity to oil saturation has been the main tool for evaluating the hydrocarbon content of oil reservoirs around the world. Borehole electrical methods are also widely used in mineral and groundwater exploration, as well as in environmental monitoring. As a result, electrical logging is today one of the largest commercial geophysical service businesses.

Methods in Geochemistry and Geophysics, Volume 44
ISSN 0076-6895, DOI: 10.1016/S0076-6895(10)44007-X
© 2010 Elsevier B.V.
All rights reserved.

As in previous chapters, we will focus on the key physical principles that determine the current and electric field when electrodes are deployed in boreholes. We will start with the simplest case when the geological formation surrounding the borehole is infinite and homogeneous.

7.1. CURRENT ELECTRODE AT THE BOREHOLE AXIS

7.1.1. Distribution of Charges and the Asymptotic Field Behavior

Suppose that a small current electrode A is situated along the axis of a borehole (Fig. 7.1A). The borehole radius is a, the conductivity of the borehole fluid (for simplicity, we will just use the term "borehole") is γ_1, and that of the surrounding medium is γ_2. Our main goal is to determine the potential and electric field within the borehole. Before formulating the boundary-value problem, let us discuss the distribution of charges and some general features of the behavior of the field. First of all, the charge on the electrode surface is

$$e_A = \varepsilon_0 \rho_1 I, \qquad [7.1]$$

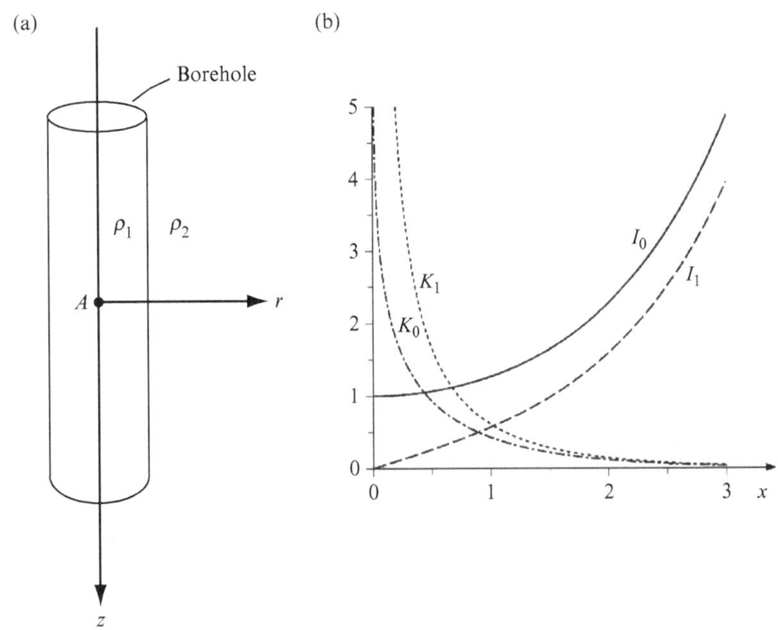

Figure 7.1 (A) Current electrode at the borehole axis. (B) Modified Bessel functions.

and it creates a primary electric field everywhere inside and outside the borehole, given by

$$E_0(p) = \frac{\rho_1 I}{4\pi L_{0p}^3} L_{0p}, \quad [7.2]$$

where $\mathbf{L}_{0p}$ is the distance from the electrode to an observation point p. The primary field induces electric charges to appear on the cylindrical interface between the borehole and the formation, creating a secondary field. As shown in Chapter 2, the density of surface charge on the interface between two conductive media is

$$\sigma(q) = 2\varepsilon_0 \frac{\rho_2 - \rho_1}{\rho_2 + \rho_1} E_n^{\text{av}}(q),$$

where $E_n^{\text{av}}(q)$ is the "average" normal component of the total electric field at point q of the interface: that is, the average of its values on either side of the interface. Recall from Chapter 2 that this average is the field at point q caused by the primary charge e_A and by the induced surface charge, excluding the charge induced at the point q itself. From this and Eq. [7.2], we expect that the surface density σ decreases rapidly with distance from the electrode. Using the boundary condition for the electric field and Ohm's law

$$E_{2n} - E_{1n} = \frac{\sigma}{\varepsilon_0} \quad \text{and} \quad j_n = \gamma E_n,$$

we can represent the surface charge density as simply

$$\sigma = \varepsilon_0(\rho_2 - \rho_1)j_n.$$

Since the normal component of current density is continuous, there is no need to take its average value in this formula. Integrating the charge density over the entire borehole surface gives for the total induced charge:

$$e_s = \varepsilon_0(\rho_2 - \rho_1)\oint j_n\,dS = \varepsilon_0(\rho_2 - \rho_1)I.$$

Thus, there are two distributions of the charge: the first is on the surface of the current electrode, while the second is distributed with changing density on the borehole surface. The total charge is

$$e = e_0 + e_s = \varepsilon_0\rho_1 I + \varepsilon_0(\rho_2 - \rho_1)I = \varepsilon_0\rho_2 I.$$

As expected from earlier chapters, the total charge coincides with the charge that would be induced on the current electrode as if it were located in a uniform medium with resistivity ρ_2, the resistivity of the infinite medium outside the borehole. The potential of the primary field at any point of a medium is

$$U_0(p) = \frac{\rho_1 I}{4\pi L_{0p}}, \qquad [7.3]$$

while the potential caused by the charges on the borehole surface is

$$U_s(p) = \frac{1}{4\pi\varepsilon_0}\int_S \frac{\sigma(q)\mathrm{d}S}{L_{qp}} = \frac{1}{4\pi}(\rho_2 - \rho_1)\int_S \frac{j_n \mathrm{d}S}{L_{qp}}. \qquad [7.4]$$

Although this last equation obviously cannot be used to calculate the field until the current j_n is known, it is nonetheless useful for understanding the physics. Consider, for example, the field behavior near and far away from the electrode. It follows from Eqs. [7.3] and [7.4] that the primary field approaching the electrode diverges if the electrode radius is sufficiently small, while the secondary field tends to a constant, because the distance L_{q0} is never less than the borehole radius. Thus, the primary field prevails near the electrode and only information about the borehole resistivity can be obtained from nearby measurements. In the opposite case when the observation point is far from the electrode, charges at the borehole surface create practically the same field as that of an elementary charge of magnitude e_s located at the current electrode (provided of course that the charge density falls off rapidly enough with distance from the electrode). A similar equivalence is observed for the gravitational field measured at sufficiently large distances from a mass distribution, which tends to the field of a point distribution of the same total mass. In our case, considering the potential or electric field far from the current electrode, one can, to a first order, replace the surface distribution of charges by an elementary charge e_s situated at the electrode. We thus have

$$U(p) = \frac{\rho_1 I}{4\pi L_{0p}} + \frac{\rho_2 - \rho_1}{4\pi L_{0p}} = \frac{\rho_2 I}{4\pi L_{0p}}, \qquad [7.5]$$

and

$$\mathrm{E}(p) = \frac{\rho_2 I}{4\pi L_{0p}^3}\mathrm{L}_{0p}, \text{as } L_{0p} \gg a. \qquad [7.6]$$

Note that this happens even if the observation point remains in the borehole. Thus, as the distance between the current electrode and observation point p increases, the field approaches that of a uniform medium with resistivity ρ_2. As shown in Chapter 2, the same result holds in the presence of several cylindrical surfaces. This situation is important in electrical logging, because when the rocks are permeable, the drilling fluid will generally flow outward from the well into the formation displacing the native fluids and altering the resistivity in a cylindrical region around the hole. This process is called "invasion," and the zone of altered resistivity is called the "invaded zone." In the next section, we will set up the full boundary-value problem and derive exact formulas that demonstrate the approximate analysis just presented.

7.1.2. Formulation of the Boundary-value Problem

We will use a cylindrical system of coordinates (r,ϕ,z), with its origin located at the center of the current electrode and the z-axis aligned with the borehole axis (Fig. 7.1A).

From the axial symmetry of the model around the z-axis and with respect to the plane $z = 0$, we assume that the potential is independent of the coordinate ϕ and is an even function of z:

$$U(r,z) = U(r, -z). \qquad [7.7]$$

The boundary-value problem for the potential consists of the following:

1. The potentials inside the borehole U_1 and within the formation U_2 satisfy Laplace's equation,

$$\nabla^2 U_1 = \nabla^2 U_2 = 0.$$

2. The potential near the current electrode tends to that of the electrode charge,

$$U_1 \to \frac{\rho_1 I}{4\pi R}, \quad \text{as}\, R = (r^2 + z^2)^{1/2} \to 0.$$

3. The potentials in the borehole and in the formation approach zero at large distances from the electrode,

$$U_1 \to 0, \quad U_2 \to 0, \quad \text{as}\, R \to \infty.$$

4. The potential and the radial component of the current density are continuous at the interface between the borehole and formation,

$$U_1 = U_2, \quad \gamma_1 \frac{\partial U_1}{\partial r} = \gamma_2 \frac{\partial U_2}{\partial r}, \quad \text{if}\, r = a.$$

7.1.3. Solution of Laplace's Equation

We need to construct a function satisfying Laplace's equation and all three boundary conditions. Laplace's equation in cylindrical coordinates with axial symmetry,

$$\frac{\partial U}{\partial \phi} = 0,$$

is

$$\frac{\partial^2 U}{\partial r^2} + \frac{1}{r}\frac{\partial U}{\partial r} + \frac{\partial^2 U}{\partial z^2} = 0. \quad [7.8]$$

We solve the equation using the method of separation of variables, assuming that its solution can be represented as the product of two functions, where each function depends on one argument only:

$$U(r,z) = T(r)\Phi(z). \quad [7.9]$$

Substituting Eq. [7.9] into Eq. [7.8] and dividing both sides by $U = T\Phi$ gives

$$\frac{1}{T}\frac{\partial^2 T}{\partial r^2} + \frac{1}{rT}\frac{\partial T}{\partial r} + \frac{1}{\Phi}\frac{\partial^2 \Phi}{\partial z^2} = 0. \quad [7.10]$$

The two terms,

$$\frac{1}{T}\frac{\partial^2 T}{\partial r^2} + \frac{1}{rT}\frac{\partial T}{\partial r} \quad \text{and} \quad \frac{1}{\Phi}\frac{\partial^2 \Phi}{\partial z^2},$$

depend separately on the arguments r and z, respectively, but sum to zero. Since r and z can vary independently, this is possible only if these terms are in fact constants of equal magnitude, but opposite sign. For convenience, we represent this constant as m^2, giving two ordinary differential equations of second order,

$$\frac{1}{T}\frac{\mathrm{d}^2 T}{\mathrm{d}r^2} + \frac{1}{rT}\frac{\mathrm{d}T}{\mathrm{d}r} = \pm m^2 \quad \text{and} \quad \frac{1}{\Phi}\frac{\mathrm{d}^2 \Phi}{\mathrm{d}z^2} = \mp m^2. \quad [7.11]$$

Solution of these equations can be constructed from known functions. To fix the sign on the right-hand side of Eq. [7.11], we use the fact that the potential is an even function of the coordinate z [Eq. 7.7] and choose the minus sign in the equation for Φ, which gives

$$\frac{\mathrm{d}^2 \Phi}{\mathrm{d}z^2} + m^2\Phi = 0. \quad [7.12]$$

This equation has two independent solutions, sin mz and cos mz, but we require only the even solution cos mz:

$$\Phi(z,m) = C_m \cos mz. \qquad [7.13]$$

The equation for $T(r)$ becomes

$$\frac{d^2T}{dr^2} + \frac{1}{r}\frac{dT}{dr} - m^2T = 0, \qquad [7.14]$$

and introducing the new variable $x = mr$ gives

$$\frac{d^2T}{dx^2} + \frac{1}{x}\frac{dT}{dx} - T = 0. \qquad [7.15]$$

Equation [7.15] can be compared with a similar equation derived in Chapter 4 [Eq. 4.43], from which it differs only in the sign of the last term. The solutions here are modified zero-order Bessel functions of the first and second kind, $I_0(x)$ and $K_0(x)$, respectively (Fig. 7.1B). We will also need their derivatives, the modified Bessel functions of first order, $I_1(x)$ and $K_1(x)$,

$$\frac{dI_0(x)}{dx} = I_1(x), \quad \frac{dK_0(x)}{dx} = -K_1(x). \qquad [7.16]$$

Their graphs are also shown in Fig. 7.1B. The asymptotic behavior of these functions for large and small arguments are important to the boundary conditions 2 and 3 above:

$$I_0(x) \to 1, \quad K_0(x) \to -\ln x, \quad I_1(x) \to \frac{x}{2}, \quad K_1(x) \to \frac{1}{x}; \quad \text{as}\, x \to 0$$

and

$$I_0(x) = I_1(x) \to \frac{\exp(x)}{(2\pi x)^{1/2}}, K_0(x) = K_1(x) \to \left(\frac{\pi}{2x}\right)^{1/2} \exp(-x); \text{as}\, x \to \infty. \qquad [7.17]$$

Modified Bessel functions are standard special functions of mathematical physics, and are described in numerous monographs with tables of values and various explicit formulas for their computation, including power series and integral representations, polynomial approximations, etc. A general solution of Eq. [7.15] can be represented as

$$T(x) = DI_0(x) + FK_0(x),$$

or

$$T(mr) = D_m I_0(mr) + F_m K_0(mr), \qquad [7.18]$$

where D_m and F_m are arbitrary constants independent on r. Using Eq. [7.9], we have

$$U(r,z,m) = [A_m I_0(mr) + B_m K_0(mr)] \cos mz, \qquad [7.19]$$

where A_m and B_m are unknown coefficients that depend on m. It is possible to show that a general solution of Laplace's equation can be constructed as a linear superposition (integral) of $U(r,z,m)$ for all positive values of m:

$$U(r,z) = \int_0^\infty [A_m I_0(mr) + B_m K_0(mr)] \cos mz \, dm. \qquad [7.20]$$

A solution satisfying particular boundary conditions is obtained by fixing the coefficients A_m and B_m.

7.1.4. Expressions for the Potential and Electric Field

We now find coefficients A_m and B_m such that $U(r,z)$ obeys the boundary conditions on the current electrode and at infinity, as well as the continuity conditions at the borehole surface. With this end in mind, we write the potential in the usual way as the sum of the primary field,

$$U_0(r,z) = \frac{\rho_1 I}{4\pi R},$$

plus the secondary field caused by the induced charge on the borehole wall. The secondary field is finite everywhere. For the representation of the secondary field, first consider the asymptotic behavior of the functions $I_0(mr)$ and $K_0(mr)$. As shown earlier, $K_0(mr)$ tends to infinity as its argument approaches zero, and therefore this function cannot describe the potential of the secondary field within the borehole where $r = 0$. Similarly, $I_0(mr)$ increases without bound as r increases, and therefore cannot characterize the field outside of the borehole. Thus, instead of Eq. [7.20], we write

$$\begin{aligned} U_1(r,z) &= \frac{\rho_1 I}{4\pi R} + \int_0^\infty A_m I_0(mr) \cos mz \, dm, \quad \text{if } r \le a, \\ U_2(r,z) &= \int_0^\infty B_m K_0(mr) \cos mz \, dm, \quad \text{if } r \ge a. \end{aligned} \qquad [7.21]$$

These two functions satisfy Laplace's equation and the two asymptotic boundary conditions: near the origin, the function U_1 tends to the potential caused by the charge on the electrode surface, whereas U_2 tends to 0 as $r \to \infty$. Also, it is possible to show that the rapid oscillations of $\cos mz$ in the integrand force both U_1 and U_2 to zero as $z \to \infty$, for any reasonable physical values of the coefficients A_m and B_m. This asymptotic behavior will be considered later in detail. Continuity of the potential and the normal component of the current density on the borehole surface can be satisfied by appropriate choice of A_m and B_m. As in Chapter 4, the derivation is simplified by expressing the primary field as a superposition of modified Bessel functions, using the following version of the Sommerfeld integral introduced earlier:

$$\frac{1}{R} = \frac{2}{\pi}\int_0^\infty K_0(mr)\cos mz\, \mathrm{d}m, \qquad [7.22]$$

where $R = (r^2 + z^2)^{1/2}$ and $r \neq 0$.

The expression for the potential inside the borehole can then be written as

$$U_1(r,z) = \int_0^\infty [CK_0(mr) + A_m I_0(mr)] \cos mz\, \mathrm{d}m, \quad \text{if } r \neq 0. \qquad [7.23]$$

Here,

$$C = \frac{\rho_1 I}{2\pi^2}.$$

Now, the conditions at the interface $r = a$ become

$$\int_0^\infty [CK_0(ma) + A_m I_0(ma)] \cos mz\, \mathrm{d}m = \int_0^\infty B_m K_0(ma) \cos mz\, \mathrm{d}m,$$

$$\gamma_1 \int_0^\infty m[CK_0'(ma) + A_m I_0'(ma)] \cos mz\, \mathrm{d}m = \gamma_2 \int_0^\infty m B_m K_0'(ma) \cos mz\, \mathrm{d}m. \qquad [7.24]$$

where

$$I_0'(x) = \frac{\mathrm{d}I_0(x)}{\mathrm{d}x} \quad \text{and} \quad K_0'(x) = \frac{\mathrm{d}K_0(x)}{\mathrm{d}x}.$$

These equations contain an infinite number of unknowns A_m and B_m, but reduce to separate equations for each value of m on using the uniqueness property of the Fourier cosine integral: that is,

$$\text{if} \int_0^\infty [F(m) - G(m)] \cos mz\, dm = 0, \quad \text{for all } z, \quad \text{then } F(m) = G(m).$$

Therefore, we have

$$\begin{aligned} CK_0(ma) + A_m I_0(ma) &= B_m K_0(ma), \\ \gamma_1\{-CK_1(ma) + A_m I_1(ma)\} &= -\gamma_2 B_m K_1(mr), \end{aligned} \qquad [7.25]$$

since $I_0'(x) = I_1(x)$ and $K_0'(x) = -K_1(x)$. The solution of the linear system [7.25] is

$$A_m = (\gamma_1 - \gamma_2)C\frac{K_0(ma)K_1(ma)}{\gamma_2 I_0(ma)K_1(ma) + \gamma_1 I_1(ma)K_0(ma)}$$

and [7.26]

$$B_m = \frac{\gamma_1 C}{ma}\frac{1}{\gamma_2 I_0(ma)K_1(ma) + \gamma_1 I_1(ma)K_0(ma)}.$$

In deriving the expression for B_m, the identity

$$I_1(x)K_0(x) + I_0(x)K_1(x) = \frac{1}{x}$$

was used. The functions $U_1(r,z)$ and $U_2(r,z)$ given by Eqs. [7.21] and [7.26] satisfy all conditions of the boundary-value problem and, by uniqueness theorem, therefore represent the potential caused by the electrode charge and the charge distributed on the borehole surface. Measurements of the potential and the vertical component of the electric field along the borehole play a role in electric logging, so let us write their expressions inside the borehole:

$$U(r,z) = \frac{\rho_1 I}{4\pi}\left[\frac{1}{(r^2+z^2)^{1/2}} + \frac{2}{\pi}(\gamma_1 - \gamma_2)\int_0^\infty \frac{K_0(ma)K_1(ma)I_0(mr)\cos mz}{\gamma_2 I_0(ma)K_1(ma) + \gamma_1 I_1(ma)K_0(ma)}\,dm\right]$$

and [7.27]

$$E_z(r,z) = \frac{\rho_1 I}{4\pi}\left[\frac{z}{(r^2+z^2)^{3/2}} + \frac{2}{\pi}(\gamma_1 - \gamma_2)\int_0^\infty \frac{mK_0(ma)K_1(ma)I_0(mr)\sin mz}{\gamma_2 I_0(ma)K_1(ma) + \gamma_1 I_1(ma)K_0(ma)}\,dm\right].$$

Also, for the radial component, we have

$$E_r(r,z) = \frac{\rho_1 I}{4\pi}\left[\frac{r}{(r^2+z^2)^{3/2}} + \frac{2}{\pi}(\gamma_1-\gamma_2)\right.$$
$$\left.\int_0^\infty \frac{mK_0(ma)K_1(ma)I_1(mr)\cos mz}{\gamma_2 I_0(ma)K_1(ma)+\gamma_1 I_1(ma)K_0(ma)}\mathrm{d}m\right]. \quad [7.28]$$

Along the borehole axis $r = 0$, these expressions reduce to

$$U(0,z) = \frac{\rho_1 I}{4\pi}\left[\frac{1}{|z|} + \frac{2}{\pi}(\gamma_1-\gamma_2)\int_0^\infty \frac{K_0(ma)K_1(ma)\cos mz}{\gamma_2 I_0(ma)K_1(ma)+\gamma_1 I_1(ma)K_0(ma)}\mathrm{d}m\right],$$

$$E_z(0,z) = \frac{\rho_1 I}{4\pi}\left[\frac{1}{z^2} + \frac{2}{\pi}(\gamma_1-\gamma_2)\int_0^\infty \frac{mK_0(ma)K_1(ma)\sin mz}{\gamma_2 I_0(ma)K_1(ma)+\gamma_1 I_1(ma)K_0(ma)}\mathrm{d}m\right],$$

and

$$E_r(0,z) = 0. \quad [7.29]$$

Here, it is appropriate to make two comments:

1. Algorithms for calculation and integration of modified Bessel functions are well known, so, computation of the potential and the electric field by Eqs. [7.27] and [7.28] is a relatively simple task.
2. The forward problem can be solved for more complicated systems of current and receiver electrodes using the principle of superposition.

7.1.5. Asymptotic Behavior of the Field

We can now formally derive asymptotic formulas for the potential and electric field that were suggested earlier by considerations of the distribution of charge on the electrode and borehole surface. Let us consider three different cases when $r = 0$.

Case 1: $z \to 0$

Equation [7.29] are simplified by introducing the new variables

$$x = ma, \quad \alpha = \frac{z}{a}, \quad \mu = \frac{\gamma_1}{\gamma_2} = \frac{\rho_2}{\rho_1}.$$

Then,

$$U(0,z) = U_0\left[1 + (\mu - 1)\alpha\frac{2}{\pi}\int_0^\infty A^*(x)\cos\alpha x\,\mathrm{d}x\right]$$

and [7.30]

$$E_z(0,z) = E_{0z}\left[1 + (\mu - 1)\alpha^2\frac{2}{\pi}\int_0^\infty xA^*(x)\sin\alpha x\,\mathrm{d}x\right],$$

where

$$A^* = \frac{K_0(x)K_1(x)}{I_0(x)K_1(x) + \mu I_1(x)K_0(x)}$$

and [7.31]

$$U_0 = \frac{\rho_1 I}{4\pi z},\quad E_{0z} = \frac{\rho_1 I}{4\pi z^2}.$$

U_0U_0 and E_{0z} are the potential and electric field on the z-axis caused by the electrode charge e_A situated in a uniform medium with the resistivity of the borehole. As z decreases, the ratio α tends to zero and, to first order in α, we have

$$U(r,0) \rightarrow U_0\left[1 + (\mu - 1)\frac{2\alpha}{\pi}\int_0^\infty A^*(x)\,\mathrm{d}x\right] \rightarrow U_0,\quad \text{if}\,\alpha \rightarrow 0.$$

Clearly, the potential approaches the potential caused by the charge e_A located on the surface of the current electrode. This result is of course not surprising since the solution for U was explicitly constructed to satisfy this condition.

Case 2: $z \rightarrow \infty$

Next, consider the case where the separation between the current electrode and an observation point increases and the parameter α tends to infinity. To obtain the asymptotic behavior of the potential, we analyze the integrand in the first equation of the set [7.30], which is the product of the functions $A^*(x)$ and $\cos\alpha x$, the first of which gradually decreases without a change of sign, while the second oscillates with a regular period. Sign reversals of $\cos\alpha x$ occur at intervals

$$\Delta x = \frac{\pi}{\alpha};$$

so, as the parameter α increases, the frequency of sign reversals increases. Since A^* is a continuous function of x, the integrals over neighboring intervals Δx

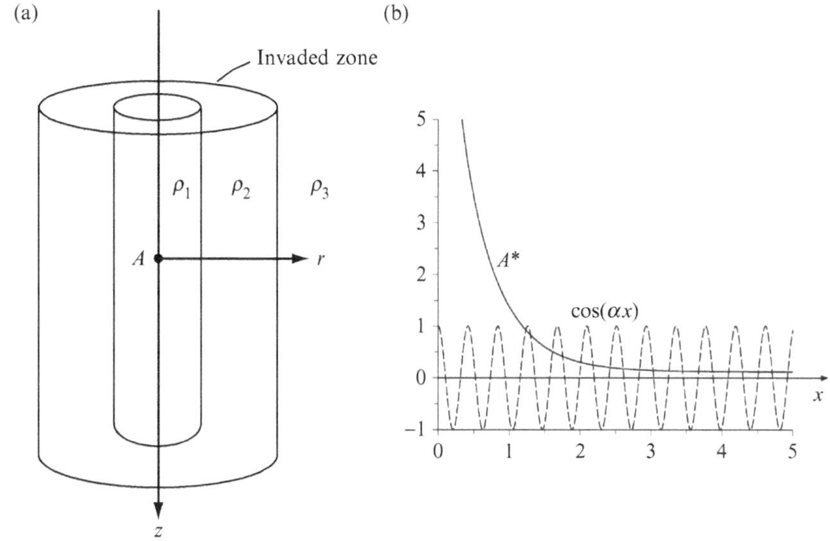

Figure 7.2 (A) Borehole surrounded by invaded zone and formation. (B) Oscillating integral to calculate field on the borehole axis.

are nearly equal in magnitude but opposite in sign and cancel each other more and more precisely as α increases. Also, since the rate of change of A^* decreases as x increases, the limit of the integral as α tends to infinity is dominated by small values of x (Fig. 7.2A). This analysis can be formalized by making the change of variable $x = x'/\alpha$ in the integral, and expanding the function A^* in powers of x, which goes to zero as α goes to infinity. We have

$$A^*(x) \to K_0(x), \quad \text{if } x \to 0,$$

and the asymptotic expression for the potential becomes

$$U(0,z) = U_0\left[1 + (\mu - 1)\frac{2\alpha}{\pi}\int_0^\infty K_0(x)\cos \alpha x \, dx\right], \quad \text{if } \alpha \to \infty.$$

Using the Sommerfeld integral [Eq. 7.22] gives

$$U(0,z) = U_0\left[1 + (\mu - 1)\frac{\alpha}{(1+\alpha^2)^{1/2}}\right] = \frac{\rho_2 I}{4\pi z}, \quad \text{if } \alpha \to \infty.$$

Thus, we have demonstrated from the solution of the boundary-value problem that the potential along the borehole axis at large distances from the current electrode approaches that of a uniform medium with the resistivity of the formation. In the same manner, we obtain

$$E_z(0,z) \to \frac{\rho_2 I}{4\pi z^2}, \quad \text{if } \alpha \to \infty.$$

Case 3: zone S

If the formation is relatively resistive, $\mu \gg 1$, there is an intermediate range of distances where the field has an interesting behavior. First, it is helpful to see that the distribution of currents in the borehole can be analyzed as follows in the limit when the formation becomes an insulator ($\rho_2 = \infty$). In this case, all the current flows along the borehole. Near the electrode, the current lines are almost radial in direction, but at larger distances they must tend to be parallel to the borehole axis. Correspondingly, at sufficiently large distances from the electrode, we can expect a nearly uniform current density across the borehole cross-section, with only a z-component. Since the total current flowing along the borehole must equal the current I emitted by the electrode, we have

$$j_z = \frac{I}{2\pi a^2},$$

where a is the borehole radius and the factor of 1/2 enters because (by symmetry) half the current flows up the borehole and the other half flows down. Ohm's law, $j_z = E_z/\rho_1$, gives the electric field inside a borehole with resistivity ρ_1 surrounded by an insulating medium:

$$E_z(r,z) = \frac{\rho_1 I}{2\pi a^2} = \frac{I}{2S}, \tag{7.32}$$

where S is the borehole conductance,

$$S = \gamma_1 \pi a^2. \tag{7.33}$$

For instance, if $a = 0.1\,\text{m}$ and $\rho_1 = 1\,\Omega\text{-m}$, we have $S \approx 3 \times 10^{-2}\,\text{m}/\Omega$. Where it applies, Eq. [7.32] describes an electric field that is uniform over the borehole cross section and constant along the z-axis. Because of its dependence on the parameter S, this interval is sometimes called the "S-zone." The voltage V measured between any two points M and N within the S-zone is

$$V = U(M) - U(N) = E_z MN = \frac{MN}{2S} I, \tag{7.34}$$

and, therefore, depends only on the separation MN. This means that the potential U behaves as a linear function of distance. We derived Eq. [7.32] from simple physical arguments in the limit $\gamma_2 = 0$. The same result can be obtained from Eq. [7.27]. Setting $\gamma_2 = 0$ gives

$$E_z = \frac{\rho_1 I}{4\pi}\left[\frac{z}{\left(r^2+z^2\right)^{3/2}} + \frac{2}{\pi}\int_0^{\infty}\frac{mK_1(ma)I_0(mr)\sin mz}{I_1(ma)}\mathrm{d}m\right],$$

which describes the field at any distance from the current electrode. The integral can be calculated using contour integration by finding poles of the function $I_1(mr)$ in the complex m-plane and applying the theorem of residues. The first term gives Eq. [7.32] provided that $z \gg a$. Certainly, the first approach is much simpler. The behavior of the radial component E_r is more complicated; in particular, it vanishes along the z-axis and at the borehole surface, and its magnitude decreases with distance z. Note that the current is accompanied by the appearance of surface charges that create different fields inside and outside the borehole. In fact, inside the borehole the electric field is directed along the z-axis and is practically uniform if $z/a \gg 1$, while outside the borehole, the charges create a strong radial field E_r. In real conditions when the surrounding medium has a finite conductivity and $\rho_2 \gg \rho_1$, there is always a finite interval, $z_1 < z < z_2$, where the electric field is described with high accuracy by Eq. [7.32]. As before, the current inside the borehole is mainly directed along the z-axis, but outside, there is a small leakage of the current into the formation. As the formation resistivity increases, this behavior is noticeable even at very large distances from the source. In the limit when the surrounding medium becomes an insulator, the upper limit of the S-zone tends to infinity.

7.2. A MEDIUM WITH TWO CYLINDRICAL BOUNDARIES

7.2.1. Expressions for the Potential and Electric Field

Next, we study a more complicated case when the borehole fluid penetrates the formation to a certain distance, creating an "invaded zone." This situation can be described by a model with two coaxial cylindrical interfaces: the inner surface at radius a_1 is the borehole wall; the outer surface at radius a_2 is the limit of the invaded zone. By analogy with the previous case, expressions for the potential at each uniform medium can be written as

$$U_1(r,z)=\int_0^{\infty}[CK_0(mr)+A_1I_0(mr)]\cos mz\,dm,\quad \text{if}\, r\leq a_1,$$

$$U_2(r,z)=\int_0^{\infty}[A_2I_0(mr)+B_2K_0(mr)]\cos mz\,dm,\quad \text{if}\, a_1\leq r\leq a_2, \qquad [7.35]$$

$$U_3(r,z)=\int_0^{\infty}B_3K_0(mr)\cos mz\,dm,\quad \text{if}\, r\geq a_2.$$

As before,

$$C=\frac{\rho_1 I}{2\pi^2},$$

to give the proper behavior of the field U_1 approaching the current electrode. Note that the expression for the potential U_2 in the invaded zone contains both Bessel functions, since its extension in the radial direction is limited. These potentials all satisfy Laplace's equation and the asymptotic at infinity, while U_1 satisfies the boundary condition on the current electrode at the origin. To ensure continuity of the potential and normal component of the current density at the interfaces, the unknown coefficients must obey the following system of linear equations:

$$\begin{aligned}
CK_0(ma_1)+A_1I_0(ma_1)&=A_2I_0(ma_1)+B_2K_0(ma_1),\\
\gamma_1[-CK_1(ma_1)+A_1I_1(ma_1)]&=\gamma_2[A_2I_1(ma_1)-B_2K_1(ma_1)],\\
A_2I_0(ma_2)+B_2K_0(ma_2)&=B_3K_0(ma_2),\\
\gamma_2[A_2I_1(ma_2)-B_2K_1(ma_2)]&=-\gamma_3B_3K_1(ma_2).
\end{aligned} \qquad [7.36]$$

Our main goal is to find the field inside the borehole and, correspondingly, we determine only the coefficient A_1. From the last two equations of the system, we have

$$A_2=MB_2, \qquad [7.37]$$

where

$$M=\frac{(\gamma_2-\gamma_3)K_0(ma_2)K_1(ma_2)}{\gamma_3I_0(ma_2)K_1(ma_2)+\gamma_2I_1(ma_2)K_0(ma_2)}. \qquad [7.38]$$

Substituting into the first two equations of the set [7.36] gives

$$\frac{CK_0(ma_1)+A_1I_0(ma_1)}{\gamma_1[-CK_1(ma_1)+A_1I_1(ma_1)]}=\frac{MI_0(ma_1)+K_0(ma_1)}{\gamma_2[MI_1(ma_1)-K_1(ma_1)]}.$$

Therefore,

$$A_1 = \frac{(\gamma_1 - \gamma_2)K_0(ma_1)K_1(ma_1) + M[\gamma_2 I_1(ma_1)K_0(ma_1) + \gamma_1 I_0(ma_1)K_1(ma_1)]}{\gamma_1 I_1(ma_1)K_0(ma_1) + \gamma_2 I_0(ma_1)K_1(ma_1) + (\gamma_1 - \gamma_2)MI_0(ma_1)I_1(ma_1)} C. \quad [7.39]$$

Thus, the potential and the vertical component of the electric field in the borehole in the presence of an invaded zone are given by

$$U_1(r,z) = \frac{\rho_1 I}{4\pi}\left[\frac{1}{R} + \frac{2}{\pi}\int_0^\infty A_1(m)I_0(mr)\cos mz\, dm\right]$$

and [7.40]

$$E_{1z}(r,z) = \frac{\rho_1 I}{4\pi}\left[\frac{z}{R^3} + \frac{2}{\pi}\int_0^\infty mA_1(m)I_0(mr)\sin mz\, dm\right].$$

As a check, consider several particular cases when the rather complicated expression for A_1 reduces to the expressions derived earlier.

1. $\gamma_2 = \gamma_3$
   In this case, $M = 0$ and Eq. [7.39] coincides with the first equation of the set [7.26].
2. $\gamma_1 = \gamma_2$
   Then, Eq. [7.39] becomes $A_1 = MC$ and, taking into account Eq. [7.38], we again arrive at Eq. [7.26] when the borehole radius is equal to a_2.
3. $a_2 \to \infty$
   Taking into account asymptotic behavior of Bessel function, we can see that $M \to 0$ and coefficient A_1 corresponds to a medium with one boundary at $r = a_1$.
4. $\gamma_2 = 0$
   If the invaded zone is perfectly insulating, the result should reduce to the case derived earlier for the S-zone. Setting $\gamma_2 = 0$ gives

$$A_1 = \frac{K_1(ma_1)}{I_1(ma_1)},$$

   and shows that the influence of the medium with resistivity ρ_3 disappears.

7.2.2. Asymptotic Behavior of the Field

We now study the asymptotic behavior of the field in the presence of invasion. First, it follows from Eq. [7.40] that the field near the current electrode tends to that in a uniform medium with the resistivity ρ_1,

$$U_1(r,z) \to \frac{\rho_1 I}{4\pi R}, \quad \text{if } R \to 0.$$

To find asymptotic expressions for large separations, we use the fact that the oscillating integrals in Eq. [7.40] will be dominated by small values of m as z goes to infinity (again, this can be formalized by expanding the integral in powers of $m' = m/z$). Using the small argument expansions of the Bessel function and preserving the leading terms, we have

$$M \to \frac{\gamma_2 - \gamma_3}{\gamma_3} K_0(ma_2)$$

and

$$A_1 = \frac{\gamma_1}{\gamma_2} K_0(ma_1) - K_0(ma_1) + \frac{\gamma_1}{\gamma_2} M.$$

Substituting these into the first equation of the set [7.40] and using the Sommerfeld integral give

$$U_1(r,z) = \frac{\rho_1 I}{4\pi}\left[\frac{1}{(r^2+z^2)^{1/2}} + \left(\frac{\gamma_1}{\gamma_2} - 1\right)\frac{1}{(z^2+a_1^2)^{1/2}} + \left(\frac{\gamma_1}{\gamma_3} - \frac{\gamma_1}{\gamma_2}\right)\frac{1}{(z^2+a_2^2)^{1/2}}\right].$$

Thus, when $z \gg a_2 > a_1 > r$, we obtain

$$U_1(r,z) \to \frac{\rho_3 I}{4\pi z}.$$

That is, the borehole and invaded zone become transparent, and the field is defined by the resistivity of the formation only. Finally, if the resistivity of the formation greatly exceeds that of the borehole and invasion zone, then there is an intermediate range of separations where the electric field E_{1z} is defined by the conductance S:

$$E_{1z}(r,z) = \frac{I}{2S} = \frac{I}{2(S_1 + S_2)},$$

where

$$S = \pi\left[\frac{a_1^2}{\rho_1} + \frac{a_2^2 - a_1^2}{\rho_2}\right].$$

In particular, when $\rho_1 \ll \rho_2$,

$$S = \frac{\pi a_1^2}{\rho_1}.$$

7.3. THE METHOD OF LATERAL SOUNDINGS

The equivalent of vertical sounding in electrical logging is called "lateral sounding." It uses the property that measurements at longer and longer electrode separations are sensitive to regions farther and farther from the borehole axis. By measuring the response as a function of electrode separation, it is possible to study the resistivity of geological formations "laterally" away from the borehole. The measurements are usually carried out with two types of devices, called normal and lateral probes.

7.3.1. Normal and Lateral Probes

The normal probe consists of one current and one voltage electrode—the source and receiver—separated by a distance L called probe length (Fig. 7.3A). The return current electrode B and the voltage reference electrode N are located far from the probe (sometimes on the earth's surface). The separation between electrodes B and N also greatly exceeds the probe length L. The lateral probe consists of two receiver electrodes M and N and one current electrode A (Fig. 7.3B), with the separation between receiver electrodes much smaller than the distance to the source, that is, $MN \ll AM$. The distance from the current electrode A to the midpoint of the receiver electrodes is the length of the lateral probe. Again, the return current electrode B is grounded far away from the other electrodes. As with arrays in surface prospecting, the normal and lateral probes

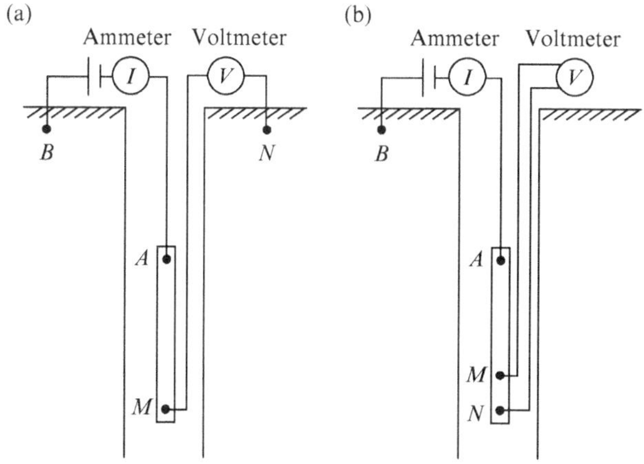

Figure 7.3 (A) Normal or potential probe. (B) Lateral or gradient probe.

measure the current I_{AB} in the source line AB and voltage V_{MN} in the receiver line MN. Then, the ratio

$$R_{MN} = \frac{V_{MN}}{I_{AB}} \qquad [7.41]$$

is calculated, giving a "resistance" measurement. By definition, R_{MN} is the resistance of the region between the two equipotential surfaces passing through the receiver electrodes. The normal probe, thus, measures the resistance of the region between the equipotential surface passing through the electrode M and the surface at zero potential located "at infinity." The lateral probe measures the resistance of the region between the equipotential surfaces passing through the voltage electrodes M and N. This suggests that a gradient probe has higher vertical resolution (i.e., resolution along the z-axis) than a normal probe. The voltage measured by a normal probe in practice coincides with the absolute electrical potential at point M, since the second voltage electrode is located far from current electrodes, and its potential $U(N)$ is nearly zero. Thus, Eq. [7.41] becomes

$$R_M = \frac{U_M}{I_{AB}}.$$

This is why the normal probe is also called a "potential probe." In contrast, the distance between the receiver electrodes of a lateral probe is relatively small, and, therefore, the measured voltage can be represented as $V_{MN} = E_z MN$, where E_z is the average electric field between the electrodes. If the distance is so short that the electric field is nearly constant between the electrodes, the voltage difference is directly proportional to the field E_z at the receiver electrodes, which is why the lateral probe is also called a "gradient probe." Thus, we have approximately,

$$R_{MN} = \frac{E_z MN}{I_{AB}}.$$

Next, consider the apparent resistivity measured by these probes, which as usual is related to the voltage in the following way:

$$\frac{\rho_a}{\rho_1} = \frac{V_{MN}(L)}{V^0_{MN}(L)}, \qquad [7.42]$$

where V^0_{MN} is the voltage measured by the same probe in a uniform medium with resistivity ρ_1. Thus, we have for the potential and gradient probes,

$$\frac{\rho_a}{\rho_1} = 4\pi L_{AM} \frac{U(M)}{I_{AB}} \quad \text{and} \quad \frac{\rho_a}{\rho_1} = 4\pi L_{A0}^2 \frac{E_z(O)}{I_{AB}}, \qquad [7.43]$$

where O is the midpoint between receiver electrodes, and L_{A0} is the length of the gradient probe. Both formulas are approximations, ignoring the positions of the other electrodes ("at infinity"), but are close enough for most applications.

7.3.2. Apparent Resistivity Curves for Lateral Soundings

Next, we consider the behavior of apparent resistivity curves for normal and lateral probes in the presence of invasion. Recall that in this case there are three independent regions in the model: the borehole with resistivity ρ_1, the invaded zone with resistivity ρ_2, and the (undisturbed) geological formation with resistivity ρ_3. It follows from Eqs. [7.40] that the apparent resistivity for the potential probe is

$$\frac{\rho_a}{\rho_1} = 1 + \frac{2\alpha}{\pi} \int_0^\infty A_1(m,\mu_1,\mu_2,p) \cos \alpha m \, dm, \qquad [7.44]$$

whereas for gradient probe it is

$$\frac{\rho_a}{\rho_1} = 1 + \frac{2\alpha^2}{\pi} \int_0^\infty m A_1(m,\mu_1,\mu_2,p) \sin \alpha m \, dm. \qquad [7.45]$$

Here,

$$\mu_1 = \frac{\rho_2}{\rho_1}, \quad \mu_2 = \frac{\rho_3}{\rho_1}, \quad p = \frac{a_2}{a_1}, \quad \text{and} \quad \alpha = \frac{L}{a_1}.$$

As in the case of vertical soundings, it is convenient to plot curves of apparent resistivity on a log–log scale, with vertical axis normalized by the resistivity of the borehole, ρ_a/ρ_1, and horizontal axis normalized by the borehole diameter, L/d_1. First, consider lateral sounding when there is no invasion $(\rho_3 = \rho_2)$. Figure 7.4A and B show curves for the normal and lateral probes. We can immediately distinguish several features of the curves that reflect the behavior of the potential and electric field along the borehole axis:

1. All curves for the normal (potential) probe have left and right asymptote corresponding to the resistivity of the borehole and formation, respectively.
2. As the formation resistivity either increases or decreases—or more precisely as the ratio μ_1 gets further from unity—the curves approach their right asymptotes at greater electrode separations.

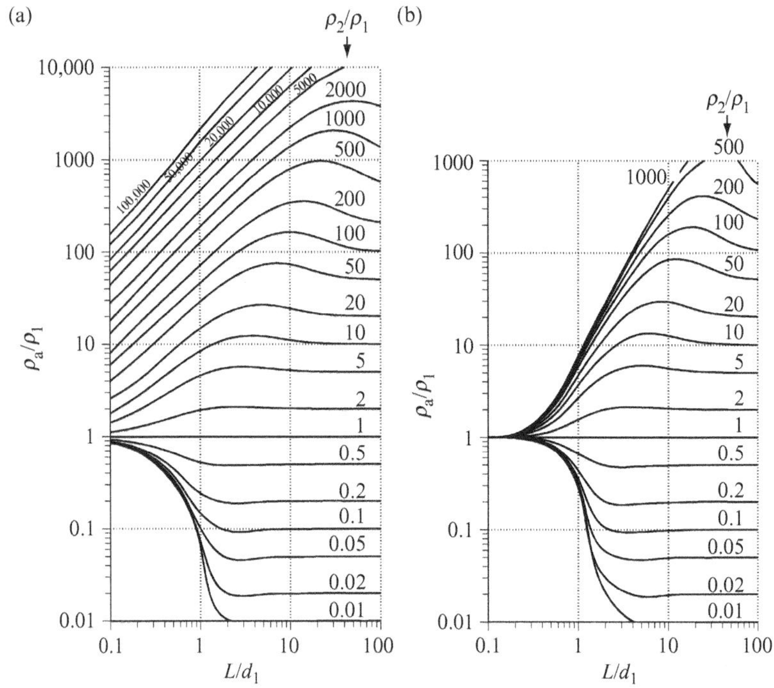

Figure 7.4 Curves of the apparent resistivity for (A) normal and (B) lateral probe ($\rho_2 = \rho_3$).

3. When the outside medium is more resistive ($\mu_1 > 1$), the curves of apparent resistivity for the potential probe first overshoot its right asymptote and reach a maximum, and then decrease gradually back to the asymptote. This indicates that there is a range of separations where the potential exceeds that in a uniform medium with resistivity ρ_2. This behavior does not occur in the case of vertical sounding.
4. The apparent resistivity curves also initially overshoot the right asymptote when the outside medium is more conductive ($\mu_1 < 1$).
5. As the parameter μ_1 gets farther from unity, the influence of charges at the borehole surface becomes noticeable at shorter electrode separations. Their effect is significant as soon as the probe length equals the borehole diameter.
6. If the formation is highly resistive ($\mu_1 \gg 1$), there is an intermediate range of separations where the apparent resistivity curve has a slope approximately equal to $63°26'$ and this interval widens as μ_1 increases. This behavior of the curve indicates that the potential is increasing almost linearly.

7. Within the intermediate zone, the potential can depend strongly on the formation resistivity even when it is very large, $\mu_1 = \rho_2/\rho_1 \gg 1$. This property is related to the fact that the normal probe measures the total resistance of the medium between the equipotential surface crossing the electrode M and infinity. In the region close to the probe, the current mainly flows through the (relatively conductive) borehole and its contribution to the measured resistance is small. At sufficiently large distances, where the current has mainly a radial direction, the influence of ρ_2 becomes dominant, and this region gives the main contribution to the measured resistance. As the ratio μ_1 increases, the position of the upper boundary of this region is located at greater distances from the normal probe.
8. If the formation is more conductive ($\mu_1 < 1$), the apparent resistivity curves also have left ($\rho_a = \rho_1$) and right ($\rho_a = \rho_2$) asymptotes, and for small values of L_{AM}/d_1, the curves are close to each other. This is understandable because the contribution of the formation to the resistance R_{MN} is relatively small.

Curves of apparent resistivity for the lateral (gradient) probe show similar behavior (Fig. 7.4B), but there is a difference between normal and lateral probes especially within the S-zone. All curves for the lateral probe practically coincide with each other in this zone and approach the curve for $\mu_1 = \infty$. Since the electric field is almost independent of distance and

$$\rho_a = 4\pi L_{A0}^2 \frac{E_z MN}{I_{AB}} = 2\pi L_{A0}^2 \frac{MN}{S},$$

we have

$$\log \rho_a = \log \frac{2\pi MN}{S} + 2 \log L_{A0}.$$

As with the normal probe, the slope of the curves in the S-zone is 63°26′.

Next, we consider apparent resistivity curves for a lateral probe in the presence of invasion for the two cases (Fig. 7.5A and B),

$$\rho_1 < \rho_2 < \rho_3 \, \text{and} \, \rho_1 < \rho_2 > \rho_3,$$

called "conductive" and "resistive" invasion. To understand the two cases and the terminology, it is helpful to recall that the drilling mud leaking out from the borehole only partly displaces the native formation fluids in the invaded zone. If the drilling mud is more conductive than native formation fluids—for example, if the mud is mixed with salt water and the rocks are filled with oil—then the conductive drilling mud will lower the resistivity of the invaded zone. The second case can arise when the drilling

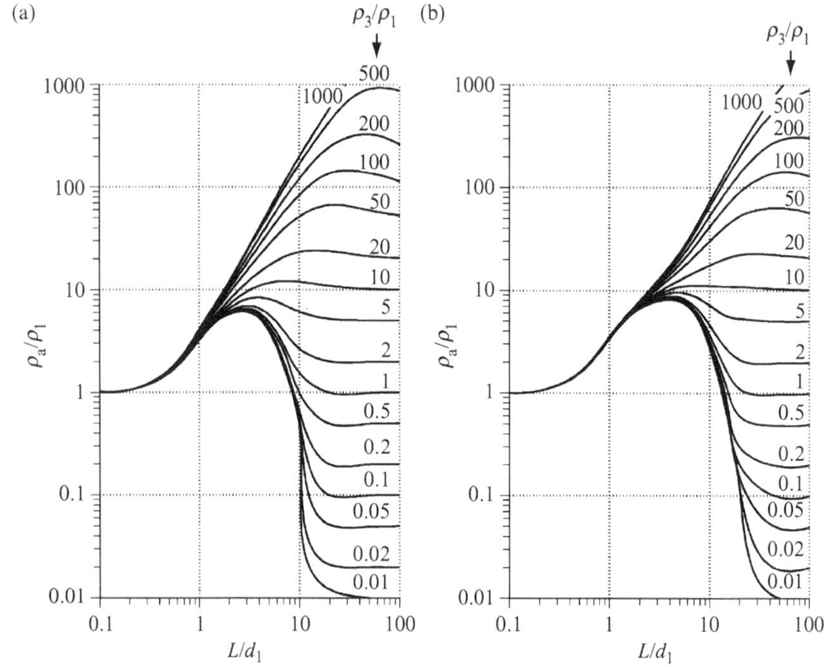

Figure 7.5 Curves of apparent resistivity for lateral probe: (A) $\rho_2/\rho_1 = 8$, $d_2/d_1 = 4$; (B) $\rho_2/\rho_1 = 8$, $d_2/d_1 = 8$.

fluid is resistive—for example, the mud is mixed with oil—but the rock pores are filled with a mixture of conductive water and resistive oil. If the drilling fluid displaces water preferentially from the pore space in the invaded zone, it can create an annular region that is more resistive than either the borehole or native formation.

With a conductive invaded zone, curves of apparent resistivity are similar to those without invasion, but have the following features:

a. The curves ρ_a/ρ_1 grow more slowly as L_{A0}/d_1 increases.

b. The maximum above the right asymptote is lower and is shifted to the right. If the radius of an invasion zone is large, then the right branch of the curve ρ_a/ρ_1 tends to that of a two-layer curve with index ρ_3/ρ_2, but the left asymptote approaches that of a two-layer curve with index ρ_2/ρ_1. Curves of apparent resistivity in the presence of a resistive invaded zone have a maximum that overshoots the right asymptote to a higher level and is displaced to the left compared to the two-layer curves.

As usual, comparison of measured and modeled curves of apparent resistivity can in principle determine the resistivity of all layers uniquely—that is, solve the inverse problem, but in real cases the inverse problem for lateral sounding is ill-posed. This means that some of parameters of a medium can be determined reliably within reasonable uncertainties, but other parameters cannot be determined to any reasonable accuracy after taking into account measurement errors. But, in fact, there are often important combinations of parameters that can be reliably determined. For instance, one such parameter in the presence of invasion is proportional to the total conductance of the borehole and invasion zone and is given by

$$S_{\mathrm{T}} = \frac{d_1^2}{\rho_1} + \frac{d_2^2 - d_1^2}{\rho_2}.$$

On the contrary, when the invasion zone has higher resistivity and sufficiently small diameter, the current mainly moves in the radial direction, and the parameter

$$U = \left(\frac{\rho_2}{\rho_1} - \frac{\rho_3}{\rho_1}\right) \ln\frac{d_2}{d_1},$$

is the second example of a stable parameter that characterizes an increase of resistance in a radial direction. Correspondingly, the equivalence by U takes place.

7.4. NORMAL AND LATERAL PROBES IN A MEDIUM WITH TWO HORIZONTAL INTERFACES

We now consider how the apparent resistivity varies along the borehole when formation layers have a finite thickness. To understand the effect of layering, we first consider a layered model with no borehole (Fig. 7.6). Note that this model is similar to the models of vertical sounding

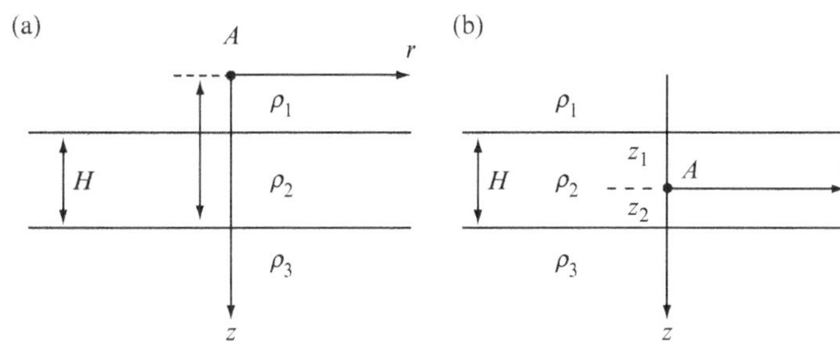

Figure 7.6 (A) Current electrode in a surrounding medium. (b) Current electrode in a layer.

considered in Chapter 4, but with two differences: first, there is no earth–air interface, and second, the measurements will be made along an axis (the borehole) that is perpendicular, not parallel, to the layers. Including the earth–air interface in the model is not difficult, but its effects on measurements deep underground are usually negligible.

7.4.1. Expressions for the Potential and Electric Field

We start with a model in which a layer of finite thickness is sandwiched between two half-spaces (Fig. 7.6). To find expressions for the potential, we reuse the results derived in Chapter 4. Let us assume that the current electrode is placed at the origin of the cylindrical coordinates and the z-axis is perpendicular to boundaries and directed downward. First, suppose that the current electrode is located above the layer. Then, from Eqs. (4.47)–(4.53), we can write

$$U_1(r,z) = \frac{\rho_1 I}{4\pi}\int_0^\infty [\exp(-m|z|) + A_1\exp(mz)]J_0(mr)\mathrm{d}m, \quad \text{if } z < z_1$$

$$U_2(r,z) = \frac{\rho_1 I}{4\pi}\int_0^\infty [A_2\exp(mz) + B_2\exp(-mz)]J_0(mr)\mathrm{d}m, \quad \text{if } z_1 < z < z_2$$

$$U_3(r,z) = \frac{\rho_1 I}{4\pi}\int_0^\infty B_3\exp(-mz)J_0(mr)\mathrm{d}m, \quad \text{if } z > z_2$$

[7.46]

Here $z_2 = z_1 + H$, where H is the thickness of the layer. It is obvious that these functions obey the boundary conditions near the current electrode and at infinity. As usual, the unknown constants are determined to provide continuity of the potential and normal component of the current density at both interfaces, giving the following system of equations:

$$\begin{aligned}
\exp(-mz_1) + A_1\exp(mz_1) &= A_2\exp(mz_1) + B_2\exp(-mz_1),\\
\gamma_1[-\exp(-mz_1) + A_1\exp(mz_1)] &= \gamma_2[A_2\exp(mz_1) - B_2\exp(-mz_1)],\\
A_2\exp(mz_2) + B_2\exp(-mz_2) &= B_3\exp(-mz_2),\\
\gamma_2[A_2\exp(mz_2) - B_2\exp(-mz_2)] &= -\gamma_1 B_3\exp(-mz_2).
\end{aligned}$$

[7.47]

The last two equations give

$$A_2 = -K_{12}\exp(-2mz_2)B_2. \qquad [7.48]$$

Substitution into the first two equations yields

$$\frac{\gamma_1 A_1 - \exp(-2mz_1)}{\gamma_2 A_1 + \exp(-2mz_1)} = N_1, \text{where}\, N_1 = -\frac{1 + K_{12}\exp(-2mH)}{1 - K_{12}\exp(-2mH)}. \quad [7.49]$$

Thus,

$$A_1 = N_2 \exp(-2mz_1), \quad \text{where}\, N_2 = \frac{\gamma_1 + \gamma_2 N_1}{\gamma_1 - \gamma_2 N_1}. \quad [7.50]$$

The remaining equations give

$$B_2 = \frac{1 + N_2}{1 - K_{12}\exp(-2mH)}, \quad A_2 = -K_{12}\frac{(1 + N_2)\exp(-2mz_2)}{1 - K_{12}\exp(-2mH)} \quad [7.51]$$

and

$$B_3 = \frac{(1 + N_2)(1 - K_{12})}{1 - K_{12}\exp(-2mH)}. \quad [7.52]$$

We can now write expressions for the apparent resistivity at different positions of the normal probe, when the current electrode is located in the lower half-space. Letting $z = L$, $r = 0$ in Eq. [7.46], we

$$\frac{\rho_a}{\rho_1} = \begin{cases} 1 + L\displaystyle\int_0^\infty A_1 \exp(mL)\mathrm{d}m, & \text{if}\, L \le z_1, \\ L\displaystyle\int_0^\infty [A_2 \exp(mL) + B_2 \exp(-mL)]\mathrm{d}m, & \text{if}\, z_1 \le L \le z_2, \\ L\displaystyle\int_0^\infty B_3 \exp(-mL)\mathrm{d}m, & \text{if}\, L \ge z_2. \end{cases} \quad [7.53]$$

The three expressions apply respectively to cases where the receiver electrode is (1) in the lower half-space with the current electrode, (2) in the middle layer, and (3) in the upper half-space. Since the model is symmetric, the same expressions apply if the current electrode is in the upper half-space. Similarly, for the gradient probe we have

$$\frac{\rho_a}{\rho_1} = \begin{cases} 1 - L^2\displaystyle\int_0^\infty mA_1 \exp(mL)\mathrm{d}m, & \text{if}\, L < z_1, \\ L^2\displaystyle\int_0^\infty m[-A_2 \exp(mL) + B_2 \exp(-mL)]\mathrm{d}m, & \text{if}\, z_1 < L < z_2, \\ L^2\displaystyle\int_0^\infty mB_3 \exp(-mL)\mathrm{d}m, & \text{if}\, L > z_2. \end{cases}$$

[7.54]

Next, suppose that the current electrode is located inside the layer (Fig. 7.6B). As before, we keep the origin of coordinates with this electrode. Unlike the previous case, the coordinate z_1 is now negative. Expressions for the potential in each region are

$$U_1 = \frac{\rho_2 I}{4\pi}\int_0^\infty A_1 \exp(mz) J_0(mr) dm, \qquad \text{if } z \le z_1,$$

$$U_2 = \frac{\rho_2 I}{4\pi}\int_0^\infty [\exp(-m|z|) + A_2 \exp(mz) + B_2 \exp(-mz)] J_0(mr) dm, \quad \text{if } z_1 \le z \le z_2, \tag{7.55}$$

$$U_3 = \frac{\rho_2 I}{4\pi}\int_0^\infty B_3 \exp(-mz) J_0(mr) dm, \quad \text{if } z \ge z_2.$$

Enforcing continuity of the potential and normal component of the current density at interfaces gives

$$\begin{aligned}
A_1 \exp(mz_1) &= \exp(mz_1) + A_2 \exp(mz_1) + B_2 \exp(-mz_1),\\
\gamma_1 A_1 \exp(mz_1) &= \gamma_2[\exp(mz_1) + A_2 \exp(mz_1) - B_2 \exp(-mz_1)],\\
\exp(-mz_2) + A_2 \exp(mz_2) + B_2 \exp(-mz_2) &= B_3 \exp(-mz_2),\\
\gamma_2[-\exp(-mz_2) + A_2 \exp(mz_2) - B_2 \exp(-mz_2)] &= -\gamma_1 B_3 \exp(-mz_2).
\end{aligned} \tag{7.56}$$

The last two equations yield

$$A_2 = -K_{12}(1 + B_2)\exp(-2mz_2), \tag{7.57}$$

and substituting this into the first two equations of the set [7.56] gives

$$A_1 = 1 - K_{12}\exp(-2mz_2) + [\exp(-2mz_1) - K_{12}\exp(-2mz_2)]B_2,$$
$$\gamma_1 A_1 = \gamma_2\{1 - K_{12}\exp(-2mz_2) - [\exp(-2mz_1) + K_{12}\exp(-2mz_2)]B_2\}.$$

This determines B_2:

$$B_2 = -\frac{K_{12}[\exp(2mz_1) - K_{12}\exp(-2mH)]}{1 - K_{12}^2\exp(-2mH)}, \tag{7.58}$$

and Eq. [7.57] gives A_2:

$$A_2 = -K_{12}\frac{\exp(-2mz_2) - K_{12}\exp(-2mH)}{1 - K_{12}^2\exp(-2mH)}. \tag{7.59}$$

Finally,

$$A_1 = -K_{12}\frac{1 - K_{12}\exp(-2mz_1)\exp(-2mH)}{1 - K_{12}^2\exp(-2mH)} \tag{7.60}$$

and

$$B_3 = \frac{(1 - K_{12})[1 - K_{12}\exp(2mz_1)]}{1 - K_{12}^2 \exp(-2mH)}. \qquad [7.61]$$

The corresponding expressions of apparent resistivity for the potential and gradient probes are

$$\frac{\rho_a}{\rho_1} = -L\int_0^\infty A_1 \exp(mL)J_0(mr)\mathrm{d}m,$$
$$\frac{\rho_a}{\rho_1} = -L^2\int_0^\infty mA_1 \exp(mL)J_0(mr)\mathrm{d}m, \quad \text{if } z < z_1.$$

Also,

$$\frac{\rho_a}{\rho_1} = \left[1 + L\int_0^\infty [A_2\exp(mL) + B_2\exp(-mL)]J_0(mr)\mathrm{d}m\right]$$

and

$$\frac{\rho_a}{\rho_1} = \left[1 - L^2\int_0^\infty [A_2\exp(mL) - B_2\exp(-mL)]J_0(mr)\mathrm{d}m\right], \quad \text{if } z_1 < z < z_2$$

and

$$\frac{\rho_a}{\rho_1} = L\int_0^\infty B_3\exp(-mL)J_0(mr)\mathrm{d}m,$$

$$\frac{\rho_a}{\rho_1} = L^2\int_0^\infty mB_3\exp(-mL)J_0(mr)\mathrm{d}m, \quad \text{if } z > z_2.$$

[7.62]

Consider now the apparent resistivity curves when the layer has either higher or lower resistivity than the surrounding medium. It is clear that the shape of these curves depends on the type of probe, on the ratio between the layer thickness and length of the probe, and on the ratio of resistivities, ρ_2/ρ_1. Examples are shown in Fig. 7.7 for the case where the probe moves upward and the receiver electrodes are located above the current electrode. In these curves, of course, the effect of the borehole is absent.

7.4.2. Layer with Higher Resistivity, $\rho_2 > \rho_1$

First, consider the gradient probe and suppose that $H > L$ (Fig. 7.7A). This case resembles the model studied earlier of a plane interface that intersects the earth's surface. Here, when the current electrode is located beneath the resistive layer, positive charge is induced at the bottom of the layer, because positive current flows out from the probe and crosses a boundary where the contrast coefficient is positive, $K_{12} = (\rho_2 - \rho_1)/(\rho_1 + \rho_2)$.

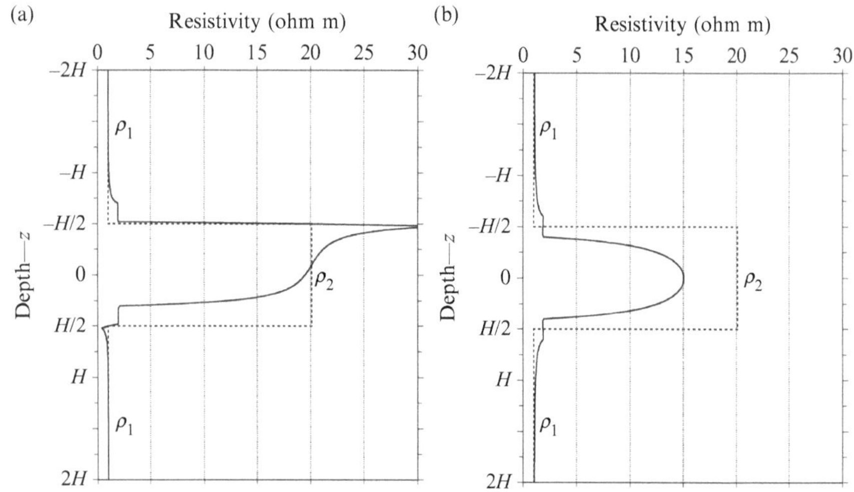

Figure 7.7 Apparent resistivity for (A) lateral and (B) normal probes, $H > L$, $\rho_2 > \rho_1$.

Similarly, negative charges are induced at the top of the layer; that is, an interface with a negative contrast coefficient. When the receiver electrode is in the lower half-space (along with the current electrode), the vertical electric field from the positive charge at the bottom interface points in the opposite direction from the primary field of the current electrode. Since the receiver is closer to the bottom interface, its effect is larger than that of the negative charge on the top interface. Therefore, the apparent resistivity decreases when the probe approaches the layer bottom, despite the fact that it is approaching a layer of higher resistivity. This well-known effect in electrical logging is called screening. When the receiver electrode crosses into the resistive layer, the apparent resistivity of a lateral probe jumps, because the electric field is discontinuous. When the electrodes are located on opposite sides of the interface, the apparent resistivity remains nearly constant, provided that $L/H < 1$. When the current electrode crosses into the layer, negative charges arise at both interfaces (positive current is flowing out from the layer into a lower resistivity medium). The negative charges on the bottom interface create an electric field that points downward above the interface and, therefore, opposes the primary field at the location of the receiver. The charges at the top of the layer create an electric field with the opposite effect. When the probe is near the bottom interface, its charges play the dominant role, but its field decreases relatively quickly. The interplay of these effects causes the ratio ρ_a/ρ_1 to increase rather rapidly and tend to ρ_2/ρ_1 as the probe moves away from the bottom interface. As the probe approaches the top interface, its negative induced charge begins to have a stronger effect, and the apparent resistivity increases even more. When the receiver electrode crosses into the upper half-space, we again observe a jump of the apparent resistivity. This is followed by a small interval, equal to the probe length L, where the apparent resistivity remains constant. Afterwards, it gradually decreases and approaches the resistivity of the upper half-space, ρ_1. Next, consider the behavior of ρ_a/ρ_1 for a normal probe when the layer thickness exceeds its length (Fig. 7.7B). It is obvious that the potential increases when approaching the bottom of the layer. Unlike the previous case, however, the apparent resistivity is continuous because the potential is continuous. When electrodes of the normal probe are located at opposite sides of the boundary, the apparent resistivity is constant and then approaches ρ_2. Afterwards, the influence of negative charges at the layer top dominates, and the apparent resistivity decreases. Obviously, the apparent resistivity curve for a normal probe must be symmetrical with respect to the middle of the layer. Now suppose that the length of the lateral and normal probes is much greater than the layer thickness, $L \gg H$. When electrodes of a gradient probe are far from the thin layer, the positive and negative charges induced at its bottom and top interfaces can be treated approximately as a double layer. The field of the double layer is small (the charges nearly cancel), and the apparent resistivity remains almost constant beneath the layer. As soon as the receiver electrode crosses into the thin layer, the charges at both interfaces cause a strong increase of the field, and the apparent resistivity curve reaches a maximum (Fig. 7.8A). When the receiver electrode crosses the top interface, the apparent resistivity again has a discontinuity and then approaches ρ_1. The curve has a complicated pattern caused by the induced charges changing sign in response to the different positions of the current electrode. When the receiver electrode M of a potential probe approaches the bottom of a thin layer, a small maximum is observed (Fig. 7.8B), because the potential of the double layer and that of the charge on the electrode A have the same sign at this position. Since the layer is thin and the density of charges at both interfaces is nearly the same, the secondary potential is nearly zero at the middle of the layer and the apparent resistivity is practically equal to ρ_1.

Of course, the curve is symmetrical with respect to middle plane.

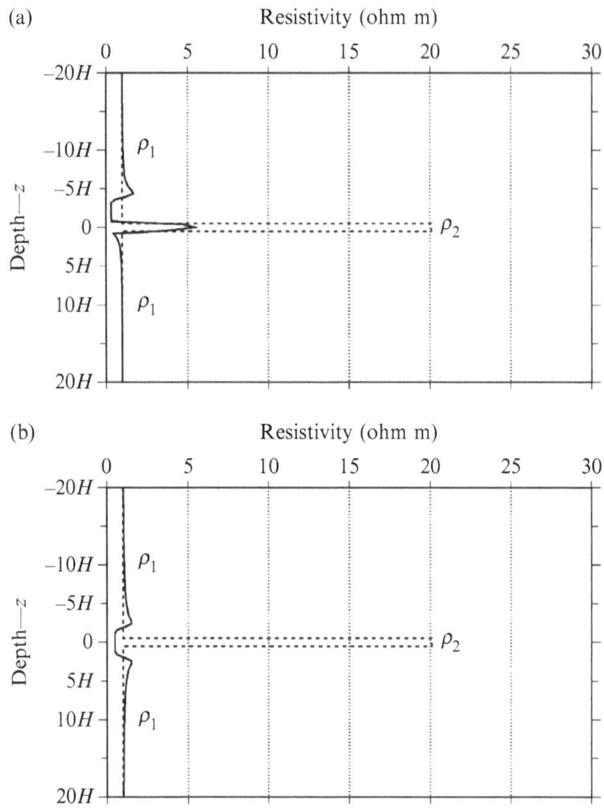

Figure 7.8 (A) Apparent resistivity curves for lateral probe. (B) Apparent resistivity curves for normal probe, $H > L$, $\rho_2 > \rho_1$.

7.4.3. Layer with the Lower Resistivity, $\rho_2 < \rho_1$

Curves of the apparent resistivity for a conductive layer are shown for gradient and potential probes in Fig. 7.9 for the case $H > L$. The patterns can be easily explained from the distribution of charges. When current electrode is beneath the layer, negative charge is induced at its bottom interface, with positive charge at the top. Therefore, in approaching the conductive layer, the apparent resistivity increases. When the receiver enters the layer, the apparent resistivity remains nearly constant over an interval equal to the probe length. Then, the apparent resistivity quickly becomes smaller and there is an interval where the apparent resistivity is close to the resistivity of the layer ρ_2. As the probe moves closer to the top interface, the influence of its positive charges becomes stronger and the apparent resistivity becomes even smaller. When the receiver electrodes intersect the top boundary,

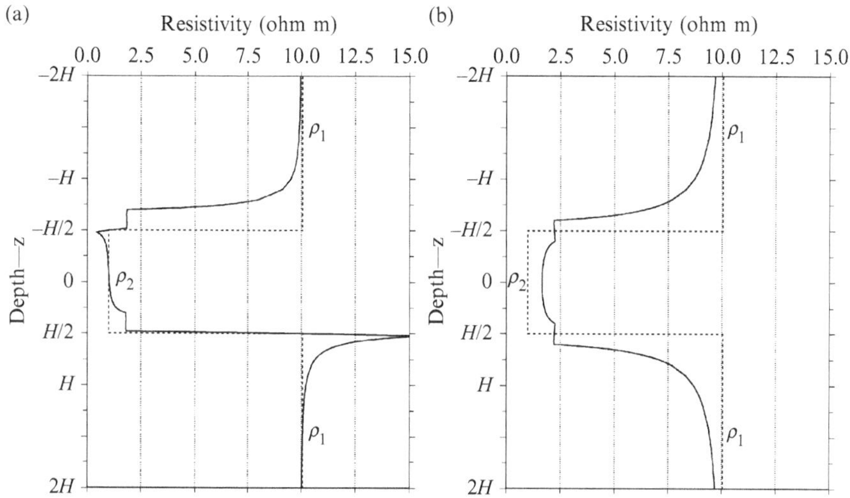

Figure 7.9 Apparent resistivity curves for (A) lateral and (B) normal probes $H > L$, $\rho_2 < \rho_1$.

there is an instant increase of ρ_a, followed by an interval with constant apparent resistivity. Finally, the apparent resistivity gradually increases to the resistivity of the upper half-space, ρ_1 (Fig. 7.9A).The curve of apparent resistivity for the potential probe is symmetrical with respect to the middle of the layer where a minimum is observed (Fig. 7.9B). Curves for a thin conductive layer are shown in Fig. 7.10A and B. In all these simulations, the effect of the borehole is absent.

Often, its influence makes only small changes in the shape of the curves. For example, when the borehole is present, there are no discontinuities in the apparent resistivity curve for a gradient probe, just intervals of rapid smooth change. In addition, there are no intervals where the apparent resistivity is exactly constant.

7.4.4. Influence of the Finite Thickness of the Layers

In logging applications, it is useful to study the dependence of the apparent resistivity on the thickness of a layer. Consider a normal or lateral probe situated at the center of a layer between two half-spaces. As the layer thickness becomes large compared to the probe length, the apparent resistivity approaches the "true" resistivity of the medium, provided, of course, that the effect of the borehole is neglected. All geological layers have, indeed, finite thickness, and it is often important to know how much the apparent resistivity measured at the center of the bed deviates from

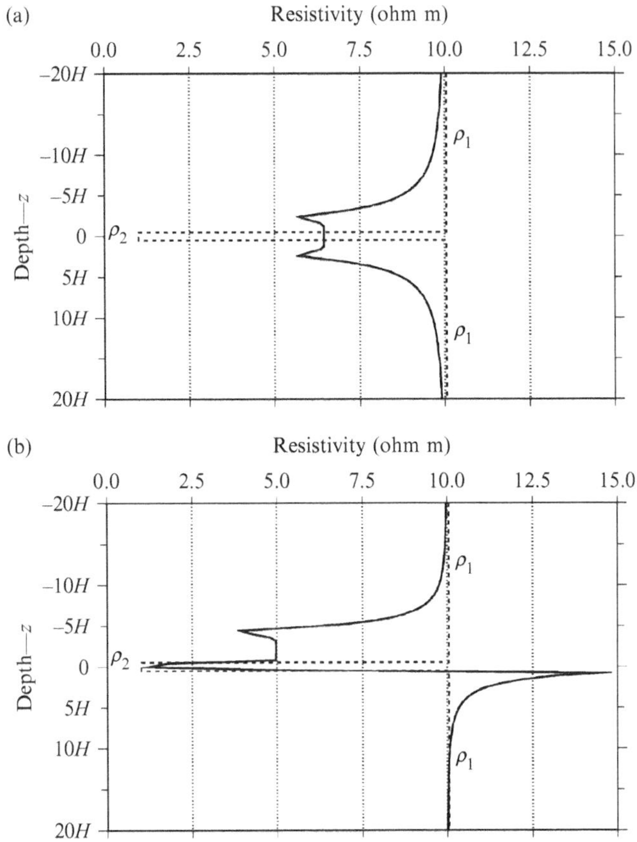

Figure 7.10 Apparent resistivity curves for (A) normal and (B) lateral probes, $H < L$, $\rho_2 < \rho_1$.

the true resistivity of a layer. This deviation is generally called the "shoulder-bed" effect and is important, for example, in the evaluation of layers for the presence of oil. Figures 7.11 and 7.12 illustrate the shoulder-bed effect by showing the apparent resistivity of a centered probe as a function of layer thickness for models with and without boreholes. The curves with borehole are calculated with a finite-element code that solves the appropriate boundary-value problem numerically; there is no closed-form analytical solution for a model with both horizontal and vertical boundaries. Clearly, the shoulder-bed effect—that is, the influence of the surrounding medium—is noticeable even for relatively thick layers. And for this reason, it is customary to correct the apparent resistivity curve to obtain the true resistivity using models similar to those in the figures.

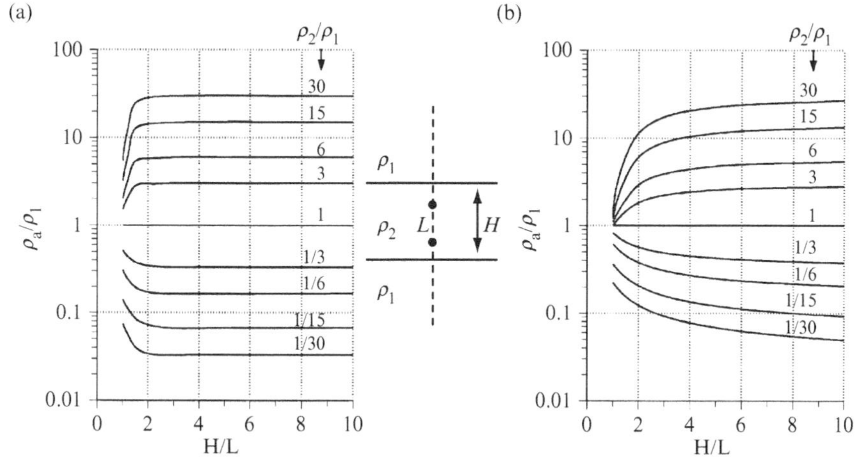

Figure 7.11 Dependence of apparent resistivity on layer thickness for (A) lateral and (B) normal probes. No borehole effect.

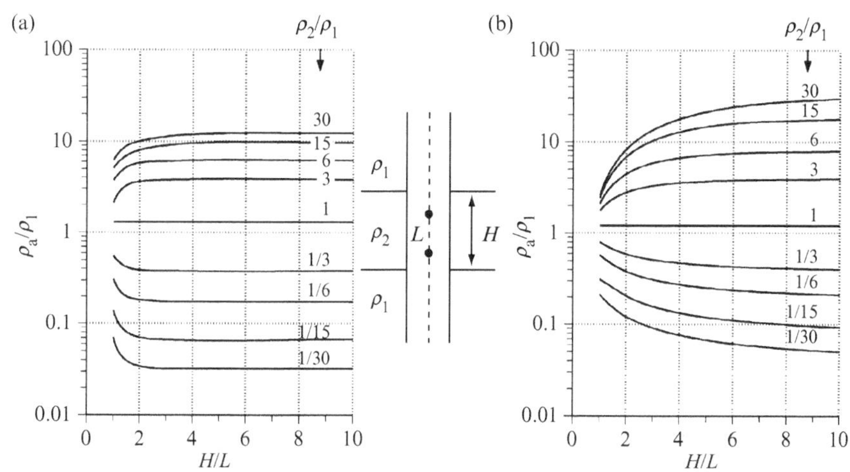

Figure 7.12 Dependence of apparent resistivity of layer thickness for (A) lateral and (B) normal probes. With borehole effect.

7.5. FIELD IN A BOREHOLE SURROUNDED BY ANISOTROPIC MEDIUM

As we know from Chapter 4, a stack of thin layers can be effectively anisotropic on the scale of electrical measurements. In this section, we will study the influence on borehole probes of electrical anisotropy characterized by different values for the normal ρ_n and transverse

ρ_t components of the resistivity tensor. If the borehole is perpendicular to the layering, then the "normal" and "transverse" resistivities apply to currents flowing parallel and perpendicular to the borehole, respectively. The degree of anisotropy is characterized by the parameter $\lambda = (\rho_n/\rho_t)^{1/2}$. We will proceed from the principle of charge conservation, which in cylindrical coordinates has the form

$$\frac{\partial^2 j_r}{\partial r^2} + \frac{1}{r}\frac{\partial j_r}{\partial r} + \frac{\partial^2 j_z}{\partial z^2} = 0. \qquad [7.63]$$

From Ohm's law, we have

$$j_r = \gamma_t E_r = -\frac{1}{\rho_t}\frac{\partial U}{\partial r} \quad \text{and} \quad j_z = \gamma_n E_z = -\frac{1}{\rho_n}\frac{\partial U}{\partial z}. \qquad [7.64]$$

So, written in terms of the potential, Eq. [7.63] becomes

$$\lambda^2\frac{\partial^2 U}{\partial r^2} + \frac{\lambda^2}{r}\frac{\partial U}{\partial r} + \frac{\partial^2 U}{\partial z^2} = 0 \quad \text{or} \quad \frac{\partial^2 U}{\partial r_1^2} + \frac{1}{r_1}\frac{\partial U}{\partial r_1} + \frac{\partial^2 U}{\partial z^2} = 0, \qquad [7.65]$$

where

$$r_1 = \frac{r}{\lambda}. \qquad [7.66]$$

Thus, we again have Laplace's equation in the coordinates r_1 and z. Using results already derived, we can write

$$U_1(r,z) = \frac{\rho_1 I}{2\pi^2}\int_0^\infty [K_0(mr) + A_m I_0(mr)]\cos mz\, dm$$

and [7.67]

$$U_2(r,z) = \frac{\rho_1 I}{2\pi^2}\int_0^\infty B_m K_0(mr/\lambda)\cos mz\, dm.$$

These expressions satisfy conditions near the current electrode A and at infinity. We use the boundary conditions at the borehole radius a to determine the unknown coefficients

$$U_1(a,z) = U_2(a,z) \quad \text{and} \quad \gamma_1\frac{\partial U_1(a,z)}{\partial z} = \gamma_t\frac{\partial U_2(a,z)}{\partial z},$$

which give the two equations

$$K_0(ma) + A_m I_0(ma) = B_m K_0(ma/\lambda)$$

and

$$\gamma_1[-K_1(ma) + A_m I_1(ma)] = -\frac{\gamma_t}{\lambda} B_m K_1(ma/\lambda). \qquad [7.68]$$

The coefficient A_m describing the field in the borehole is therefore

$$A_m = \frac{sK_0(ma/\lambda)K_1(ma) - (1/\lambda)K_0(ma)K_1(ma/\lambda)}{sI_1(ma)K_0(ma/\lambda) + (1/\lambda)I_0(ma)K_1(ma/\lambda)}, \qquad [7.69]$$

where $s = \rho_t/\rho_1$. Setting $\lambda = 1$, we obtain the expression for A_m obtained earlier for an isotropic model. First, consider the potential on the borehole axis. From Eqs. [7.67], we have

$$U_1(0,L) = \frac{\rho_1 I}{4\pi L} + \frac{\rho_1 I}{2\pi^2}\int_0^\infty A_m \cos mL\, dm, \qquad [7.70]$$

where L is the length of the potential probe. When the probe is small, the borehole dominates and the effect of anisotropy in the formation is negligible. To study long probes, we use the asymptotic expansion in which Eq. [7.70] is mainly defined by small values of m. The expansions for the modified Bessel functions give two leading terms,

$$A_m \approx \frac{sK_0(ma/\lambda)(1/ma) - (1/ma)K_0(ma)}{1/ma} = sK_0(ma/\lambda) - K_0(ma). \qquad [7.71]$$

The terms can be evaluated analytically with the Sommerfeld integral,

$$\frac{\rho_1 I}{2\pi^2}\int_0^\infty A_m \cos mL\, dm \to s\frac{\rho_1 I}{4\pi(L^2 + a^2/\lambda^2)^{1/2}} - \frac{\rho_1 I}{4\pi(L^2 + a^2)^{1/2}}, \quad L \to \infty.$$

Since $L \gg a$, this simplifies to $(s-1)(\rho_1 I/4\pi L)$. Substituting into Eq. [7.70] yields

$$U_1(r,z) \to \frac{\rho_t I}{4\pi L}, \quad \text{if } L \gg a. \qquad [7.72]$$

Thus, in the limit the field is defined by the transverse resistivity only, and is independent of the coefficient of anisotropy. In fact, this holds as soon as the probe length is several times larger the borehole radius. For example, consider the function $\rho_a/\rho_a(s,\lambda,L/a)$, shown in Fig. 7.13. As expected, there is only a narrow, intermediate range of separations where the influence of anisotropy is noticeable.

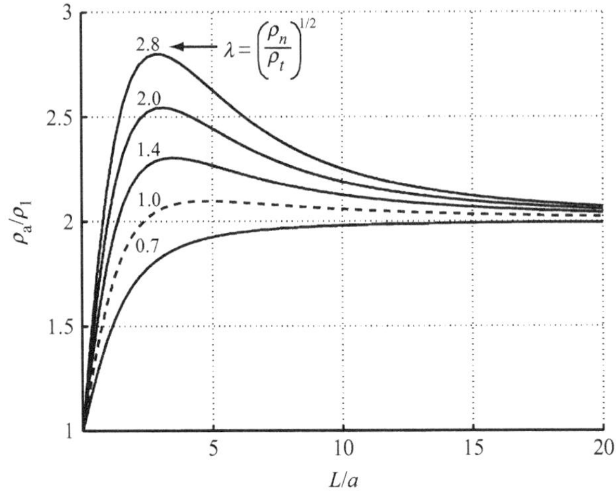

Figure 7.13 Apparent resistivity curves for normal probe. λ is the index of curves.

7.6. RESISTANCE R_{MN} AND ITS MEANING

7.6.1. Resistance R_{MN} as a Sum of Two Terms

We showed earlier that, as the formation resistivity ρ_1 decreases, the apparent resistivity curves for normal and lateral probes (on the borehole axis) approach their asymptotes only at relatively long probe lengths. In other words, to reduce the effects of the borehole and invaded zone in such cases, it is necessary to use very long probes. As usual, this effect can be understood from the distribution of induced charges: the density of induced charges on the cylindrical boundaries decays slowly with distance away from the current source, and only at large distances does their field approach that of an equivalent point charge at the current electrode. Let us discuss this subject from the point of view of the resistance R measured by the normal or lateral probe. Recall that R is the resistance of the region confined by equipotential surfaces passing through the receiver electrodes. In general, this region is not uniform and includes parts of the borehole, invaded zone, and formation. First, consider the distribution of equipotentials and electric field lines when invasion is absent (Fig. 7.14). The former are almost parallel to the borehole axis through an intermediate range of distances, whereas in the near and far zones they are quasi-circular. Focus now on the region confined by equipotential lines passing through the receiver electrodes M and N (Fig. 7.15A). The portion inside the borehole has resistance R_1 is equal to

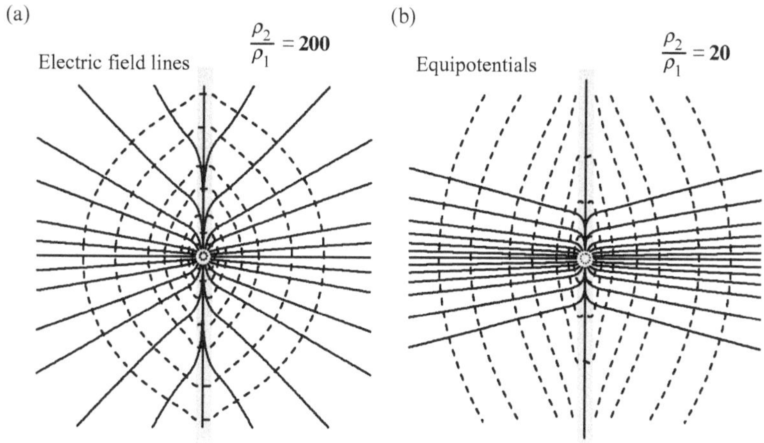

Figure 7.14 Distribution of equipotential lines and vector lines of electric field.

$$R_1 = \rho_1 \frac{MN}{\pi a^2} = \frac{MN}{S_1},$$

since the electric field and current flow is practically parallel to the borehole. In this case, the resistance is defined by only ρ_1 and the borehole geometry. The corresponding voltage between the receiver electrodes is

$$V_{MN} = I_1 R_1,$$

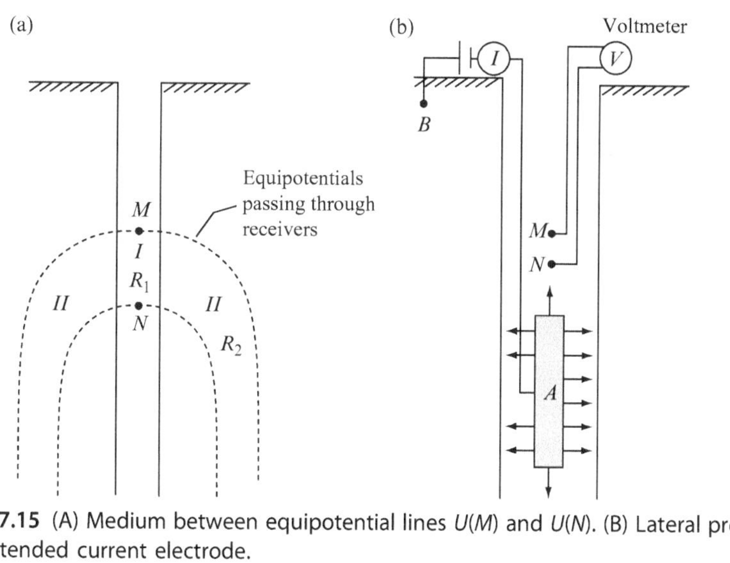

Figure 7.15 (A) Medium between equipotential lines $U(M)$ and $U(N)$. (B) Lateral probe with extended current electrode.

where I_1 is the current going through borehole (region 1). This voltage can also be written as

$$V_{MN} = I_2 R_2,$$

where I_2 is the current through region 2 of this configuration, and R_2 is its resistance. By definition, the total current from the electrode A is the sum

$$I = I_1 + I_2.$$

Applying a rather intuitive approach, assume that the internal and external portions 1 and 2 are connected in parallel, that is,

$$\frac{1}{R} = \frac{1}{R_1} + \frac{1}{R_2}, \qquad [7.73]$$

where the second term obviously depends on the resistivity of the medium surrounding the borehole. In the near and intermediate zones, the first term plays the dominant role; that is, $R_2 \gg R_1$, but in the far zone $R_2 \ll R_1$. Note that, unlike R_1, the resistance R_2 will generally depend on the current distribution. Therefore, in order to reduce the borehole effect, it is necessary to create a current field such that the resistance of region 2 will be smaller than the resistance of region 1 and most of the current will flow through region 2:

$$R_2 \ll R_1.$$

Using this and the previous equalities, the same result can represented slightly differently. Since

$$I_1 R_1 = I_2 R_2 = (I - I_1) R_2 = (I - I_2) R_1,$$

Equation [7.73] becomes

$$\frac{1}{R} = \frac{1}{R_2} \frac{I}{I - I_1} \quad \text{or} \quad \frac{1}{R} = \frac{1}{R_1} \frac{I}{I_1}. \qquad [7.73a)$$

This relation shows that when the current I flowing through region 1 (the borehole path between receiver electrodes) is reduced, the influence of the surrounding medium becomes larger. An elementary current electrode in a uniform medium initially creates a uniform distribution of current in all directions. In the presence of a conductive borehole, the secondary field tends to concentrate the current in the borehole (toward the path of least resistance) and the current I_1 becomes stronger. In effect, the measurement becomes more sensitive to the borehole effect. Thus, one way to avoid this effect is to create a primary electric field whose direction is mainly perpendicular to the borehole. There are several ways to accomplish this task: one is based on the use of elongated current electrodes; another uses a system of electrodes that actively directs, or "focuses," the primary electric field in the direction perpendicular to the borehole. The second type of device is called a "laterolog" after a design proposed by Henri Doll in 1948, which became a standard device for logging oil wells for more than 50 years. Before we discuss how to realize this goal, let us use Eq. [7.73a] to demonstrate the relation between the depth of investigation of the lateral and Schlumberger sounding and the distance of observation point from the current electrode. As an example, suppose that the gradient probe is located at the borehole axis and the invasion zone is absent. Near the current electrode, the medium bounded by equipotential surfaces $U(M)$ and $U(N)$ is practically inside the borehole and, therefore, $I = I_1$. As follows from Eq. [7.73a] the measured resistance R is almost equal to R_1, which depends on the borehole resistivity ρ_1. With an increase of the distance from the current electrode, the leakage of the current into a formation becomes more noticeable, and at large distances $I_1 \approx 0$. This means that the measured resistance is close to R_1, and we obtain information about the resistivity of the surrounding medium ρ_2. The same behavior is observed in the presence of the invasion zone as well as in the case of a horizontally layered medium. Now, we return to the main subject and show that there are at least two ways that allow reduction of the influence of the borehole and invasion zone, using relatively short probes. As was pointed out, one of them is based on measuring grounding resistance of the elongated current electrode, while the other applies a system of current electrodes which performs a focusing of the primary electric field. Now, we consider devices that implement these approaches, but at the beginning outline one possible probe, which is still based on measuring the voltage between the receiver electrodes.

7.6.2. Lateral and Normal Probes with an Extended Current Electrode

Suppose that a lateral or normal probe located on the borehole axis uses a long metal tube as the current electrode A (Fig. 7.15B). Since the metal surface is almost equipotential, the electric field is directed radially along its lateral surface and, therefore, flows mainly into the formation. There is a small leakage of the current into the borehole at the top and bottom of the electrode, but this will generally be much smaller than in the case of a spherical or a short cylindrical electrode. For this reason, we expect

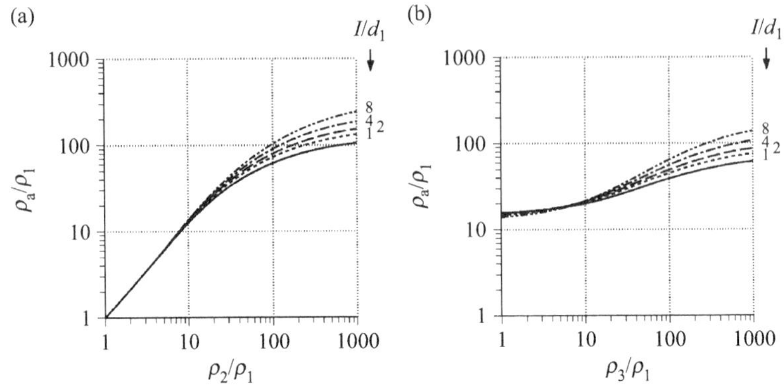

Figure 7.16 Apparent resistivity curves for conventional lateral probe (dashed line) and for the probe with extended current electrode (solid line), $L/d_1 = 10$. (A) Borehole and formation. (B) Invaded zone is present, $\rho_2/\rho_1 = 20$, $d_2/d_1 = 4$.

that the influence of the borehole will be reduced in this arrangement, especially at distances comparable with the electrode length. When the distance between the receiver electrodes and the current electrode is much greater than the probe length, the current electrode should behave as a point source. For illustration, consider the apparent resistivity curves for the lateral probe for different values of the ratio $l/d_1 = 1,2,4$ (Fig. 7.16), where l is the length of the current electrode and the distance between the receiver electrodes and the nearest point of the current electrode stays constant, $l/d_1 = 10$. As the electrode increases in length, the effect of the borehole and invaded zone should decrease; in fact, the contribution of these regions tends to some constant. Correspondingly, by carrying out soundings with either lateral or normal probes having extended current electrode of different lengths and separations, it may be possible to improve the radial response of this method. A version of this latter approach is used in a modern electrical logging device called the high-resolution laterolog array (HRLA).

7.7. MEASUREMENTS OF GROUNDING RESISTANCE

So far, we have considered normal and lateral probes where the current and receiver lines are separated. We now discuss a series of probes that measure the grounding resistance of a current electrode. It turns out that such a transition allows one to greatly reduce the influence of a conductive borehole and invaded zone. In fact, laterolog probes can be viewed as

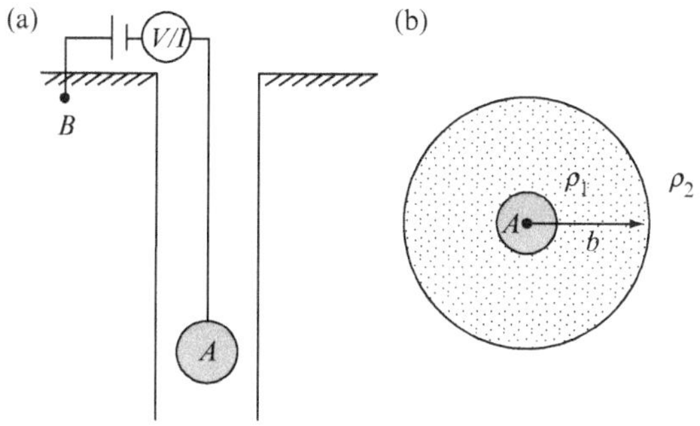

Figure 7.17 (A) Single electrode probe measuring grounding resistance. (B) Grounding resistance of a medium with one spherical interface.

measuring the grounding resistance of extended current electrodes. The simplest two-electrode array that measures the grounding resistance is shown in Fig. 7.17A. From Ohm's law, the current in this array is

$$I = \frac{\Xi}{R_\mathrm{i} + R_\mathrm{g}},$$

where Ξ is the electromotive force, R_g is grounding resistance of the electrode A which depends on the resistivity of the surrounding medium, and R_i is the combined resistance of the wire, current source, ammeter, and electrode B. The latter can be treated as a constant during a measurement. First, suppose that $R_\mathrm{i} \ll R_\mathrm{g}$.
Then,

$$I \approx \frac{\Xi}{R_\mathrm{g}},$$

that is, the current is directly proportional to the grounding conductivity. In contrast,
when $R_\mathrm{i} \gg R_\mathrm{g}$, we have

$$I = \frac{\Xi}{R_i(1 + R_\mathrm{g}/R_\mathrm{i})} \approx \frac{\Xi}{R_\mathrm{i}} - \frac{\Xi}{R_\mathrm{i}^2} R_\mathrm{g} \quad \text{or} \quad \Delta I \approx C R_\mathrm{g},$$

since Ξ/R_i is a constant and can be compensated. Thus, unlike the previous case, the current is proportional to the grounding resistance of the surrounding medium. Note that this method of logging was used in mining geophysics and in the early years of oil exploration. To see the importance of measuring grounding resistance, consider the simplest possible example of a spherical electrode of radius a surrounded by a medium with one spherical interface of radius b (Fig. 7.17B). As shown in Chapter 2, the grounding resistance in this configuration is

$$R = \frac{\rho_1}{4\pi a}\left(1 - \frac{a}{b}\right) + \frac{\rho_2}{4\pi b} = \frac{\rho_2}{4\pi a}\left[\frac{\rho_1}{\rho_2}\left(1 - \frac{a}{b}\right) + \frac{a}{b}\right].$$

Assuming that $\rho_1 \ll \rho_2$ and radius b is only several times larger than a, we have

$$R \approx \frac{\rho_2}{4\pi b},$$

that is, the grounding resistance becomes directly proportional to the resistivity of the external medium. Of course, this result cannot be applied directly to the case where a spherical electrode is placed on the borehole axis because the currents will no longer be purely radial. The example nevertheless illustrates the efficiency of this approach in reducing the influence of a conductive medium surrounding the probe. For instance, the apparent resistivity measured by a very short lateral probe would be mainly defined by the borehole resistivity. Now let us discuss how actual laterolog probes work.

7.8. THE LATEROLOG 1

7.8.1. The Laterolog 1 in a Uniform Medium

First, consider a probe that consists of an elongated metal shell of length l (electrode) with external radius b (Fig. 7.18A). Near the middle of the electrode, there is a small contact that feeds current to the electrode from an external source. Since all portions of the metal current electrode must be at (nearly) the same potential, the induced charge at its surface will be distributed in such a way that its electric field is mainly perpendicular to the shell. Of course, there will also be some leakage of current along the borehole. The probe measures the potential U of the electrode along with the current I, from which the grounding resistance R is calculated:

$$R = \frac{U}{I}. \qquad [7.74]$$

By definition, this is the resistance of the region between the equipotential surface of the electrode and infinity (zero potential). To find the relation between the grounding resistance and the resistance of the surrounding

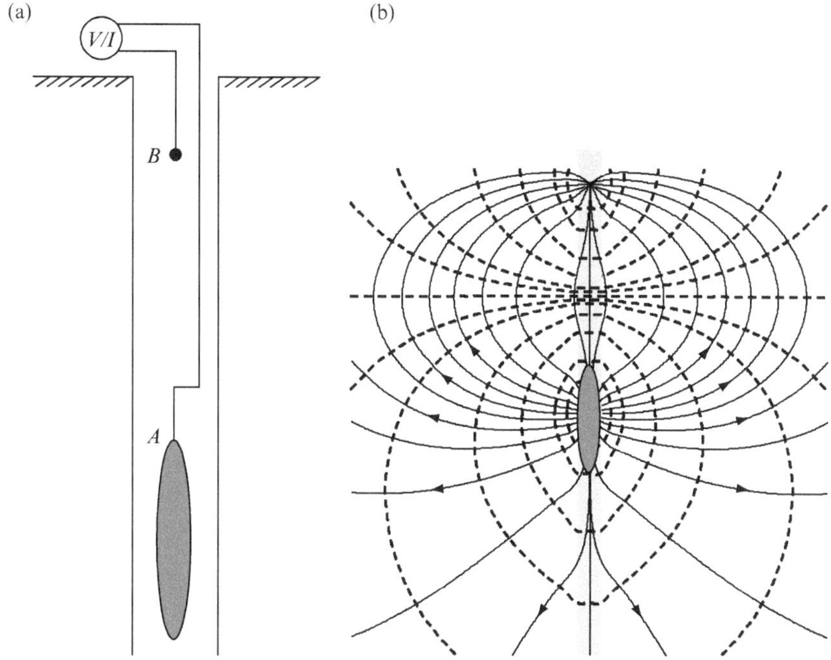

Figure 7.18 (A) Laterolog 1. (B) Equipotential lines and vector lines of electric field.

medium, we assume that the long probe can be treated as an elongated spheroid with semi-axes a and b, with $a = (l/2) \gg b$. Using the results of Section 3.5, we have for a uniform medium

$$R = \frac{\rho}{8\pi c} \quad \ln \frac{a + (a^2 - b^2)^{1/2}}{a - (a^2 - b^2)^{1/2}}, \qquad [7.75]$$

where $c = (a^2 - b^2)^{1/2}$. Assuming that the electrode is highly elongated $(a \gg b)$, we have

$$(a^2 - b^2)^{1/2} \approx a - \frac{b^2}{2a},$$

which gives (see Eq. [3.78])

$$\frac{U}{I} = R = \frac{\rho}{4\pi a} \quad \ln \frac{2a}{b}. \qquad [7.76]$$

From this equation, it follows that the formula for the apparent resistivity is

$$\rho_a = k\frac{V}{I},$$

where the probe coefficient is

$$k = \frac{4\pi a}{\ln(2a/b)}. \qquad [7.77]$$

7.8.2. A Medium with Ellipsoidal Interfaces

To determine the relationship between grounding resistance and the parameters of the surrounding medium, it is necessary to solve a more complicated model problem than in the case when the field is caused by an elementary source. This is because the distribution of charge on the surface of an extended electrode depends on the geoelectric section, which is not known in advance. Thus, we cannot just use the formulas for the potential of a point charge along the borehole axis and apply the principle of superposition. To understand this situation better, we will reexamine the model problem studied in Chapter 3 of a nonuniform medium whose interfaces are ellipsoidal surfaces. Although unrealistic for the borehole geometry, it nevertheless illustrates some important aspects of grounding resistance. Recall that, in this model, the coordinate surface $\xi_1 =$ constant in ellipsoidal coordinates is the boundary between media with resistivity ρ_1 and ρ_2. Before solving for the potential and current through the electrode, let us discuss the distribution of surface charges. As usual, the total charge on the electrode surface is

$$e_A = \varepsilon_0 \rho_1 I, \qquad [7.78]$$

while at the interface ξ_1 we have for the density of charge:

$$\sigma = \varepsilon_0(\rho_2 - \rho_1) j_\xi.$$

Thus, the total surface charge at the interface is

$$e_1 = \varepsilon_0(\rho_2 - \rho_1)I, \qquad [7.79]$$

while the total charge in the medium is

$$e = e_A + e_1 = \varepsilon_0 \rho_2 I,$$

which coincides with the charge on an electrode located in a uniform medium with resistivity ρ_2. Next, we write expressions for the potential in each medium. As shown in Chapter 3, the potential in this configuration depends on only one coordinate, the quasi-radial coordinate ξ, and the relevant solutions of Laplace equation are Legendre functions of zero order:

$$U_1(\xi) = A_0 + A_1 Q_0(\xi), \quad \text{if } \xi_0 \le \xi \le \xi_1$$

and [7.80]

$$U_2(\xi) = B_1 Q_0(\xi), \quad \text{if } \xi \ge \xi_1.$$

These functions obey Laplace's equation and the condition at infinity. The boundary conditions at the interface (continuity of the potential and normal current) determine the unknown constants through the system:

$$U_0 = A_0 + A_1 Q_0(\xi_0), \quad A_0 + A_1 Q_0(\xi_1) = B_1 Q_0(\xi_1),$$
$$\gamma_1 A_1 Q'_0(\xi_1) = \gamma_2 B_1 Q'_0(\xi).$$

We have

$$A_0 = \frac{[(\rho_2/\rho_1) - 1] Q_0(\xi_1) U_0}{D}, \quad A_1 = \frac{U_0}{D}, \quad B_1 = \frac{\rho_2}{\rho_1}\frac{1}{D}U_0, \qquad [7.81]$$

where

$$D = Q_0(\xi_0) + \left(\frac{\rho_2}{\rho_1} - 1\right) Q_0(\xi_1). \qquad [7.82]$$

Thus, potential at any point has now been expressed in terms of the potential at the probe surface and the medium parameters. Next, we find a relationship between the current I and potential U_0, which will in turn determine the resistance of the electrode. First, let us calculate the normal component of the current density at the surface of the probe. As follows from Section 3.5,

$$j_\xi = -\gamma_1 \frac{\partial U_1}{h_\xi \partial \xi} = -\gamma_1 \frac{U_0}{h_\xi D} Q'_0(\xi_0).$$

Thus, from Chapter 3, we have

$$I = \frac{4\pi c U_0}{\rho_1 D}, \qquad [7.83]$$

and the grounding resistance of the probe is

$$R = \rho_1 \frac{D}{4\pi c} = \frac{\rho_1}{4\pi c}\left[Q_0(\xi_0) + \left(\frac{\rho_2}{\rho_1} - 1\right) Q_0(\xi_1)\right], \qquad [7.84]$$

which in a uniform medium coincides with Eq. [7.76].

7.8.3. Geometric Factors of Uniform Regions

We can represent Eq. [7.84] in the form

$$R = \frac{1}{4\pi c}[\rho_1[Q_0(\xi_0) - Q_0(\xi_1)] + \rho_2 Q_0(\xi_1)], \qquad [7.85]$$

which conveniently shows that the grounding resistance is a sum of two resistances, $R_1 = \rho_1 N_1$ and $R_2 = \rho_2 N_2$, connected in series. Here, the terms

$$N_1 = Q_0(\xi_0) - Q_0(\xi_1) \quad \text{and} \quad N_2 = Q_0(\xi_1) \qquad [7.86]$$

are geometric factors for each uniform region of the configuration. Since $Q_0(\xi)$ tends to zero if $\xi \to \infty$, the function N_2 can also be written as

$$N_2 = Q_0(\xi_0) - Q_0(\infty).$$

In general, the geometric factor for any uniform region is written in the form

$$N_n = Q_0(\xi_{n-1}) - Q_0(\xi_n). \qquad [7.87]$$

In particular, when an intermediate layer is present, we have

$$R = R_1 + R_2 + R_3, \qquad [7.88]$$

or

$$R = \frac{1}{4\pi c}\{\rho_1[Q_0(\xi_0) - Q_0(\xi_1)] + \rho_2[Q_0(\xi_1) - Q_0(\xi_2)] + \rho_3 Q_0(\xi_2)\}.$$

Then, it follows from Eq. [7.77] that the apparent resistivity is

$$\rho_a = \frac{1}{\ln(2a/b)}(\rho_1 N_1 + \rho_2 N_2 + \rho_3 N_3). \qquad [7.89]$$

This formula clearly shows that the apparent resistivity is the sum of the resistivities of each uniform piece of the medium weighted by a factor that depends only on geometry. Such a simple result was not obvious in advance. In fact, as we know, the density of charge induced at interfaces itself depends on the distribution of resistivity (which determines the current paths), and therefore it is natural to expect that the formula would mix different resistivities in individual terms. This model is an exception because of its high degree of symmetry: the presence of interfaces does not distort the geometry of equipotential lines of the primary electric field. Moreover, charges on a coordinate surface ξ = constant do not create an electric field inside the surface. The electric field in each uniform region, thus, becomes directly proportional to its resistivity. In actual conditions, the geometry of the total field strongly differs from that of the primary field, and equipotentials will cross boundaries between different media, including the borehole wall and the limits of the invaded zone. In general cases, the concept of a pure geometrical factor multiplying the resistivities of each uniform region to give the apparent resistivity no longer holds. Instead, one has to solve the boundary-value problem for Laplace's

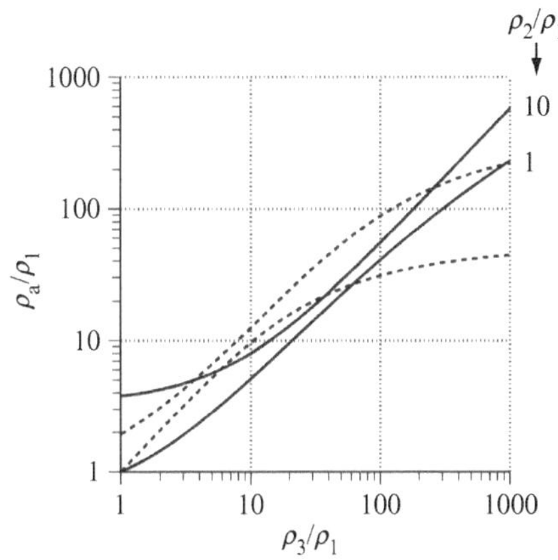

Figure 7.19 Apparent resistivity curves for the laterolog (solid line) and gradient probe of the same length (dashed line), $L/d_1 = 10$, $d_2/d_1 = 4$.

equation in the actual configuration, usually using numerical methods. Examples of such modeling are shown in Fig. 7.19. Comparison of the apparent resistivity curves for a laterolog with those of a gradient probe of the same length shows that measuring grounding resistance of the extended electrode (Laterolog 1) allows one to reduce the influence of the borehole and a low-resistivity invaded zone.

7.9. LATEROLOG 3

We now show that it is possible to improve the response of a logging probe in the vertical and radial directions by considerations based on the distribution of equipotential surfaces and electric field lines near the borehole axis (Fig. 7.18B). It is clear that there is a region near the central part of an extended probe where the electric field lines are nearly perpendicular to the borehole. Let I_0 be the current flowing from this part of the probe. Of course, at large distances from the borehole, these lines will eventually bend. The ratio

$$R = \frac{U}{I_0} \qquad [7.90]$$

approximately defines the grounding resistance of a cylindrical disc of a finite thickness, whose inner annular surface surrounds the central part of the probe and whose outer annulus is at infinity. By analogy with Eq. [7.88], R is the sum of resistances of the borehole, invasion zone, and formation, respectively. It is obvious that, as the borehole resistivity ρ_1 decreases (the borehole fluid becomes more conductive), the contribution of the borehole resistance becomes smaller, whereas the influence of formation decreases when the resistivity in the intermediate (invaded) zone increases. A simple probe, called the Laterolog 3, for measuring the current I_0 and grounding resistance R consists of three electrodes (Fig. 7.20A). The central electrode is shorter compared to the long outer electrodes, called "guard electrodes." At the reference point C, the current I moves directly to the guard electrodes, which are held at the same potential. There is a known but small resistance R_s between C and the midpoint of each guard electrodes; this means that the potential of the central and lateral electrodes is practically the same. The probe measures the potential U as well as the voltage $V = I_0 R_s$ which determines I_0. Near the central electrode, the current lines are nearly radial, and this segment of the borehole contributes little to the grounding resistance. At somewhat larger radial distances, the current lines tend to become horizontal because of the symmetric action of charges on the guard electrodes. Finally, at larger distances from the probe, the field lines start to bend vertically and move toward the electrode N at the earth's surface (Fig. 7.20B). Intuitively, we expect that parts of the medium far away from the probe give a relatively small contribution to the grounding resistance, which is mainly defined by region

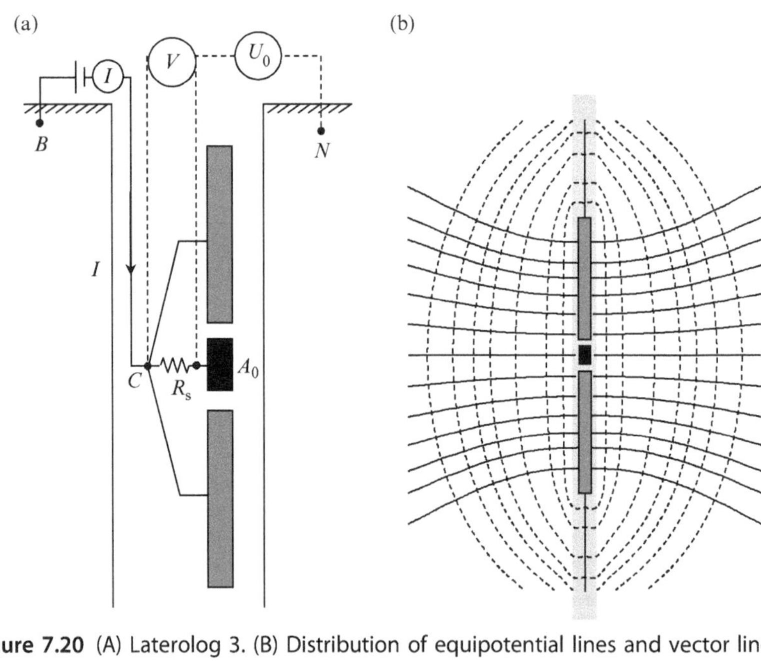

Figure 7.20 (A) Laterolog 3. (B) Distribution of equipotential lines and vector lines of electric field.

in the shape of a cylindrical disc whose height is comparable to length of the central electrode. Note that there are situations where the effect of distant, highly resistive regions is significant. Figure 7.21 compares the response of the Laterolog 3 to that of the gradient probe. It may proper to point out that, while the radial response of the Laterolog 3 is only slightly better than that of a single extended electrode (Laterolog 1), there is substantial improvement in its vertical resolution.

A few other observations are appropriate. By definition, the grounding resistance is given by

$$R = \frac{1}{I_0}\int_a^{\infty} \mathbf{E}\cdot d\mathbf{l}, \qquad [7.91]$$

where a is any point of the electrode A_0, and $\mathbf{E}$ is the electric field caused by all charges in the model, including those on the surface of the central and guard electrodes, as well as induced charges on boundaries between media with different resistivities. Similar expressions can be written for the resistances of each uniform region of the model, for instance, borehole, invaded zone, and formation. Since the field $\mathbf{E}$ at any point in the model itself depends on the distribution of resistivity throughout the model, it is in general impossible to write the grounding resistance as a sum of individual resistances multiplied by pure geometric factors.

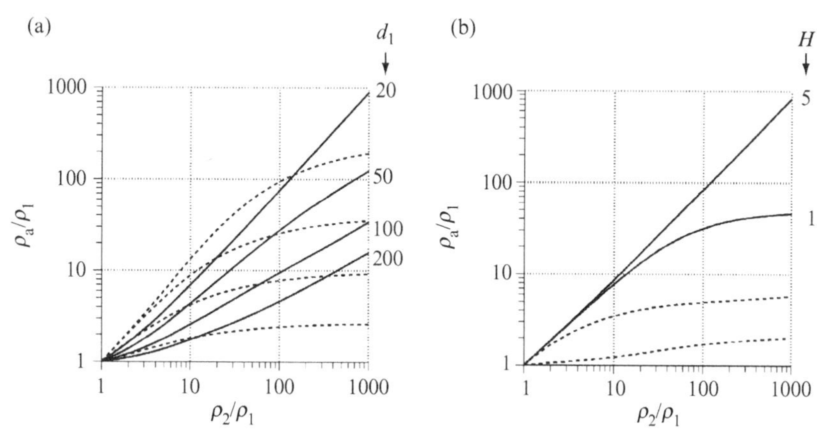

Figure 7.21 (A) Radial responses of Laterolog 3 (solid lines) and gradient probe (dashed lines); d_1 is diameter of cylindrical surface in inches, the probe length $L=100$ in. (b) Vertical responses of Laterolog 3 and gradient probe, located at the middle of layer; H is layer thickness in feet, L is the probe length in inches; ρ_1, ρ_2 are resistivity of the surrounding medium and a layer.

7.10. LATEROLOG 7

Our final example of an electrical probe that focuses current perpendicular to the borehole is the Laterolog 7. Instead of long guard electrodes, this device uses a system of elementary electrodes to perform the measurement (Fig. 7.22A). Three current electrodes are arranged with a central electrode A_0 flanked symmetrically by auxiliary electrodes A_1 and A_2. A pair of receiver electrodes M_1N_1 is located between the central current electrode and A_1; a second pair M_2N_2 is located between the central electrode and A_2. The currents on the three source electrodes can be controlled independently, and of course the charges on these electrodes are the source of the primary electric field. The concept of the Laterolog 7 is as follows. First, imagine that the electrode pair A_0 and A_1 carry equal (positive) charges, and recall from Chapter 1 that this arrangement creates an electric field that is perpendicular to the line connecting the charges at its midpoint. This electric field will generate a current flow that is also perpendicular to the line joining the electrodes. Operating such a probe in a borehole will, however, induce additional charges at interfaces in the surrounding medium; and these secondary charges will distort the equipotentials and lines of current. The Laterolog 7 compensates for this

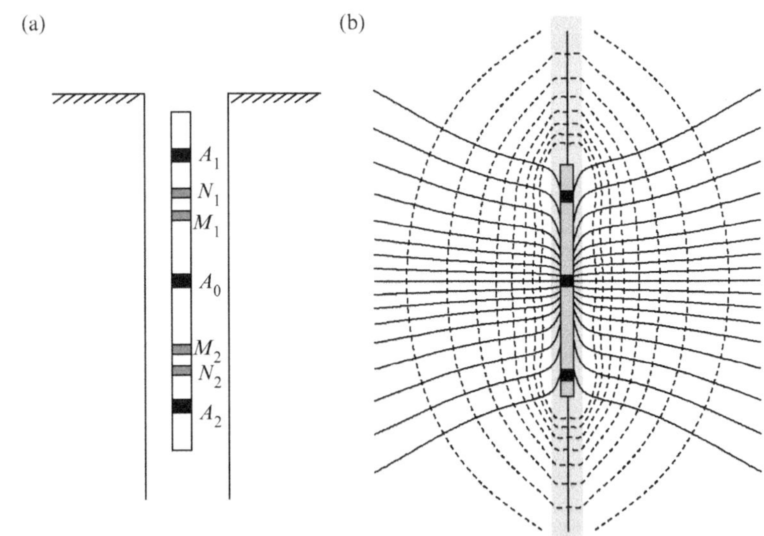

Figure 7.22 (A) Laterolog 7. (B) Distribution of equipotential lines and vector lines of electric field.

distortion by dynamically adjusting the three currents A_0, A_1, A_2 to cancel the potential difference between the pair of receiver electrodes M_1N_1 and between the pair M_2N_2 (Fig. 7.22A). That is, during the measurement, we have

$$V(M_1N_1) = V(M_2N_2) = 0. \qquad [7.92]$$

"Dynamic" adjustment means that conditions [7.92] are enforced at each position of the tool along the borehole before recording the measurement at that position. The conditions themselves imply that the average vertical component of the electric field between each pair of receiver electrodes is zero; this means, in turn, that there is no net current flowing vertically in the borehole at those points. In other words, the net current density is forced in the radial direction in the region between the pairs of monitoring electrodes. Two quantities are measured to determine the resistance of the medium to the forced current flow. One is the current I_0 through electrode A_0, the other is the potential at one of the receiver electrodes, for instance M_1, with respect to an electrode N located far away from the probe ("at infinity"). As usual, the resistance of the measurement is defined by

$$R = \frac{U(M_1)}{I_0}, \qquad [7.93]$$

and can be viewed as the grounding resistance of the current electrode A_0 of the probe. To convert the resistance to apparent resistivity, we need an expression for the resistance in a uniform medium. Since the voltage V_{M_1N1} is zero and the distance between the receiver electrodes is small, we can say that the vertical component of the electric field at the midpoint O between these electrodes is zero. This yields the following condition:

$$\frac{\rho_1}{4\pi}\left[\frac{I_0}{(A_0O)^2} - \frac{I_1}{(A_1O)^2} + \frac{I_1}{(A_2O)^2}\right] = 0,$$

since (positive) charges on electrodes A_0 and A_2 create fields that reinforce each other, and are opposite to field of (positive) charge on electrode A_1. This provides a relationship between currents in a uniform medium:

$$I_1 = \alpha I_0,$$

where

$$\alpha = \left[\left(\frac{A_0O}{A_1O}\right)^2 - \left(\frac{A_0O}{A_2O}\right)^2\right]^{-1}.$$

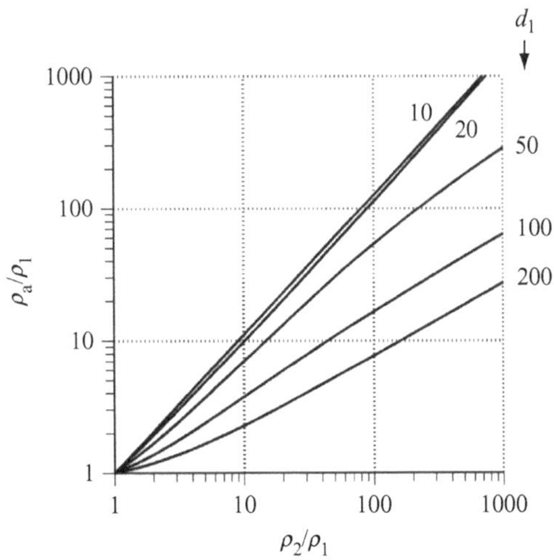

Figure 7.23 Radial responses of Laterolog 7; d_1 is diameter in inches of the cylinder with resistivity ρ_1, and ρ_2 is resistivity of a surrounding medium. Length of the probe is 80 in.

Therefore, the potential at the point M when the probe is surrounded by a uniform medium is

$$U_0(M) = \frac{\rho_1 I_0}{4\pi}\left(\frac{1}{A_0M} + \frac{\alpha}{A_1M} + \frac{\alpha}{A_2M}\right),$$

and the resistance to the current I_0 is equal to

$$R_0 = \frac{\rho_1}{4\pi}\left[\frac{1}{A_0M} + \alpha\left(\frac{1}{A_1M} + \frac{1}{A_2M}\right)\right]. \qquad [7.94]$$

Thus, for the apparent resistivity we have

$$\rho_a = 4\pi\left[\frac{1}{A_0M} + \alpha\left(\frac{1}{A_1M} + \frac{1}{A_2M}\right)\right]^{-1} R. \qquad [7.95]$$

Figure 7.23 shows typical apparent resistivity curves for Laterolog 7.

7.11. ELECTRIC LOGGING THROUGH CASING

A simple but surprising application of the concepts developed so far shows that it is possible to measure the electrical resistivity of geological layers penetrated by a borehole even when the hole is lined with metal casing. In this section, we develop the theory of such measurements.

7.11.1. Three Zones Along the Borehole Axis

We can start from the model studied earlier consisting of the borehole, an intermediate annular region, and a homogeneous external region. In the earlier model, the intermediate region was a model for an invaded zone; here, it will be used to model a highly conductive metal casing lining the hole. Suppose that a current electrode A is placed on the borehole axis (Fig. 7.24A). Electric charges arise at all interfaces, and the electric field **E** caused by these charges strongly depends on the distance from the electrode A. In the near zone, the field **E** has both vertical and radial components and charges appear on the internal and external surfaces of the casing. Usually, this zone extends to a distance of 10–15 borehole radii away from the electrode. But if the current electrode is connected to the casing, this zone practically vanishes. In the intermediate zone, the electric field and current have some important features. In particular, considering the apparent resistivity curves in a medium with cylindrical interfaces, we found that the vertical component of the field **E** is almost constant within this S-zone. Finally, at greater distances in the range called the far zone, the influence of the borehole and casing disappear, and the field is the same as that in a uniform medium with resistivity of the formation outside

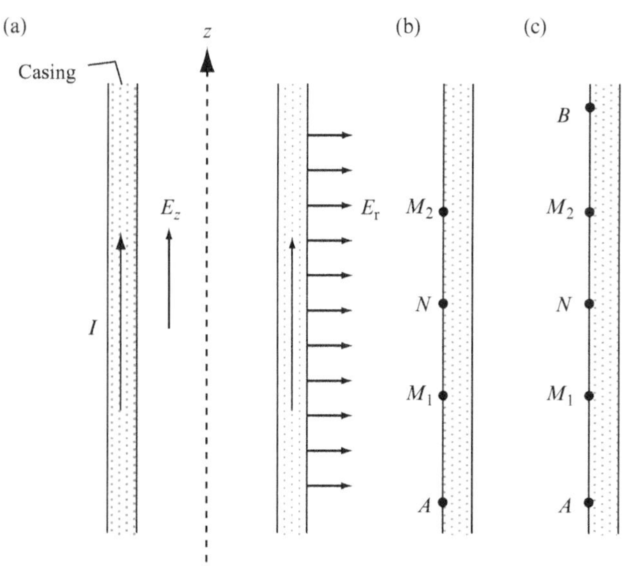

Figure 7.24 (A) Distribution of electric field in borehole, casing, and formation. (B) Probe AM_1NM_2. (C) Measuring of casing conductivity and its change.

casing. We focus on the field behavior in the intermediate zone, because the logging through casing can be performed reliably only within this zone.

7.11.2. Intermediate Zone

The remarkable features of the field in the intermediate zone that allow logging through casing are the following:

1. Inside the borehole and casing, the field **E** is directed along the z-axis; that is, the radial component E_r is negligible. Correspondingly, charges are practically absent on the internal surface of the casing.
2. At each cross section of the borehole and casing:

$$E_z(r, z) = C(z). \qquad [7.96]$$

   If $r \leq r_{max}$, where $C(z)$ is a function of z only, r_{max} is the radius of the external surface of the casing, and z is the distance from the electrode A. In other words, for each cross section the electric field is the same. At the same time, the current flows mainly inside the casing because its conductivity is many orders higher than that of the borehole.
3. Charges on the external surface of the casing mainly generate a radial component of the electric field E_r within the formation. Thus, the current density in the formation near the borehole also has mainly a component j_r in the radial direction. Of course, at relatively large distances from the borehole axis, all field lines will be directed toward the current electrode B located far away.
4. If the formation is conductive, there is leakage of current from the casing into the surrounding medium. This means that the current I flowing through the casing diminishes with distance z.
5. Calculations of the field at the borehole axis for this configuration shows that the vertical component of the electric field within the intermediate zone can be written as

$$E_z \approx \frac{I}{2S} \exp\left(\frac{-z}{\rho S}\right)^{1/2}. \qquad [7.97]$$

   In as much as measurements are performed when $(z/\rho S) \ll 1$, this equation can be reduced to

$$E_z \approx \frac{I}{2S} - \frac{Iz}{2S^{3/2}(\rho)^{1/2}}. \qquad [7.98]$$

6. Since the electric field outside the casing has mainly a radial component E_r, charges do not arise on horizontal interfaces between layers with

different resistivities. In reality, of course, surface charges appear at these interfaces at distances far from the borehole; but their influence is correspondingly small. This radial flow greatly improves the vertical response of this method. Such behavior of the electric fields and current density suggests that there is an intermediate zone where the casing can be treated as a transmission line. This analogy allows one in a very simple manner to define quantities that can be used to determine the resistivity of a surrounding medium in the presence of extremely conductive casing.

7.11.3. Distribution of Potential, its Derivatives, and Current Along the Transmission Line

Start with the current electrode A placed at the origin (Fig. 7.24). Let the potential along the casing at depth z be $U(z)$. From Ohm's law, the differential change of the potential $\mathrm{d}U$ over the vertical distance $\mathrm{d}z$ is

$$\mathrm{d}U = -\frac{I(z)\mathrm{d}z}{S}, \qquad [7.99]$$

where $I(z)$ is the current flowing vertically through the casing cross section at depth z, and S is the casing conductance, so that $\mathrm{d}z/S$ is the resistance of the casing element. The negative sign appears in Eq. [7.99] since our convention is that positive current flows from the electrode, so that the potential is dropping in the direction of current flow,

$$\mathrm{d}U = U(z + \mathrm{d}z) - U(z) \;\text{ and }\; U(z + \mathrm{d}z) < U(z).$$

The amount of current in the casing decreases with distance from the electrode because of leakage into the medium. We can represent this in the transmission-line model by adding a radial current path through the surrounding medium in parallel with the casing current. Consequently, within the interval $\mathrm{d}z$, the change $-\mathrm{d}I(z)$ in the casing current, caused by leakage, is related to potential U by

$$U = -\frac{T}{\mathrm{d}z}\mathrm{d}I, \qquad [7.100]$$

where T is the resistance per unit length $\mathrm{d}z$ of the medium for the radial current path. In writing Eq. [7.100], we assume that the radial current leaks off "to infinity" where the potential is zero. In particular, any coaxial cylindrical surface of sufficiently large radius has practically zero potential. We can write Eqs. [7.99] and [7.100] as the coupled set of first-order equations,

$$\frac{dU}{dz} = -\frac{I(z)}{S} \text{ and } \frac{dI}{dz} = -\frac{U(z)}{T}. \qquad [7.101]$$

Differentiation of these equations leads to uncoupled second-order equations for the current and potential,

$$\frac{d^2 I}{dz^2} = \alpha^2 I \text{ and } \frac{d^2 U}{dz^2} = \alpha^2 U, \qquad [7.102]$$

where

$$\alpha = \left(\frac{1}{ST}\right)^{1/2}. \qquad [7.103]$$

Solutions of Eqs. [7.102] have the form

$$A \exp(\alpha z) + B \exp(-\alpha z). \qquad [7.104]$$

Thus, the distributions of potential and current within the intermediate zone are defined by the parameter α. Its value is very small, usually within the range

$$10^{-3}\,\text{m}^{-1} < \alpha < 10^{-2}\,\text{m}^{-1}.$$

Since the current in the borehole is negligible, S is the casing conductance,

$$S = \gamma_c 2\pi a \ \Delta a, \qquad [7.105]$$

where γ_c and Δa are the conductivity and thickness of the casing, and a is its radius. The formula assumes that $\Delta a \ll a$. The most useful parameter is T, which is related to the resistivity of the surrounding formation. The theory of transmission lines does not provide a formula for T, which has to be found from a separate analysis. One way is by solving the full boundary-value problem for the field; in the case where the casing sits in a uniform medium of resistivity ρ, this yields

$$T = \rho = \frac{1}{\gamma}. \qquad [7.106]$$

7.11.4. Apparent Resistivity and Geometry of the Probe

Suppose that near depth z the conductivity γ of the surrounding medium does not change in the radial direction. Then, according to Eqs. [7.102] and [7.106], it is given by

$$\gamma = S\frac{U''(z)}{U(z)}, \qquad [7.107]$$

where $U''(z)$ is the second derivative of the potential with respect to depth z. Thus, when the conductivity does vary, we can define an apparent conductivity and apparent resistivity for this configuration by

$$\gamma_a = S\frac{U''(z)}{U(z)} \quad \text{and} \quad \rho_a = \frac{U(z)}{SU''(z)}. \qquad [7.108]$$

It follows from Eq. [7.107] that the resistivity of the formation can be measured with an electrical probe in a borehole, even when the borehole is lined with metal casing. The calculation requires three quantities, namely, the potential U, its second derivative U'' along the casing, and the casing conductance S. The second derivative U'' can be estimated using the usual finite-difference approximation,

$$\Delta^2 U \approx U''(z)(\Delta z)^2, \qquad [7.109]$$

where

$$\Delta^2 U = U(z - \Delta z) + U(z + \Delta z) - 2U(z). \qquad [7.110]$$

The error in this approximation is second order in Δz if this interval is sufficiently small and U is smooth. The function $\Delta^2 U$ can be written as

$$\Delta^2 U = [U(z - \Delta z) - U(z)] - [U(z) - U(z + \Delta z)],$$

or

$$\Delta^2 U(z) = V_1 - V_2, \qquad [7.111]$$

where V_1 is the voltage differences between points $z - \Delta z$ and z, and V_2 is the voltage between z and $z + \Delta z$. Therefore, an approximation for the second derivative of the potential is

$$U''(z) = \frac{\Delta V}{(\Delta z)^2}, \qquad [7.112]$$

where

$$\Delta V = V_1 - V_2. \qquad [7.113]$$

In practice, the quantity measured is ΔV, which is the difference of voltages going in opposite directions from a point z along the casing. Thus, determination of the second derivative $U''(z)$ by Eq. [7.112] requires three electrodes M_1, N, and M_2, placed at equal distances from each other (Fig. 7.24B),

$$M_1N = M_2N = \frac{l}{2}, \qquad [7.114]$$

where l is the distance between electrodes M_1 and M_2. These three electrodes form the main part of the probe of the length l, which can move along the borehole, while the current electrode remains in the same position. Substitution of Eq. [7.112] into Eq. [7.108] gives

$$\gamma_a = \frac{4S}{l^2}\frac{\Delta V}{U(N)} \quad \text{or} \quad \gamma_a = K\frac{\Delta V}{U(N)}, \qquad [7.115]$$

where

$$K = \frac{4S}{l^2} \qquad [7.116]$$

can be called the probe coefficient. Thus, in order to compute the apparent conductivity, it is necessary to know the casing conductance S and to measure the voltage difference ΔV, as well as the potential of the middle electrode N. In principle, instead of measuring ΔV, it is also possible to determine voltages V_1 and V_2 separately and then calculate ΔV. Replacing the second derivative $U''(z)$ by the function ΔV assumes that the probe length l is so small that the electric field E_z decreases linearly between electrodes M_1 and M_2. In other words, the approximation implies that the function $U''(z)$ does not change between the electrodes. Also, the use of Eq. [7.115] supposes that the parameter S is known and does not change between electrodes M_1 and M_2. It is clear that if these conditions are met, Eqs. [7.108] and [7.115] are equivalent.

7.11.5. Vertical Response of the Probe

The analysis above implicitly assumes that the medium surrounding the casing has only cylindrical boundaries. Next, consider the case when the probe is placed against a layer of finite thickness. This situation can be handled approximately in the transmission-line model by allowing the parameter α to vary. This is natural because α can be expressed in terms of $U''(z)$ and $U(z)$ at the same point, independent of its behavior at other points. Let us illustrate this by assuming that the probe is located inside the casing opposite a layer of thickness H. Taking into account Eq. [7.104], we can represent the potential along the casing as

$$U(z) = A\exp(\alpha z) + B\exp(-\alpha z),$$

where A and B are unknown and depend on several factors including the behavior of the parameter α as a function of z. The same is true for the second derivative

$$U''(z) = \alpha^2[A\exp(\alpha z) + B\exp(-\alpha z)].$$

The ratio U''/U, thus, allows one to eliminate the influence of unknowns and obtain the parameter α and, therefore, the conductivity γ of the layer. It is obvious that this also holds when the probe M_1NM_2, which approximates the second derivative, is placed against the casing opposite a layer whose thickness H is greater than the probe length, that is, $l < H$. Certainly, this is a remarkable feature of the measurement and illustrates its high vertical resolution. One can say that the presence of the casing makes the probe act like a laterolog with guard electrodes of enormous extension.

7.11.6. Influence of Changes in Casing Conductance

Until now, we have assumed that the casing conductance S does not change between the receiver electrodes of the probe. In real conditions, S does vary, although usually not more than a few percent. These small changes mean that the resistances of the casing between the electrodes M_1N and M_2N,

$$R_1 = \int_{M_1}^{N} \frac{dz}{S(z)} \quad \text{and} \quad R_2 = \int_{N}^{M_2} \frac{dz}{S(z)}, \qquad [7.117]$$

are different. Therefore, the voltages V_1 and V_2 can differ from each other even when leakage current is absent. For this reason, the difference

$$\Delta V = V_1 - V_2$$

is, in general, defined by two factors: leakage of current from the casing, which depends on the formation resistivity, and changes in the casing resistance between the receiver electrodes. The latter represents "noise" and has to be removed. Consider for this purpose an array consisting of the probe M_1NM_2 and two current electrodes A_1, B_1, situated symmetrically with respect to the electrode N (Fig. 7.24C). In this configuration, the difference of voltages V_1^* and V_2^* appears only when resistances R_1 and R_1 are not equal. In reality, there is still a leakage of current, which also causes a voltage difference but can be neglected since

$$\frac{\Delta V^l}{\Delta V^*} \approx \alpha l <<< 1. \qquad [7.118]$$

Here, ΔV^l is the voltage due to current leakage with array $A_1M_1NM_2B_1$. Thus, by measuring voltages V_1^* and V_2^* with the probe $A_1M_1NM_2B_1$, it is possible to compensate for changes in the casing resistance. In fact, we have

$$\frac{V_1^*}{V_2^*} = \frac{R_1}{R_2}.$$

Therefore, the voltage difference ΔV, caused directly by leakage of current into the formation, is given by

$$\Delta V = V_1 - \frac{V_1^*}{V_2^*} V_2. \qquad [7.119]$$

For this reason, an array for measuring formation resistivity through the casing normally consists of two parts, AM_1NM_2 and $A_1M_1NM_2B_1$. The second part performs a correction for changes of casing conductance simultaneously with the measurement of formation resistivity. Of course, the values of resistances R_1 and R_2 are very close to each other, but this small difference still creates a signal comparable to or much greater that that caused by the leakage current, and, therefore, must be reduced for an accurate measurement of formation resistivity. For illustration, let us estimate the size of both signals. As follows from Eq. [7.98], the voltages due to current leakage and differences of casing conductance are

$$\Delta V = \frac{I}{8S} \frac{l^2}{(S\rho)^{1/2}} \quad \text{and} \quad \Delta V^* = \frac{I\Delta S}{4S^2} l,$$

respectively. Thus,

$$\frac{\Delta V^*}{\Delta V} \approx \frac{2\Delta S}{S\alpha l}.$$

Letting $\alpha = 3 \times 10^{-3}\ \text{m}^{-1}$, $l = 1$ m, and $\Delta S/S = 0.02$ gives

$$\frac{\Delta V^*}{\Delta V} > 10,$$

that is, the noise greatly exceeds the useful signal.

7.11.7. Determination of Casing Conductance

The probe $A_1M_1NM_2B_1$ plays another important role. As follows from Eqs. [7.115] and [7.116], the determination of formation resistivity requires knowledge of the casing conductance S. This is usually calculated by measuring either voltage V_1^* or V_2^* and current I^*. Indeed, by definition we have

$$V_1^* \approx \frac{I^*}{S} M_1N \quad \text{or} \quad V_2^* \approx \frac{I^*}{S} M_2N.$$

Thus,

$$S \approx \frac{I^* M_1 N}{V_1^*} \approx \frac{I^* M_2 N}{V_2^*}, \qquad [7.120]$$

where I^* is the current in the line A_1B_1. In principle, the conductance S can be determined by other methods, including an inductive measurement with low-frequency alternating currents.

7.11.8. Influence of Finite Length of Casing

At the beginning, it was assumed that the casing is infinitely long. To study the influence of a finite length of casing, consider the distribution of the potential U and its derivatives along the casing of finite length L. Also, we suppose that the current electrode A is placed at one end of the casing, while at the opposite end of the casing the current vanishes:

$$I(0) = I_0 \quad \text{and} \quad I(L) = 0. \qquad [7.121]$$

This condition implies that all current I_0 eventually flows into the surrounding medium through the surface of the casing. Initially, we will assume that the surrounding medium is uniform. Then, Eqs. [7.104] and [7.121] give

$$A + B = I_0 \quad \text{and} \quad A\exp(\exp\alpha L) + B\exp(-\alpha L) = 0.$$

Thus,

$$I(z) = \frac{I_0}{\sinh(\alpha L)} \sinh[\alpha(L - z)]. \qquad [7.122]$$

Correspondingly, the total radial current I_r flowing into the formation between the receiver electrodes M_1N or M_2N is

$$I_r(z) = -\frac{\partial I(z)}{\partial z}MN = \frac{\alpha I_0 MN}{\sinh(\alpha L)}\cosh[\alpha(L-z)], \qquad [7.123]$$

where $M_1N = M_2N = MN$, and z is the midpoint of each pair of electrodes. Since $U = I_r T$, we have

$$U(z) = \frac{T\alpha I_0}{\sinh(\alpha L)}\cosh[\alpha(L-z)], \quad E_z(z) = \frac{I_0}{S\sinh(\alpha L)}\sinh[\alpha(L-z)]$$

and

$$\Delta^2 U = \frac{I_0 T}{\sinh(\alpha L)}\alpha^3(MN)^2\cosh[\alpha(L-z)], \qquad [7.124]$$

where z is the coordinate of electrode N and, as before, $\Delta^2 U$ is the voltage difference measured with the probe M_1NM_2. It follows from Eqs. [7.124] that, regardless of L and α, the ratio $\Delta^2 U/U$ is directly proportional to formation conductivity. Also, we have

$$\Delta^2 U = \frac{I_r}{S}MN. \qquad [7.125]$$

Now, consider two special cases, $\alpha L \ll 1$ and $\alpha L \gg 1$.

Case 1: $\alpha L \ll 1$

Since the parameter α is very small ($\alpha \approx 10^{-3}\,\mathrm{m}^{-1}$), the condition $\alpha L \ll 1$ is valid even for a relatively long casing. Using the expansions

$$\sinh x \approx x + \frac{x^3}{6}, \quad \cosh x \approx 1 + \frac{x^2}{2},$$

we obtain

$$I(z) \approx I_0\left(1 - \frac{z}{L}\right), \qquad [7.126]$$

showing that the current decreases linearly with distance. Correspondingly, the leakage current is uniformly distributed along the casing. In fact, from Eq. [7.123], it follows that the current I_r between each pair of receiver electrodes is

$$I_r = \frac{I_0 MN}{L}, \qquad [7.127]$$

and is a function of geometrical parameters only. From Eqs. [7.124], we see that the field

$$E_z(z) \approx \frac{I_0}{S}\left(1 - \frac{z}{L}\right) \quad [7.128]$$

decreases linearly with distance in the same manner as current I but the potential U is almost constant:

$$U = \frac{T}{L} I_0, \quad [7.129]$$

and is defined by the resistivity of the medium and casing length. Thus, measuring the potential $U(z_N)$ provides information about the resistivity of the medium, and there is no need to determine the other functions, except I_0. Finally, for the function $\Delta^2 U$, we have

$$\Delta^2 U = \frac{I_0}{SL}(MN)^2, \quad [7.130]$$

which, unlike the potential, is independent of the formation resistivity. Note that in the limiting case of an ideal conductor, surface charges are distributed in such way that the field inside the casing is equal to zero. Therefore, the potential is constant and the voltage difference $\Delta^2 U$ vanishes, but there is still leakage of current into the formation. Thus, if the parameter αL is very small, the casing is similar to the laterolog which measures the grounding resistance.

Case 2: $\alpha L \gg 1$

In this case, Eqs. [7.123] and [7.124] give

$$\begin{aligned} I(z) &= I_0 \exp(-\alpha z), \quad I_r(z) = \alpha I_0 (MN) \exp(-\alpha z), \\ U(z) &= \alpha T I_0 \exp(-\alpha z), \quad E_z = \frac{I_0}{S} \exp(-\alpha z), \end{aligned} \quad [7.131]$$

$$\Delta^2 U = \frac{I_0}{S} \alpha (MN)^2 \exp(-\alpha z), \quad \text{if } z/L \ll 1 \quad [7.132]$$

Thus, all functions decay exponentially with distance from the electrode A. It is obvious that the casing can be treated as infinitely long if $\alpha L > 1$. It is also a simple matter to evaluate the ratio of the radial and vertical components of the electric field in the formation surrounding a borehole. By definition, the z-component of the field is a continuous function, and, therefore, outside the casing, we have

$$E_z(a) = \frac{I_0}{S} \exp(-\alpha z).$$

The area of the casing surface between electrodes M and N is equal to $2\pi aMN$, so that the radial component of the current density is

$$j_r = \frac{\alpha I_0}{2\pi a}.$$

Therefore,

$$E_r = \frac{\rho\alpha I_0}{2\pi a}\exp(-\alpha z) \quad \text{and} \quad \frac{E_r}{E_z} = \frac{\rho\alpha}{2\pi aS} >>> 1.$$

As expected, the electric field around the borehole points mainly in the radial direction.

7.12. SELF-POTENTIAL METHOD

7.12.1. Introduction

The self-potential method is one of the oldest methods of borehole geophysics. It has applications in detecting permeable zones intersected by the borehole and in understanding the clay mineralization of rock layers. In electrical logging, the self-potential method is also called the "spontaneous-potential" method or simply "SP." Several different effects give rise to naturally occurring potentials in boreholes. The two most important are (1) the electrokinetic effect, where conducting fluid flows under pressure through a porous medium (studied in Chapter 6), and (2) electrochemical effects caused by the different mobilities of cations and anions in the pore fluids of different types of rocks, especially porous sandstones and impermeable shales. It is generally accepted that electrochemical effects dominate the "SP anomalies" commonly observed in well logs. The full mechanism for the development of electrochemical self-potentials in boreholes is complicated and took many years to unravel. Here, we give a highly simplified description. Consider the case where the natural formation water has a higher concentration of salt than the borehole fluid. This concentration difference will cause positive sodium (Na^+) and negative chlorine (Cl^-) ions to diffuse from the formation fluid into the borehole. Eventually, either the concentration difference vanishes or equilibrium is reached where the diffusion force is balanced by an electric force that appears if the ionic motion creates net volume or surface charges. It is, in fact, the latter situation that prevails in most boreholes, but the process by which equilibrium is reached is different in different types of rocks. For example, in regions where the borehole penetrates porous sandstone, charge separation occurs because negative chlorine ions (Cl^-) have higher mobility in water than positive sodium ions (Na^+). The negative ions, therefore, diffuse into the borehole at a faster rate, creating a net separation of negative and positive charge near the borehole wall. The final equilibrium configuration can be treated as double layer of charge on the borehole wall, with net negative charge on the inner wall of the borehole and net positive charge on its outer surface. The dipole moment of this double layer points from the negative to the positive layer—that is, away from the borehole. As we know from Chapter 2, there is a potential jump across the double layer, which in this case is called a "liquid-junction" potential. Because of the negative charge accumulation on the inside of the borehole, the self-potential measured in boreholes is usually lower opposite porous sandstone than surrounding impermeable layers. A different process operates in regions where the borehole penetrates a shale layer. Shales are generally impermeable to fluid flow because their pore space is extremely narrow and not well connected. But ions in solution can still diffuse through shale layers, transported along the surface of clay minerals that line the pore space of shales. These clay minerals generally have a net negative charge on their surface (through elemental substitutions in the crystal structure) and preferentially attract and transport positive ions from the pore fluid. Negative ions in the borehole fluid are repulsed by clay minerals and remain trapped in the pore space. Thus, when a shale layer is penetrated by a borehole filled with fluid having a lower concentration of salt, positive (Na^+) ions preferentially diffuse from the shale layer into the borehole. When equilibrium is reached, there is a net accumulation of positive charge near the inner surface of the borehole wall, balanced by a layer of negative charge on its outer surface. In this case, the moment of the double layer points toward the borehole axis. Its potential is generally called a "membrane potential." Membrane potentials can also develop at horizontal interfaces between porous sandstone and impermeable shale layers near the borehole.

7.12.2. Self-potential at the Borehole Axis

Figure 7.25 shows a typical configuration of naturally occurring charges surrounding a porous sandstone layer, flanked by impermeable shale beds and intersected by a borehole.

To calculate the self-potential field, the different systems of charge can be replaced by double layers. The charge distribution consists of two plane double layers S_{21} and S_{23} of infinite extension, located at the bottom and

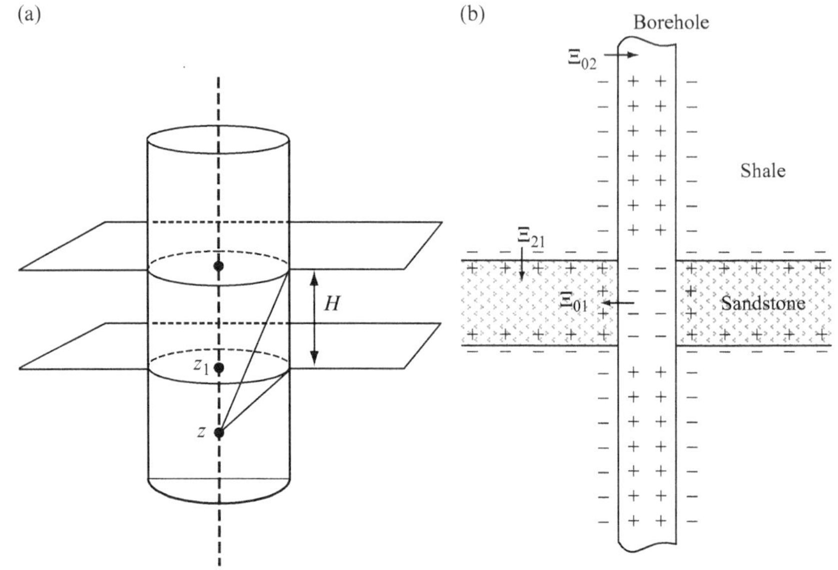

Figure 7.25 (A) Model of a medium illustrating calculation of a solid angle. (B) Distribution of double layer.

top of the layer, and cylindrical double layers S_{01}, S_{02}, and S_{03}, located at the borehole surface against the medium beneath the layer, the layer itself, and the medium above the layer. The potential at the observation point M is equal to the sum of potentials caused by each double layer:

$$U = U_{21} + U_{23} + U_{01} + U_{02} + U_{03}. \quad [7.133]$$

As was shown in Chapter 1,

$$U = \frac{\omega}{4\pi}\Xi, \quad [7.134]$$

where ω is the magnitude of the solid angle subtended by the double-layer surface as viewed from the observation point M, and Ξ is the voltage or electromotive force of the double layer. We will use a cylindrical system of coordinates r,φ,z with origin O at the middle of the layer along the borehole axis z. Also, we will use the following notations: h, d, and z are the layer thickness, borehole diameter, and distance from the observation point to the origin, respectively; $\Xi_{21},\Xi_{23},\Xi_{01},\Xi_{02},\Xi_{03}$ are the electromotive forces of the corresponding double layers—that is, the potential difference crossing the layer,

$$\Xi_{ik} = U(i) - U(k).$$

The solid angles subtended by an infinite plane and closed surface are equal to $\pm 2\pi$ and either 4π or 0, respectively, depending on the position of the observation point with respect to this surface. Then, the solid angles for each double layer can be represented as

$$\begin{aligned} &\omega_{21} = -2\pi + \omega_p, \quad \omega_{23} = 2\pi - \omega_q, \\ &\omega_{01} = 4\pi - \omega_p, \quad \omega_{02} = \omega_p - \omega_p, \quad \omega_{03} = \omega_q, \end{aligned} \qquad [7.135]$$

where ω_p and ω_q are the solid angles subtended by the borehole cross sections with coordinates $z = \pm z/2$, respectively. Substituting Eqs. [7.134] and [7.135] into Eq. [7.133], we obtain

$$\begin{aligned} U &= -\frac{2\pi - \omega_p}{4\pi}\Xi_{21} + \frac{2\pi - \omega_q}{4\pi}\Xi_{23} + \frac{4\pi - \omega_p}{4\pi}\Xi_{01} + \frac{\omega_p - \omega_q}{4\pi}\Xi_{02} + \frac{\omega_q}{4\pi}\Xi_{03} \\ &= \frac{1}{4\pi}\left[\omega_p(\Xi_{21} - \Xi_{01} + \Xi_{02}) - \omega_q(\Xi_{23} - \Xi_{03} + \Xi_{02}) + \frac{\Xi_{23} - \Xi_{21}}{2} + \Xi_{01}\right]. \end{aligned} \qquad [7.136]$$

Using the relations between the solid and plane angles, we have

$$\omega_p = 2\pi(1 - \cos\beta_p) = 2\pi\left\{1 - \frac{z - (h/2)}{\left[(z - (h/2))^2 + (d^2/4)\right]^{1/2}}\right\}$$

and

$$\omega_q = 2\pi(1 - \cos\beta_q) = 2\pi\left\{1 - \frac{z + (h/2)}{\left[(z + (h/2))^2 + (d^2/4)\right]^{1/2}}\right\},$$

where β_p and β_q are the angles shown in Fig. 7.25B. Correspondingly, instead of Eq. [7.136] we have

$$U = \left\{\frac{2z + h}{\left[(2z + h)^2 + d^2\right]^{1/2}} - \frac{2z - h}{\left[(2z - h)^2 + d^2\right]^{1/2}}\right\} \times \frac{\Xi_{21} - \Xi_{01} + \Xi_{02}}{2} + \Xi_{01}$$

or

$$U = \left\{\frac{2z^* + h^*}{\left[(2z^* + h^*)^2 + 1\right]^{1/2}} - \frac{2z^* - h^*}{\left[(2z^* - h^*)^{1/2} + 1\right]^{1/2}}\right\} \times \frac{\Xi_{21} - \Xi_{01} + \Xi_{02}}{2} + \Xi_{01} \qquad [7.137]$$

where

$$z^* = \frac{z}{d}, \quad h^* = \frac{h}{d}.$$

This equation shows that the potential U tends to a constant equal to Ξ_{01} at large distances from the layer; that is

$$U(\infty) \to \Xi_{01}.$$

Since the potential at the observation point M is defined with respect to the potential at infinity (usually zero, but not in this model), we have

$$\Delta U = U(z^*,h^*) - U(\infty) = -\frac{1}{2}F(z^*,h^*)\Xi_s, \qquad [7.138]$$

where

$$F(z^*,h^*) = \frac{2z^* + h^*}{[(2z^* + h^*)^2 + 1]^{1/2}} - \frac{2z^* - h^*}{[(2z^* - h^*)^2 + 1]^{1/2}} \qquad [7.139]$$

is a function characterizing the change of ΔU along the borehole axis, and

$$E_s = \Xi_{01} - \Xi_{02} - \Xi_{21} \qquad [7.140]$$

is the algebraic sum of electromotive forces at the layer interfaces surrounding the borehole. As is seen from Fig. 7.26, the curves illustrating

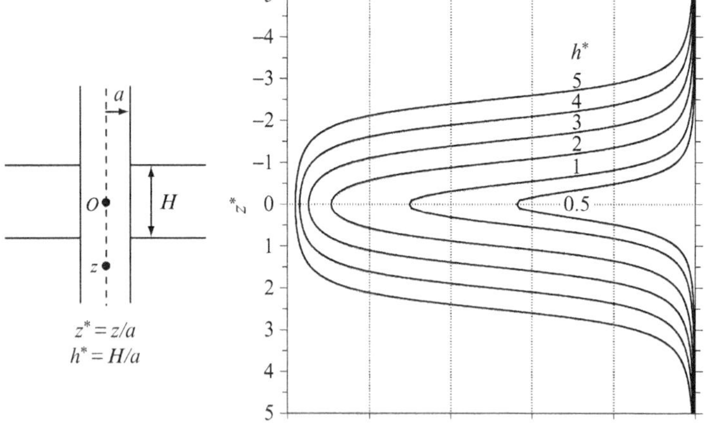

Figure 7.26 Distribution of potential on the borehole axis. Index of the curves is h^*.

the dependence of the function $\Delta U/\Xi_s$ on the coordinate z^* are symmetrical with respect to the origin. The index of the curves is h^*. Let us note that, in spite of the fact that this solution is an approximate one, since it does not take into account the change of resistivity, its analysis is useful for understanding the behavior of the spontaneous potential at the borehole axis.

BIBLIOGRAPHY

1. Alpin LM. *Theory of electric logging*. Moscow: OHTI; 1938.
2. Alpin LM. *The method of the electric logging in the borehole with casing*. US patent No. 56026.
3. Alpin LM. *Theory of dipole soundings*. Moscow: Nedra; 1950.
4. Alpin LM. *Theory of field*. Moscow: Nedra; 1966.
5. Bursian VR. *Theory of electromagnetic fields applied in electrical methods*. Moscow: Nedra; 1972.
6. Charkevich AA. *Spectra and analysis*. Moscow: GIFML; 1962.
7. Doll HG. The Laterolog: a new resistivity logging method with electrodes using an automatic focusing system: petroleum transactions. *AIME* 1951;**192**:305–16.
8. Kaufman AA. *Geophysical theory and methods: part A*. New York, NY: Academic Press; 1992.
9. Keller QV, Frischnecht FC. *Electrical methods in geophysical prospecting*. Oxford: Pergamon Press; 1966.
10. Oristaglio M, Dorozynski A. *A sixth sense: the life and science of H. Doll*. Parsippany, NJ: The Hammer Company; 2007.
11. Svetov BS. *Foundation of geolectric*. Moscow: LKI; 2008.
12. Yakubovsky YV. *Electroprospecting*. Moscow: Nedra; 1973.
13. Zabarovsky AI. *Electroprospecting*. Moscow: GNTI; 1943 p. 12.
14. Ellis DV, Singer JM. *Well logging for earth scientists*. 2nd ed. The Netherlands: Springer; 2007.

INDEX

D

E

F

G

H

I

J

K

L

M

N

O

P

V

W

FSC
www.fsc.org
MIX
Papier | Fördert gute Waldnutzung
FSC® C083411

Kolibri 360 GmbH
Ferdinand-Jühlke-Straße 7
99095 Erfurt, Deutschland
produktsicherheit@kolibri360.de